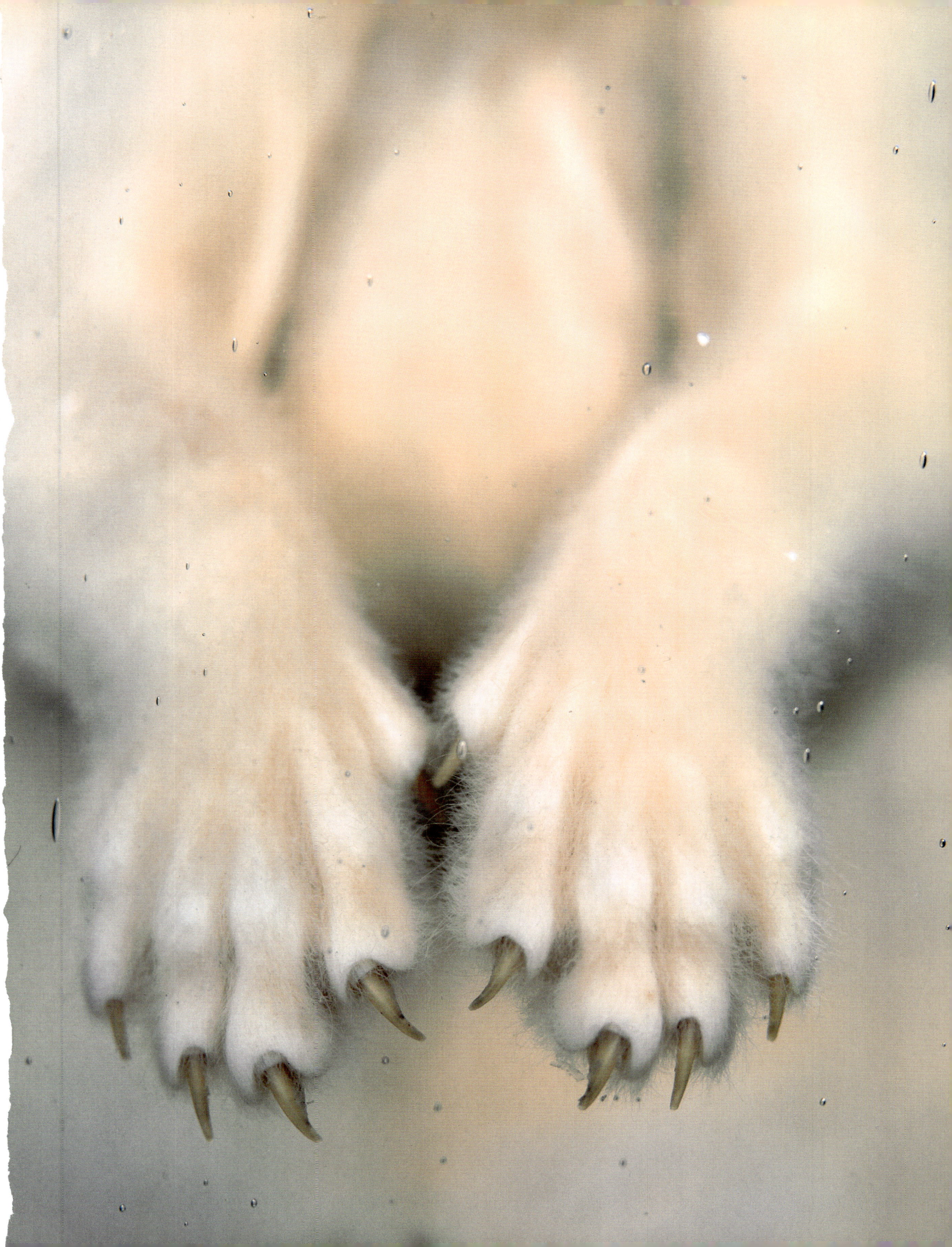

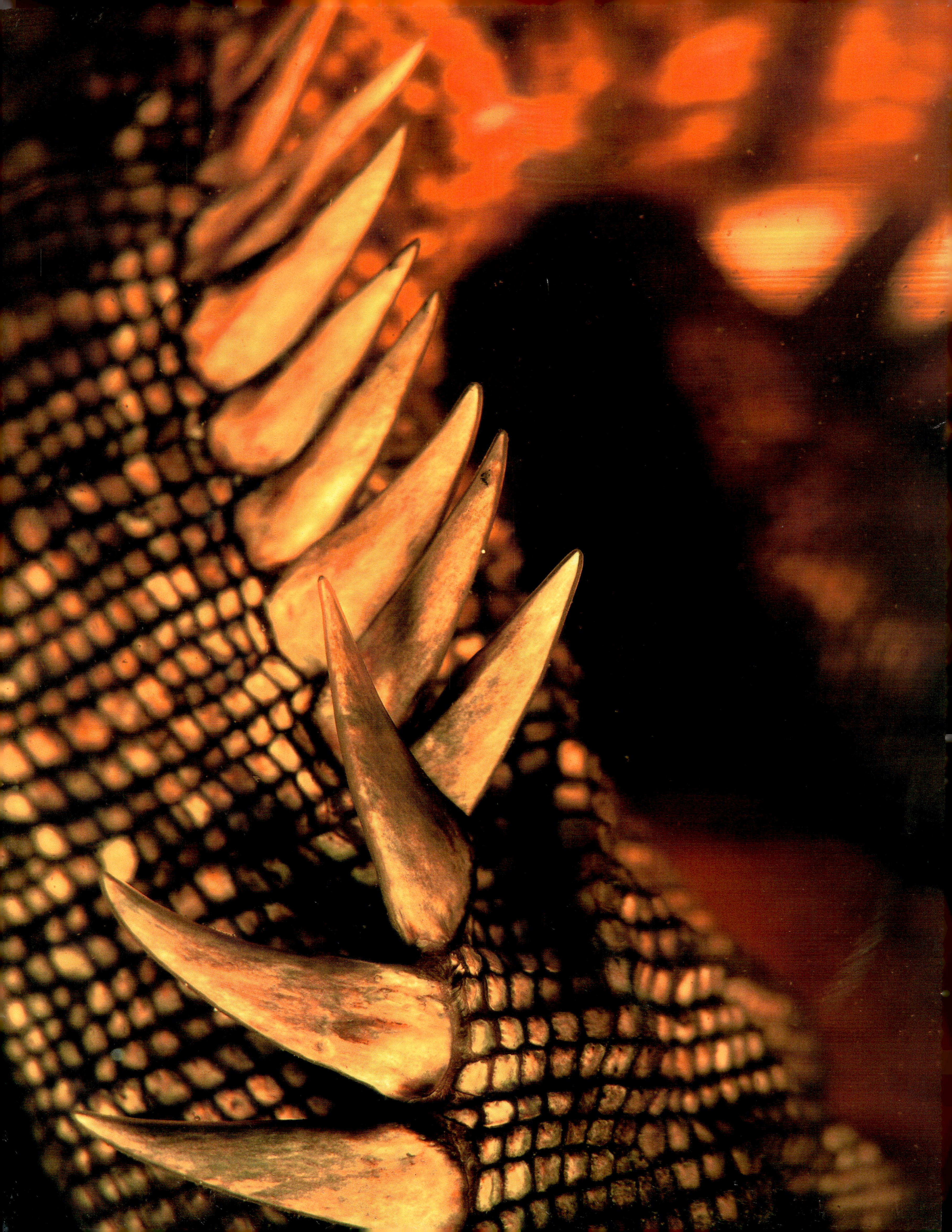

CONSERVING

DANIEL AND GEO FUCHS

EDITION REUSS

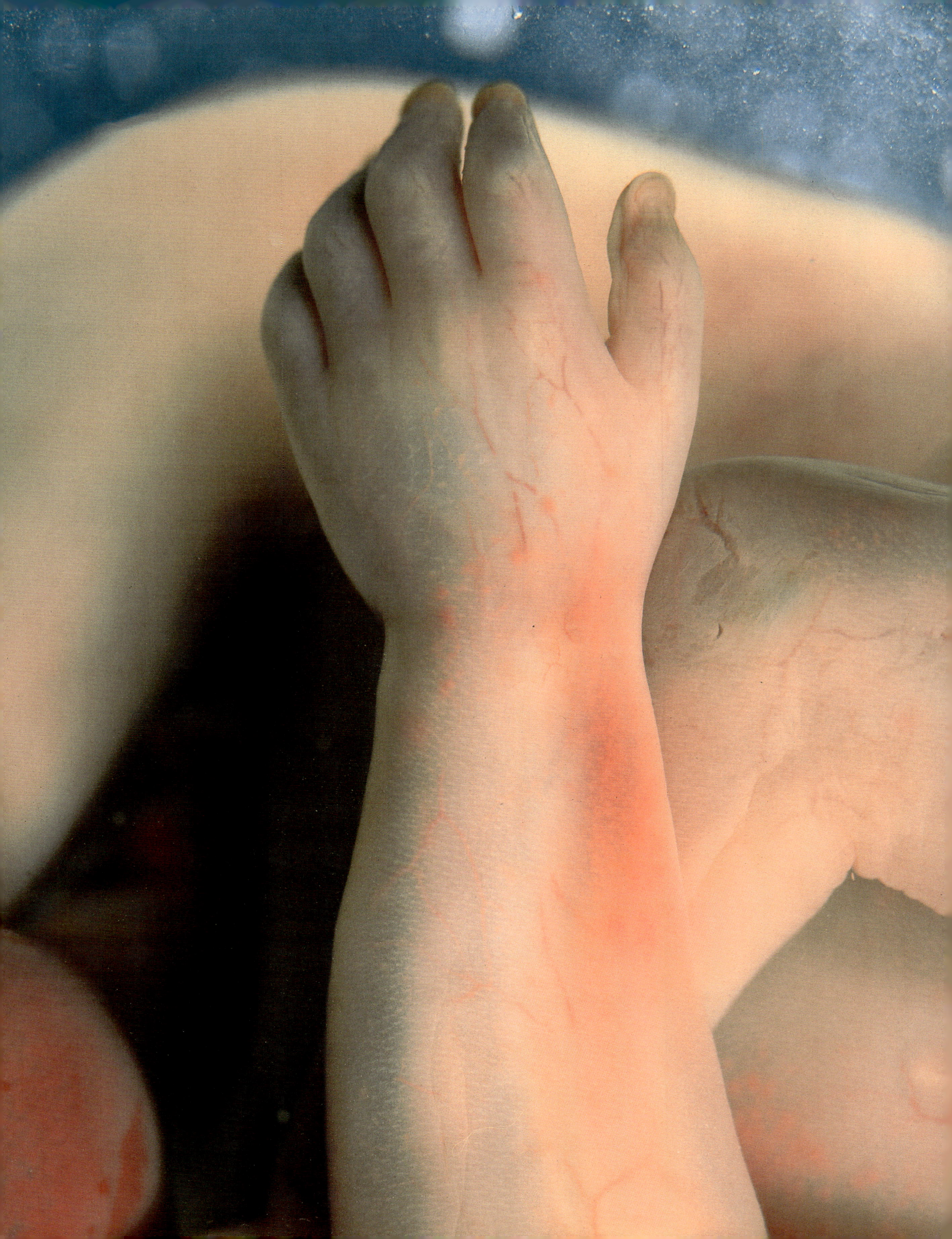

CONTENTS

CONSERVING — zu dem Zyklus von Daniel und Geo Fuchs

Die Bilder dünken die gelungensten,
in denen das absolut Gleichzeitige
wie ein Zeitverlauf scheint,
der den Atem anhält.
ADORNO

Ein roter Dämon grinst hinter einer unsichtbaren Grenze aus Glas; ein träumendes Kind, das nie geboren wurde, fast lächelnd, als störe die Zeit nicht den Schlaf der Vergänglichkeit; ein Kampffisch, früher, nun schon hinter seinem transparenten letzten Gang. Bärentatzen, die sich aneinander pressen, um sich der verlorenen Wärme zu versichern. Ein Mädchengesicht, drapierte Spitze, die umhüllt. Ein Gesicht vor dreihundert Jahren, in die Zeitlosigkeit gebannt. „Und unsere Gesichter, mein Herz, vergänglich wie Fotos." [1]

Arbeiten von Daniel und Geo Fuchs. Fotografien. Auch (Ab-)Bilder gefundener Exponate in naturhistorischen Sammlungen. Fisch, Tier, Mensch. Konservierte Fische, Tiere, Menschen. Mehr als das: Das künstlerische Arbeiten in dieser Form wird zur Dokumentation beginnender Freundschaft zwischen Dargestelltem und Darstellendem. Auseinandersetzung, Reibung. Grenzüberschreitung. Kann Tod überwunden werden? Wo beginnt Würde, wo die Pflicht des Künstlers, sie zu verletzen, ohne sie zu zerstören? Die Bilder haben Daniel und Geo Fuchs zuerst zugelassen. Über die gewahrte Mauer zwischen Aus- und Eingeschlossenen hinweg haben sowohl die Bilder sie gefunden als auch sie die Bilder.

Ein Weg von Jahren. Ängste, Veränderungen. Wer sich in die dunkleren Gefilde begibt, findet gerade in der „natura morta" Fragmente der eigenen Vergänglichkeit. Wenn „natura morta" durch menschlichen Eingriff vor dem endgültigen Verfall bewahrt werden kann, bleibt Hoffnung auch für uns? Wer eine konservierte Hand in seinen Händen hält, um sie für das Motiv in Licht zu tauchen, wird stummer Zeuge einer anderen Dimension. Zeigen. So führt das im bildlichen Sinn sichtbar Gemachte zum Bewusstsein von etwas, das durch unsensibles Vorzeigen verloren ginge. Wie Bedeutung nicht durch den Zusatz von Sichtbarem, sondern durch dessen Entzug entsteht.

Natürlich auch: Bilder einer Ausstellung. „Conserving Fish". Die Fotobilder an den Wänden. Die Exponate, Vorläufer und gefundenes Motiv zugleich, als Leihgaben. Fischkörper in der schützenden Hülle des Glases, in Flüssigkeit konserviert, fundiert analysiert, wissenschaftlich präpariert. Sie zeichnen für den Beginn der Auseinandersetzung der beiden Künstler. Ein kleines Mädchen, das vergleicht, begreift, erfährt und plötzlich sieht. Der Fisch in dem Glas ist seit hundert Jahren tot, auf der Fotografie lebt er. Wo beginnt das Danach? Vor der Entstehung? Das Staunen, die Gabe, etwas zum ersten Mal zu sehen ... das Auge des Kindes hat mit der Seele gesehen.

Die Frage des Sehens. Wir. Beobachtete Betrachter. Der von Licht durchdrungene Kopf, die Hand, die bloße Hand, die feine Führung der Linie eines ungeborenen Profils. Der Schlaf ruht tief in diesem friedlichen Gesicht. Dort, unter der Nasenspitze das zarte Grün als Flaum. Und das Abbild, das Bild (ver)birgt eine Geschichte. Vielleicht gibt es ein Erkennen. Nichts ist festgelegt.

In diesen Fotografien, in diesen „Subjekten", aufgenommen mit und durch die trennende, wenngleich auch transparente Wand, ist Bewegung. Sie erzählen Vergangenheit weiter und gestalten sie neu. Die Geschichten streben in eine Zukunft, nach vorne. Oder besser: Zeit hat sich in ihnen verfangen, und die Bilder von Daniel und Geo Fuchs haben sie zum Stocken gebracht. Und wir, die uns durch Sehen nähern, haben daran teil.

Laut Maurice Blanchot besteht das Wesen des Bildes darin, ganz außen zu sein, ohne Intimität, und dennoch unzugänglicher und rätselhafter als die innere Vorstellung; ohne Bedeutung und doch zugleich eine Herausforderung der Unergründlichkeit jedes möglichen Sinns.

In diesen Bildern ist Herausforderung, die irritiert. Außerhalb ihrer technischen Brillanz, der Leichtigkeit des Lichts, der Unschärfe der unter sich zerfließenden Farben des Hintergrunds liegt eine eigentümliche Schärfe der Wahrnehmung: das Auge des Fisches – von untrüglicher, matter Trübheit. Auf den ersten Blick ein totes Auge. Dann ist plötzlich, je länger der Blick den Gegenstand aus seiner Gegenständlichkeit löst, ihn abstrahiert, Bewegung. Nicht Stillstand des Moments, sondern Stillsein im Jetzt. Das Fühlen liegt manchmal vor dem Wissen, das Ahnen vor der Anwesenheit des Auges. Das Rätsel wehrt sich gegen den Übergriff der Ratio. Dahinter ein – unser? – Geheimnis. Man kann gehen und vergessen wollen. Die Zeit, den Tod, die Geschichte. Es gibt Bilder, die bleiben, weil sie bleiben wollen. Sie haben sich in den Betrachter eingeschlichen.

Den Werken liegt eine merkwürdige Präsenz zu Grunde. Ohne anmaßendes Kalkül, mit dem Wissen der Brisanz des Thematisierten begegnen sie den Dargestellten mit staunendem Respekt, schaffen eine Atmosphäre des Innehaltens, einer wärmenden Ruhe, respektieren Anmut und Würde. Sie rücken die bestehende Grenze nicht um eines Effektes willen weiter und lassen so Entfaltung zu. Bilder können Raum schaffen, einen aufnehmen, Assoziationen wecken. Nichts greift an oder legt fest, wenn es und wir gewahrt bleiben.

Trotzdem Fremdheit. Auch: Einsamkeit. Wer hinter eine Grenze gerät, kennt zwei Wirklichkeiten. Die Zeit, der Tod. Grenzerfahrungen. Auch wenn die Bilder uns nicht fremd sind, sie bleiben fremd – und sie bleiben einsam. Von dem Moment an, da sie ihrer Welt aus Glas entnommen würden, dieser real existierenden Welt ohne Schutz überlassen – sie zersetzten sich, zerfielen. Wir, die darum wissen. Im Gegensatz dazu die Bilder dieser Trilogie, deren jetzige geheimnisvolle Heimat den Tod nicht kennt. Er liegt hinter ihnen, und sie sind neu erfunden.

JANA MARKO

---

1 Buchtitel. John Berger. *Und unsere Gesichter, mein Herz, vergänglich wie Fotos.* Carl Hanser Verlag. München und Wien 1986, S.14

Scheintod: CONSERVING — Fotografien von Daniel und Geo Fuchs

In ihren fotografischen Werken betreten Geo und Daniel Fuchs sehr private Welten. Sie nähern sich Situationen, die sie als „Grenzsituationen" beschreiben – Situationen, denen andere lieber ausweichen; Orten und Gegenständen, die normalerweise vor dem Blick der Öffentlichkeit verborgen sind. Sie dringen in Lagerräume von Museen und wissenschaftlichen Sammlungen ein, in denen Tiere und menschliche Körperteile – konserviert in mit Alkohol oder Formaldehyd gefüllten Glasbehältern – in einem scheinbaren Schwebezustand lagern. Sie setzen sich nicht nur mit dem institutionellen Drang zum Bewahren und Sammeln auseinander, sondern auch mit ihren eigenen Reaktionen auf diese Objekte und das, wofür sie stehen. „Anfang der 90er Jahre erfuhren wir von wissenschaftlichen Sammlungen, in denen Tiere – Fische, Säugetiere und Reptilien – in Alkohol konserviert aufbewahrt werden. Vor allem der Gedanke konservierter Fische, die in ihrem Element – Flüssigkeit – wieder eine Lebendigkeit erlangen müssten, faszinierte. Bereits bei der ersten Besichtigung einer Sammlung eines naturhistorischen Museums erkannten wir, dass sich die Fische im Tod eine bemerkenswerte Lebendigkeit bewahrt hatten, die genau unser Thema – Leben im Tod zu finden – traf ... Tod und Konservierung sind für uns in dem Sinne verwandt, dass wir die Konservierung als eine andere Form des „Weiterlebens" betrachten. Manche Exponate sind über 300 Jahre alt."[1]

Daniel und Geo Fuchs haben ihr Projekt „Conserving" genannt, ein Titel, der eine doppelte Bedeutung hat. Die Objekte, die sie fotografieren, sind selbst konserviert und bewahrt. Durch den Akt des Fotografierens erschaffen die Künstler eine weitere Ebene der Dokumentation und des Bewahrens. Sie errichten diesen normalerweise verborgenen Objekten eine neue Arena und holen sie in das Umfeld der seriösen Kunst- und Dokumentarfotografie, wo sie sich vom Blick des Wissenschaftlers stark unterscheiden. „Conserving" zeigt, dass die Fotografie in der Lage ist, ein neues Publikum für bisher im Verborgenen gehaltene Sujets und Ideen zu schaffen.

In den späten 80er und frühen 90er Jahren begannen sich eine Reihe europäischer und amerikanischer Fotografen intensiv für das Thema Tod zu interessieren. Rudolf Schäfers Porträts von Gesichtern toter Menschen (die 1987 in einer ostdeutschen Leichenhalle entstanden sind) erregten sofort Aufmerksamkeit. Gleiches galt für die eindrucksvolle Serie, die Andres Serrano in den Leichenhäusern von New York City aufgenommen hat. Der englische Farbdokumentarist Nick Waplington, der in den frühen 90er Jahren mit seinen rauen Kommentaren zum Leben zweier Familien in einer Wohnsiedlung in Nottingham von sich reden machte, schuf danach eine Serie nachgestellter Situationen, die sich mit unnatürlichem Tod und Selbstmord auseinander setzen. Die Gründe für dieses neu erwachte Interesse an der Fotografie des Todes bleiben unklar und sind vielleicht auch zu vielfältig, um in eine einzige schlüssige Theorie gefasst zu werden. Seit den 80er Jahren entwickelten Künstler, die mit Fotografie arbeiten, ein großes Interesse an den „objets morte" und begannen Situationen „aus dem wirklichen Leben" in einer Weise umzusetzen, die eine stark subjektive und gezielte Sicht der realen Welt widerspiegelte. Die Fotografie des Todes, so wie man ihn in einer Leichenhalle oder

im Museum erleben kann, spielte bei dieser neuen Art, Fotografie einzusetzen, eine wichtige Rolle. Die Toten waren Teil unserer Welt und dennoch außerhalb von ihr – dem Künstler/ Fotografen in einer einzigartigen und unnachahmlichen Weise als stille und unbewegliche Objekte zur Verfügung.

Die neue fotografische Auseinandersetzung mit dem Thema Tod stellte eine Herausforderung an den Status quo dar und fügte sich in die postmoderne Methodologie der 80er und 90er Jahre ein. Die Parameter der Fotografie verschoben sich, als die Künstler damit begannen, die Fotografie nicht mehr nur zur Darstellung zu verwenden, sondern die Darstellung selbst zu erforschen. Cindy Sherman konstruierte ein Theater der Verkleidung und Fantasie, mit dem sie Vorstellungen rund um die weibliche Identität erkunden wollte. Martin Parr setzte grelle Farbfotografie ein, um bittere Dokumentationen über das britische Straßenleben zu schaffen, und stellte damit die Idee der klassischen Reportage auf den Kopf. Thomas Ruff ging im Genre der Porträtfotografie Aufsehen erregende neue Wege. Grenzwälle, die die Fotografie umgaben, wurden durchbrochen, und die wachsende Bedeutung der fotografischen Theorie (im Gegensatz zur Geschichte der Fotografie, die man zunehmend als öde und prozessbesessen empfand) machte ein neues Publikum auf die intellektuellen und politischen Dimensionen des fotografischen Bildes aufmerksam. In zunehmendem Maße begannen Künstler, die Fotografie als geeignetes und leicht zugängliches Werkzeug zu benutzen, um Ideen auszudrücken, die sie in den 60er und 70er Jahren wahrscheinlich durch Malerei, Bildhauerei und Illustration vermittelt hätten.

Innerhalb dieses postmodernen Kontextes sind Daniel und Geo Fuchs' Werke zu sehen. Sie nehmen sich das Recht, in Bereichen zu arbeiten, die zuvor mit einem Tabu belegt waren. Ihre Arbeit ist weit davon entfernt, dokumentarisch zu sein, sondern ein hoch subjektiver Prozess, der persönliche Geschichte und tiefe Anteilnahme mit einbezieht. Die Welt, so scheinen sie sagen zu wollen, ist dazu da, um betrachtet und studiert zu werden, denn sie bietet sich einer Vielzahl unterschiedlicher Blicke dar. Ihr unablässiger Wunsch ist es, an etwas zu arbeiten, das sie selbst als Grenzerfahrung beschreiben. Ihr persönliches Engagement, mit dem sie sich dem Thema widmen, und ihre Distanzierung von Sensationslüsternheit oder Voyeurismus resultieren in einer äußerst anspruchsvollen und konsequenten Vorgehensweise, die vom Betrachter Aufmerksamkeit und Verständnis verlangt. Die Bilder der „Conserving"-Serie haben durchgehend eine sehr spezielle Ästhetik, die Teile unserer Welt, die man sonst für makaber oder abstoßend halten würde, in faszinierende Objekte verwandelt. Die toten Kreaturen und Menschen, die sie fotografieren, werden zu Ikonen, zu fast ehrwürdigen Objekten – durch die Konservierungsflüssigkeit unsterblich gemacht. Lange tot und doch immer noch Teil unserer heutigen Welt, befinden sie sich in einem seltsamen Schwebezustand zwischen Leben und Tod und haben vielleicht einen mystischen Status erlangt.

„Conserving Fish" ist vielleicht die durchgeistigtste und abstrakteste der drei „Conserving"-Serien. Das Schwimmen in Flüssigkeit erlaubt den Fischen, einen Zustand einzunehmen, den man als eher natürlich bezeichnen kann. Wir sehen sie durch das

Glas wie durch die Fenster eines Aquariums, und nur eine gewisse Trübheit der Farben und ein Mangel an Lebhaftigkeit deuten auf ihren Scheintod hin. Hier und da erhaschen wir einen Blick auf ein Glasgefäß oder ein Identifikationsetikett, der uns zu der Erkenntnis führt, dass es sich hier um eine taxonomische Studie handelt und nicht um die Betrachtung zwar gefangener, aber doch lebender Tiere. Obwohl „Conserving Fish" die scheinbar am wenigsten „schockierende" Serie der Trilogie ist, ist sie von ausgesprochener Melancholie geprägt. Schuppen, die glänzen sollten, sind matt und ohne Schimmer, lebhafte Augen sind leblos, flinke Fischschwärme verharren regungslos, von Tod und Zeit für immer zum Erstarren gebracht.

„Conserving Animals" vermittelt eine ganz andere Erfahrung. Schon auf den ersten Blick erkennen wir, dass dies eine Studie der Sterblichkeit ist. Klauen und Pfoten sind von den Körpern getrennt und treiben in der Konservierungsflüssigkeit. Ausdruckslose Augen starren durch die Flüssigkeit, zu denen wir keinen Blickkontakt aufnehmen können. Manchmal ist da eine Geste, die auf Leben hindeutet, ein Augenreiben, ausgestreckte Arme und Beine, eine Umarmung, aber das sind nur Zufallsgesten, flüchtige Momente einer Lebenskraft, die sich dem Tod widersetzt. Diese Tiere schlafen einen unruhigen Schlaf, in Flüssigkeit verbannt, ihrem natürlich Element entrissen und den Blicken der Neugierigen und Experten ausgesetzt. Sie sind zu Objekten geworden, Gegenständen ohne Selbstgefühl.

Wenn wir unsere eigene Spezies betrachten (in der Serie „Conserving Humans"), wird die Verweigerung der Privatsphäre fühlbar. Gesichter toter Menschen sind nicht ohne Ausdruck oder Persönlichkeit. Ein junger Mann, ein Baby, eine alte Frau, selbst ein Embryo gehörten einmal zu unserem Stamm, hatten Gedanken und Gefühle, fällten Entscheidungen, gingen Beziehungen ein, existierten innerhalb unserer Gesellschaft. Doch hier, wo sie zu Musterexemplaren der Wissenschaft werden, als Museumsobjekte dienen, sind sie von Stammesritualen, den Übergangsriten ausgeschlossen. Wie Goethes „Faust" sind sie auf groteske Weise unsterblich geworden. „Conserving Humans" ist der Höhepunkt der „Conserving"-Serie, die uns vom Abstrakten zum Realen führt. „Conserving Humans" wird unvermeidlich als Studie des Grotesken betrachtet werden, obwohl Daniel und Geo Fuchs wenig Interesse am Grotesken habe. Sie blicken auf die Verstümmelungen, die gespaltenen Schädel, die Skelette, die tätowierte Haut fast wie Zeugen, Berichterstatter der Zeit, in der wir leben. Selbst haben sie über „Conserving Humans" geschrieben, dass sie die Arbeit an der Serie als große Herausforderung angesehen haben, „denn wir wollten die Menschen so darstellen, dass sie eine Schönheit behalten, dass ein Leben sichtbar bleibt, dass das Wissen über den Tod überwiegt und über den Tod hinaus die Würde der Menschen gewahrt bleibt."[2]

Die Arbeit an „Conserving" hat zu einem neuen, weit reichenden fotografischen Projekt zum Thema Tod geführt. Die Künstler haben sich über die wissenschaftlichen Institutionen und Museen hinausgewagt und die seltsame, stille Welt der Sektionsräume betreten, wo nichts ihren Blicken entgeht, vom Leichnam bis hin zu den Schnitt- und Sezierinstrumenten. Hier, wo der menschliche Körper, obwohl immer noch als Individuum erkennbar, zum Objekt geworden ist, seiner Intimsphäre beraubt. Sie beschäftigen sich mit den Spuren, die der Körper zurücklässt, ein Stück Fleisch, ein Haar. Ihr Interesse an diesen Vorgängen hat nichts Morbides, denn wo Neugier sich anschickt, das Groteske und Absurde zu enthüllen, behauptet sich das Künstlerauge. Die Bilder, die wir sehen, sind weder drastisch noch entsetzlich, sondern öffnen sich unserer Vorstellungskraft und unserem Verstand und einer Vielzahl von Möglichkeiten. „Nach unserer Beschäftigung mit dem Thema Tod in einer mehr abstrakten Form war es für uns ein logischer Schritt, uns mit dem realen Tod zu konfrontieren. Ohne das schützende Glas und die Flüssigkeit werden die Situationen für uns und die daraus resultierenden Bilder viel direkter. Tod und Zerfall des menschlichen Körpers werden etwas sehr Natürliches. Der Körper erscheint nur als eine leere Hülle."[3] Man spürt, dass ihr Wissensdurst, ihr Verlangen nach Antworten alles verzehrend und tief persönlich geworden ist.

In den drei Serien, aus denen „Conserving" besteht, bleiben Daniel und Geo Fuchs den Prinzipien treu, die ihr Werk bestimmen. Wenn wir die Fische, Säugetiere und Menschen ansehen, die sie fotografiert haben, fühlen wir ihre innige Verbundenheit mit den Subjekten. Obwohl sie unangenehme Situationen betrachten, spüren wir, dass ihr Blick besorgt, sogar gütig ist. „Der Mensch ist vom Gedanken an den Tod gleichermaßen fasziniert und abgestoßen – dem letzten Geheimnis des Lebens. Da das Wissen um das eigene Sterben immer mit uns lebt, gehen Tod und Leben immer nebeneinander her. Das ist auch unser Ziel – diese beiden Zustände in einem Bild zu vereinen."[4] Wir können ihre Verwundbarkeit spüren, das persönliche Risiko, das sie eingegangen sind, als sie beschlossen, ihre künstlerische Laufbahn diesen verborgenen und oft umstrittenen Terrains zu widmen. Sie haben sich selbst zu Antiquaren gemacht, zu Forschern in einem fremden Land, die gewissenhaft unbequeme Wahrheiten aufzeichnen und Fotografien erschaffen, die Diskussionen auslösen, Fragen aufwerfen und wesentlich zu unserem Verständnis der Rolle der Fotografie in der Dokumentation unserer Welt beitragen werden.

VAL WILLIAMS

1 Zitat aus einem Interview mit den Künstlern 1999

2 Ibd.

3 Zitat aus dem Interview mit Francesca Alfano Miglietti für *Virus Magazine*, Mailand, Italien

4 Zitat aus einem Interview mit den Künstlern 1999

CONSERVING – On a Cycle of Photos by Daniel and Geo Fuchs

The most successful images seem to be
those in which absolute simultaneity
appears like a temporal sequence
that is holding its breath.
ADORNO

A red demon grins behind an invisible boundary made of glass. A dreaming, unborn child almost smiling, as though time cannot disturb the sleep of transience. A fish of prey, once, now beyond its last transparent bout. Bear paws pressed together to recapture the warmth they have lost. A girl's face wrapped in lace, from three hundred years ago, exiled into timelessness. "And our faces, my heart, ephemeral as photos."[1]

Artworks by Daniel and Geo Fuchs. Photographs. Also: pictures of specimens found in natural history collections. Fish, beast, human. Conserved fishes, animals, people. More than that: this form of artistic work becomes a documentation of a friendship between the portrayed and the portrayer. Confrontation, friction, crossing frontiers. Can death be overcome? Where does dignity begin? Where does it become the artist's duty to violate dignity without destroying it? Daniel and Geo Fuchs were the first to accept the images. Crossing the existing wall between the fenced out and the fenced in, the artists found the images as much as the images found the artists.

A path that took many years. Anxieties, changes. Anyone venturing into darker fields, especially in *natura morta*, will find fragments of his own transitoriness. If human intervention can preserve *natura morta* from ultimate decay, is there still hope left for us? Anyone who holds a mummified hand in his own hands and lifts it into the light to be photographed mutely witnesses another dimension. Signs. That which has been rendered visible in the sense of an image leads us towards becoming aware of that which would have been lost if it had been insensitively displayed. Just as meaning can arise not by adding more of that which is visible, but by removing some of it.

Also, of course: pictures at an exhibition. "Conserving Fish." The photographic images on the walls. The precursors and the discovered motifs alike, the bodies of the fish displayed here on loan in protective glass containers, conserved in liquid, accurately analyzed and scientifically preserved. These specimens define the beginning of the two artists' process of confrontation. A little girl compares, comprehends, experiences and suddenly sees. Though the fish in the glass container has been dead for a hundred years, it lives in the photograph. Where does "afterwards" begin? Before the origin? To marvel, the gift of seeing something for the first time.... The child's eye saw with her soul.

The question of seeing. We. Observed observers. The head penetrated by light. The hand, a mere hand, and the delicate lines in the profile of an unborn face. Slumber rests profoundly in this peaceful face. There, beneath the tip of the nose, the faint green of downy fuzz. The likeness, the photo, encloses and discloses a story. Perhaps there is an insight. Nothing is fixed or certain.

There is movement in these photographs, in these "subjects" photographed with and through a dividing yet transparent wall. They recount the past even as they refigure it anew. The stories strive forwards into a future. Or better yet: time became entangled in them. Daniel and Geo Fuchs' photos have brought time to a halt. And we, who approach them through seeing, participate in this.

According to Maurice Blanchot, the essence of an image lies in its being utterly outside, devoid of intimacy, yet more inaccessible and more puzzling than any inner notion. Meaningless and yet a challenge to the inscrutability of any possible meaning.

There is an irritating challenge in these pictures. Beyond their technical brilliance, beyond the brightness of the light or the blur of the colors of the background faded out of focus – beyond all this there is a peculiar clearness of perception: the eye of the fish with its unmistakable dimness. At first glance, it seems to be a dead eye. But then, after one's gaze has dwelt long enough on it and, in doing so, has abstracted the object, suddenly there is movement. Not the immobility of the moment, but a stillness in the here and now. Feeling sometimes comes before knowing. A vague presentiment may precede the presence of the eye. The riddle resists reason's attempts to encroach upon it. Behind it lies a mystery. Is it, perhaps, our mystery? One can go and want to forget. Time, death, history. Some pictures remain because they want to remain. They've crept into their viewer.

These artworks have an unexpected presence. Without arrogant calculation, fully aware of the controversial nature of their theme, these artworks encounter what they portray with an attitude of awed respect. They create an atmosphere which prompts the beholder to pause and hold his breath. They breathe an aura of warming peace, respectful grace and dignity. Though they redraw the existing limits, they do so not for the sake of sensationalism, and thus they allow for evolution. Pictures can create space, can absorb their viewer and evoke associations. Nothing attacks or imposes itself if they and we remain inviolate.

There is strangeness nonetheless. Also: loneliness. Anyone who ventures beyond a boundary knows two realities. Time, death. Boundary experiences. Although the images are by no means alien to us, they remain strangers – and they remain lonely. The moment they are removed from their world of glass and left unprotected in this real, existing world, in that very moment they would begin to decompose, to decay. We know this. The pictures in the trilogy are just the opposite. Their present, enigmatic home knows no death. Death lies behind them. They have been reinvented.

JANA MARKO

---

1. The quotation is a translation of the German title of John Berger's book *Und unsere Gesichter, mein Herz, vergänglich wie Fotos*. Published by Carl Hanser Verlag, Munich and Vienna, 1986, p.14.

Suspended Animation: CONSERVING – Photographs by Daniel and Geo Fuchs

In their photographic works, Geo and Daniel Fuchs enter very private worlds. They are intent on approaching situations which they describe as "borderline," situations which others might well avoid, places and things which are normally hidden from public view. They penetrate the storerooms of museums and scientific institutions, where preserved animals and even human beings are suspended in glass cases filled with formaldehyde or alcohol, and confront not only the institutional compulsion to conserve and collect, but also their own responses to these objects and what they represent. "In the early Nineties we learned of the scientific collections in which animals, fish, mammals and reptiles are conserved in alcohol. We began to think that the fish especially must somehow regain a new form of life, back in their own liquid element. When we first visited a collection at a natural history museum, we noticed that the fish still had a remarkable liveliness which fit our theme exactly: to find life in death.... For us, death and preservation are related in the sense that we see preservation as another form of continued existence. Some of the specimens are more than 300 years old."[1]

Daniel and Geo Fuchs have entitled their project "Conserving," a title which, of course, has a dual meaning. The objects which they photograph have themselves been conserved and preserved, but by the act of photography, they have created another layer of documentation and preservation. They have also created another arena for these normally hidden objects, taken them into the surroundings of serious art and documentary photography, where they become available to a gaze quite different from that of the scientist or the mortician. "Conserving" illustrates photography's capability to create a new audience for subjects and ideas which have hitherto been made obscure.

In the late Eighties and early Nineties, the subject of death became of great interest to a number of European and American photographic artists. Rudolf Schäfer's portraits of dead people (taken in an East German mortuary in 1987) attracted immediate attention, as did Andres Serrano's powerful series taken in the morgues of New York City. English color documentarist Nick Waplington, known in the early Nineties for his raucous commentary on the lives of two families living in a Nottingham housing estate, subsequently produced a series of constructed photographs dealing with unnatural death and suicide. The reasons for this new interest in the photography of death remain unclear, perhaps too multiple to be assembled as a single coherent theory. Artists using photography certainly developed a strong interest in *objets morts* from the Eighties onwards and had begun to use "real life" situations in a way which reflected a highly subjective and directed view of the real world. The photography of death, as seen in the mortuary or the museum, played an important role in this new way of using photography. The dead were of our world yet outside it, available to the artist/photographer as still and silent subjects in a way which was unique and inimitable.

This new photography of the dead challenged the status quo and thus fit exactly into the postmodernist methodology of the Eighties and Nineties. The parameters of photography were shifting as artists began to use photography not to represent, but to explore representation itself. Cindy Sherman constructed a theater of disguise and fantasy to explore ideas around female identity. Martin Parr used bright color photography to make acerbic documentaries about street life in Britain and turned the idea of classic reporting upside down. Thomas Ruff made startling revisions within the genre of photographic portraiture. Boundaries around photography were breaking down, and the growing importance of photographic theory (as opposed to what was increasingly seen as the dull, process-fetishising process of photographic history) alerted a new audience to the intellectual and political dimensions of the photographic image. Increasingly, artists began to use photography as a convenient and accessible tool to express ideas which, in the Sixties and Seventies, might well have been conveyed through painting, sculpture and illustration.

It is within this postmodernist history that Daniel and Geo Fuchs have worked, taking as a given their right to work within areas which might once have been seen as taboo and insisting that their work, far from being a documentary, is a highly subjective process involving personal history and a deep emotional involvement. The world, they seem to insist, is there to be looked at and studied, and it has become available to a multitude of gazes. Their continuing wish to work on what they describe as the borderline of experience, their personal commitment to the subject, their distancing of themselves from the sensationalist or the voyeuristic: all these have resulted in a highly demanding and consistent practice which demands attention and understanding from the viewer. Their work, throughout the "Conserving" series, has a very particular aesthetic which makes beautiful objects out of elements of our world which might otherwise be thought of as macabre or distasteful. The dead creatures and people who they photograph, immortalized by the preserving fluid in which they are suspended, become iconic, almost reverential objects. Long dead yet still a part of our contemporary world, they have perhaps achieved a mystical status, hovering between death and life in a peculiar state of limbo.

"Conserving Fish" is perhaps the most ethereal and abstract of the three "Conserving" series. That the fish are floating in liquid allows them to inhabit what one might call a more "natural" state. We see them through glass, as we would through the windows of an aquarium, and it is only a certain muteness of color, a lack of vivacity, which gives clues to their suspended animation. Here and there, we see a glimpse of a glass bottle or an identifying label, leading us to the realization that this is a study in taxonomy rather than of a captive, but nevertheless living wild creature. Although ostensibly the least "shocking" of the three series in "Conserving," "Conserving Fish" has a distinct melancholy. Scales which should shine are dull and without luster; lively eyes are lifeless; swift-moving schools are motionless, stilled forever by death and time.

"Conserving Animals" is an altogether different experience. Even at first glance, we are aware that this is a study in mortality: claws and paws are detached from bodies and float in preservative liquid. Unseeing eyes stare through the fluid and we cannot connect with their gaze. Sometimes there is a gesture which seems to indicate life, a rubbing of the eyes, an extending of arms and legs, an embrace, but these are merely accidents, moments of vigor defied by death. These animals sleep an uneasy sleep, marooned in liquid, out of their natural element, abandoned to the gazes of the curious or the expert. They have become subjects, objects, without a sense of self.

When we look at our own kind (in the "Conserving Humans" series), the denial of privacy is acute. Dead faces are not without expression or personality: a young man, a baby, an old woman, even an embryo, were once part of our tribe, experienced thoughts and emotions, made decisions, forged relationships, existed within our society. Yet here, becoming specimens for science or exhibits in a museum, they are excluded from tribal rituals and rites of passage. Like Faustus, they have become grotesquely immortal. "Conserving Humans" is the culmination of the "Conserving" series, and takes us from the abstract to the real. "Conserving Humans" will inevitably be seen as a study of the grotesque, yet Daniel and Geo Fuchs have little interest in *grotesqueries*. They gaze on the deformed, the maimed, the split-open skulls, the skeletons and the tattooed skin almost as witnesses, recorders of the time we live in. Writing of "Conserving Humans," they have observed that the making of this series proved to be "a great challenge. We wanted to depict the people in a way that preserves their beauty, that renders their life visible and makes them seem alive even in death. We wanted the people we photographed to retain a certain dignity"[2]

The work which Daniel and Geo Fuchs produced during the making of "Conserving" led them into a new, wide-ranging and ongoing photographic project about death. The artists have ventured beyond the museum of the scientific institution and entered the strange, quiet world of the mortuary, where nothing escapes their gaze — from the corpse to the instruments of cutting and dissecting. Here, where the human body, though still recognizable as the individual it once was, has become an object, its privacy vanishes. They are concerned with the traces which the body leaves behind — a scrap of flesh, a hair. Their interest in these signs has no morbidity of its own, for when curiosity shows signs of revealing the grotesque and the absurd, the artist's eye is reasserted and the photographs which we see are not graphic or horrific, but available to our imaginations and intellects and thus to a multitude of interpretations. "After our preoccupation with a more abstract form of death, it was a logical step for us to confront real death. Without the protecting glass and liquid, our situation and the pictures which result are both far more direct.... Death and the decay of the human body become very natural. The body seems merely an empty shell."[3] One can sense that their need to know, to have answers, has become all-consuming and deeply personal.

In the three series which make up "Conserving," Daniel and Geo Fuchs have remained true to the rationale which informs their work. When we look at the fish, animals and people who they have photographed, we feel their intimate connection with their subjects. Though the Fuchses look at uncomfortable situations, we can feel that their gaze is concerned, even kindly: "People are fascinated and alarmed by death and dying — the last secrets of life. The knowledge of our own inevitable deaths is always with us. In that sense, life and death go side by side. This is our intention: to show both in our photographs."[4] We can sense their vulnerability, the personal risk which they have taken by choosing to devote their careers to studying this hidden and often disputed territory. They have positioned themselves as antiquarians, explorers in a foreign land, faithfully making records of truths which are uncomfortable, producing photographs which will engender debate, provoke questions and contribute substantially to our understanding of photography's role in documenting our world.

VAL WILLIAMS

1. Quoted from a statement by the artists, 1999.
2. Ibid.
3. Quoted from an interview with Francesca Alfano Miglietti in "Virus Magazine", Milan, Italy.
4. Quoted from a statement by the artists, 1999.

CONSERVING FISH

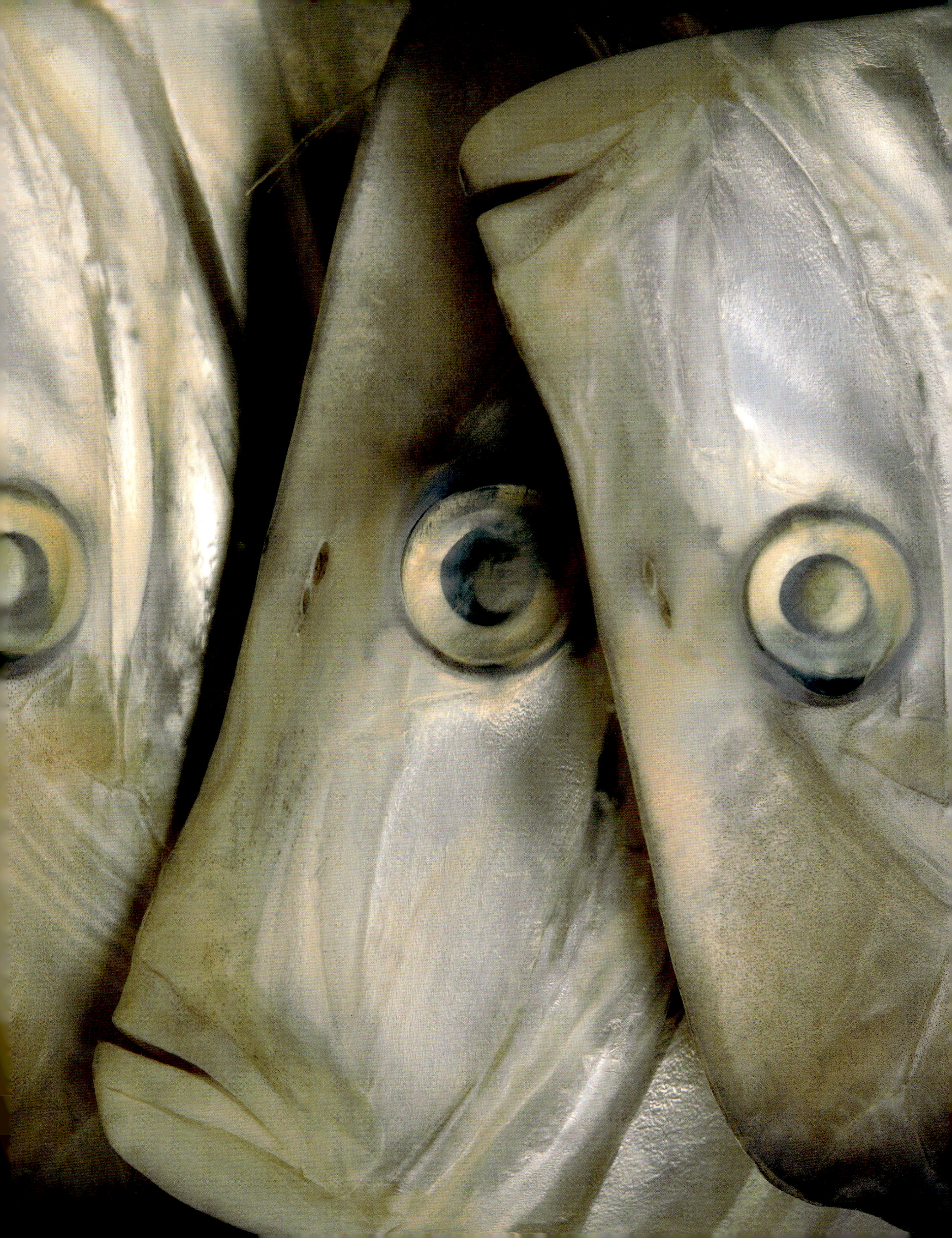

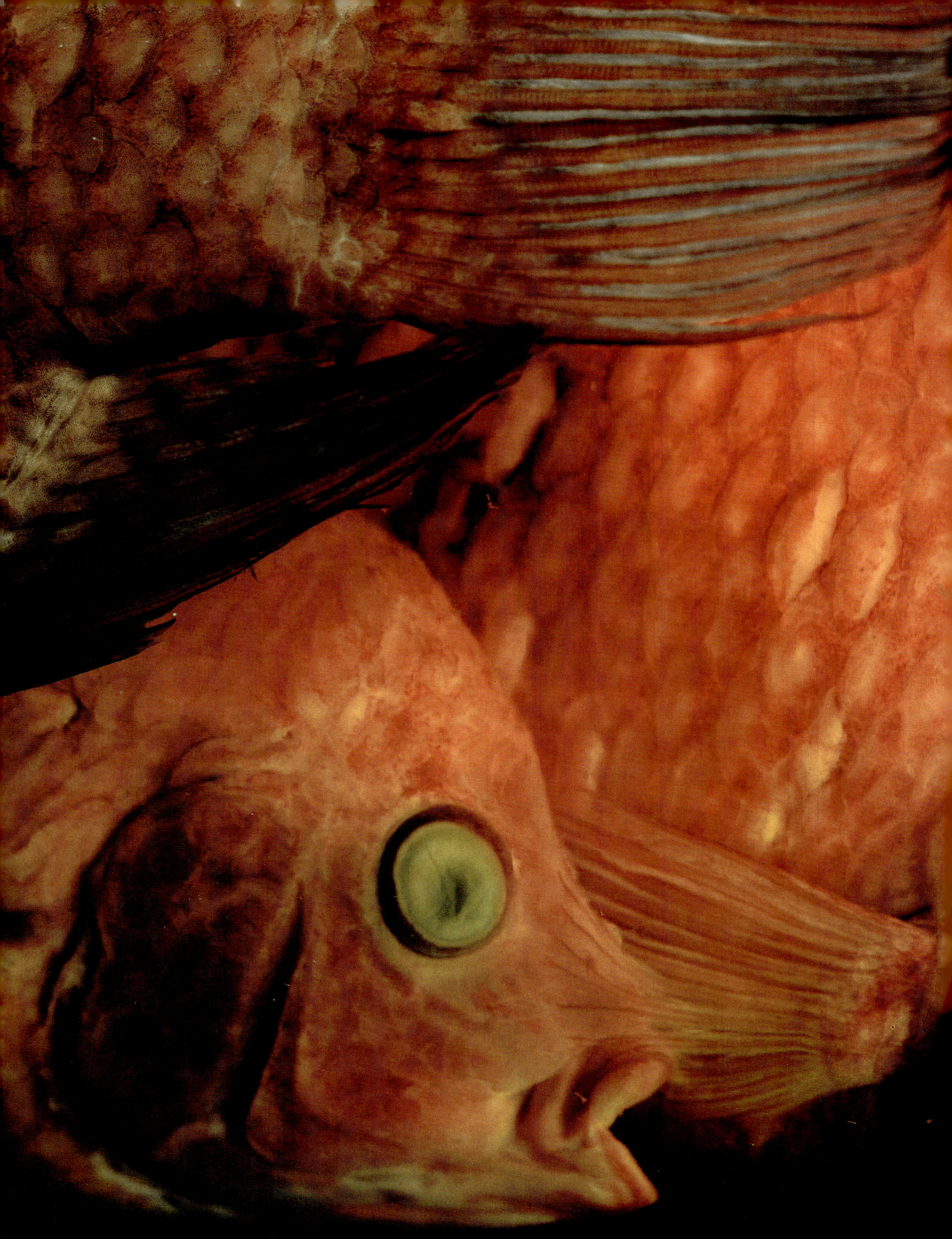

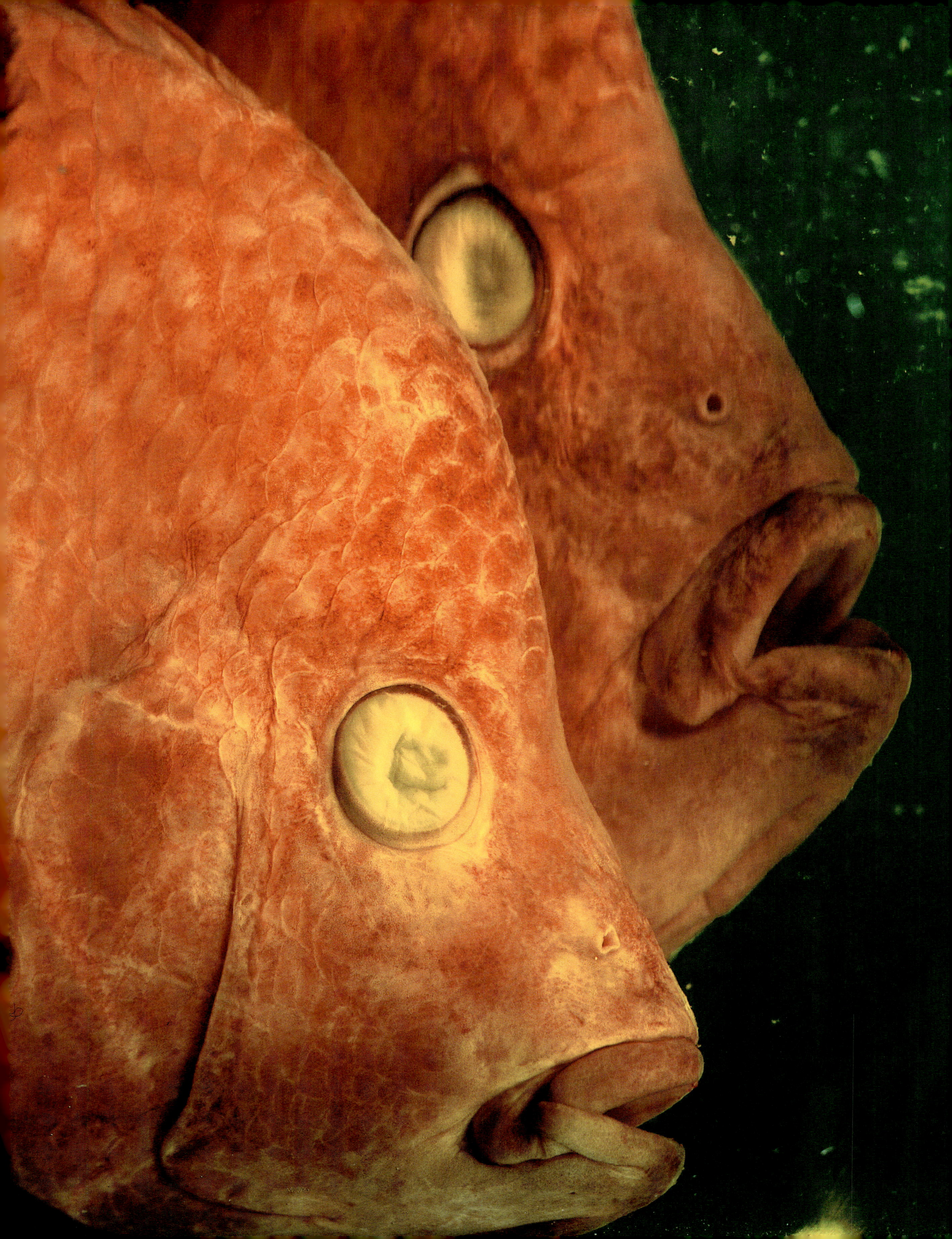

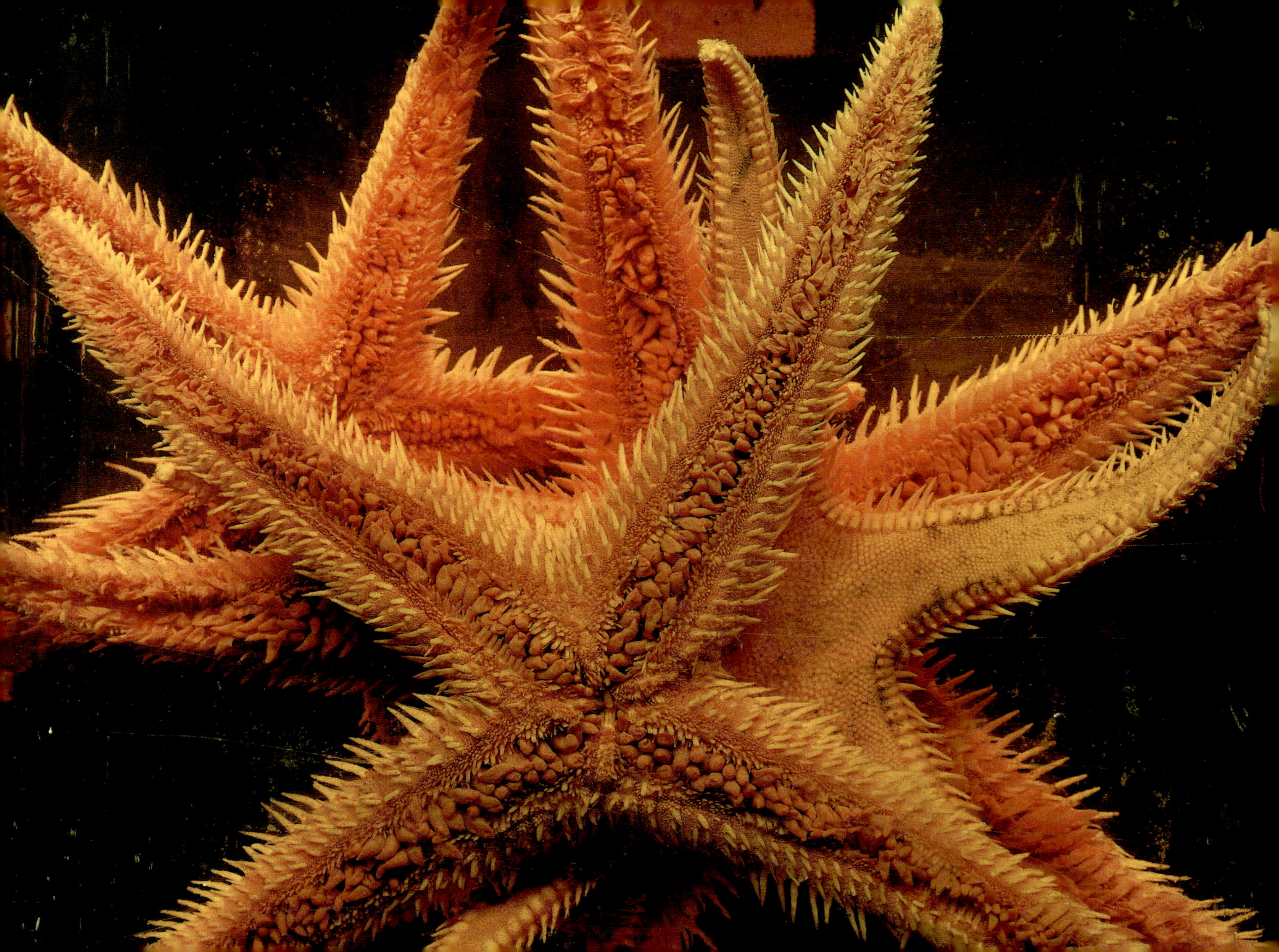

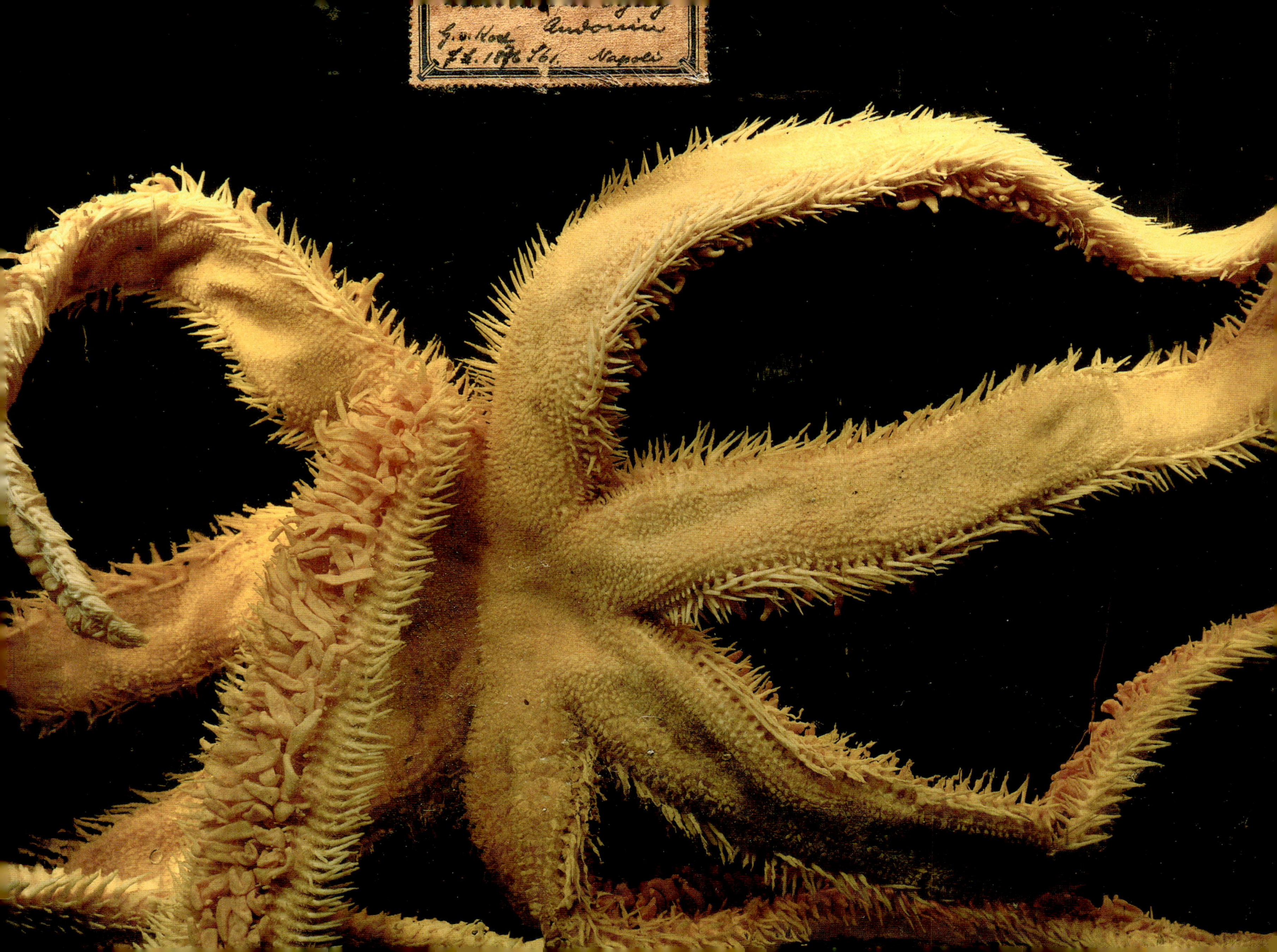
Napoli

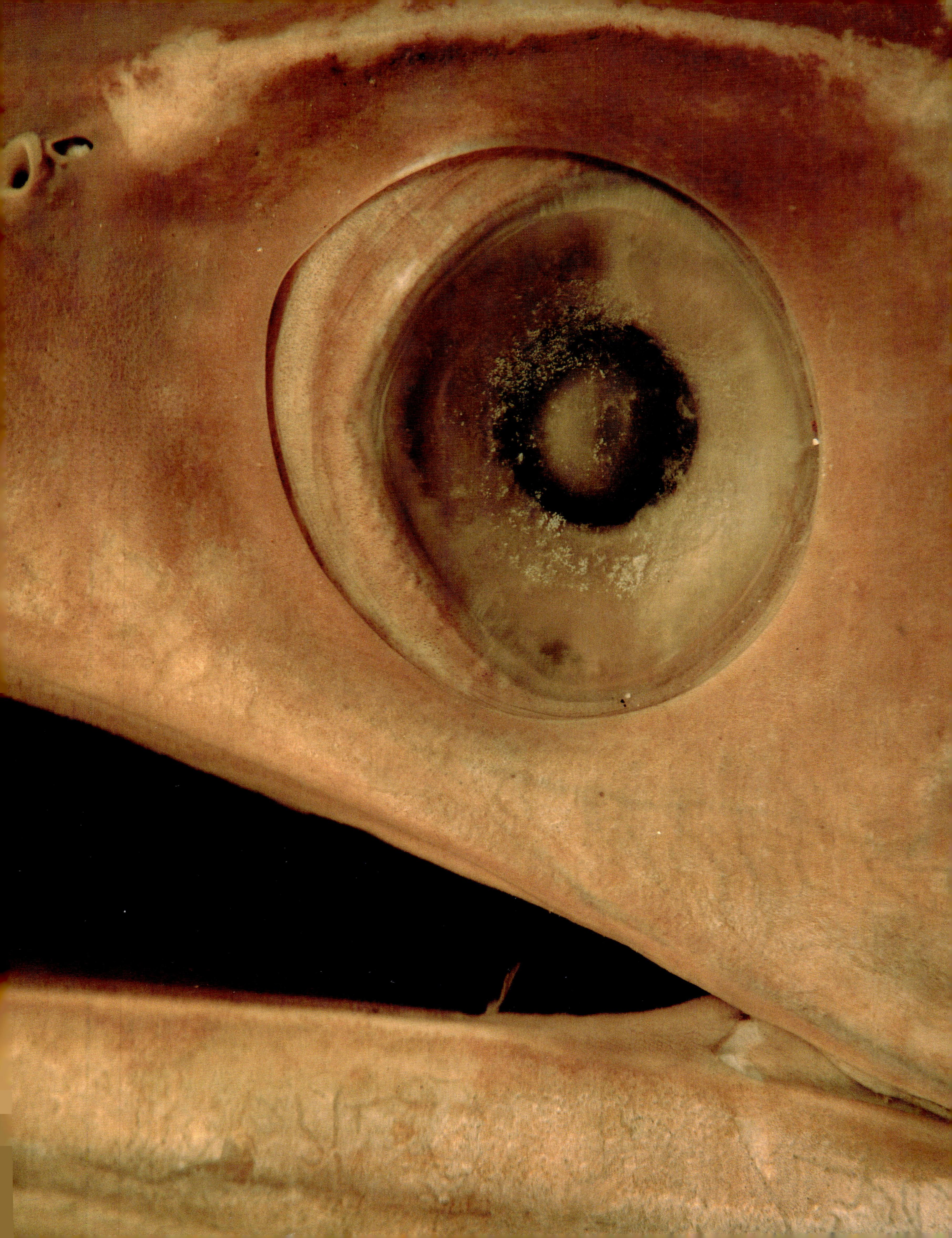

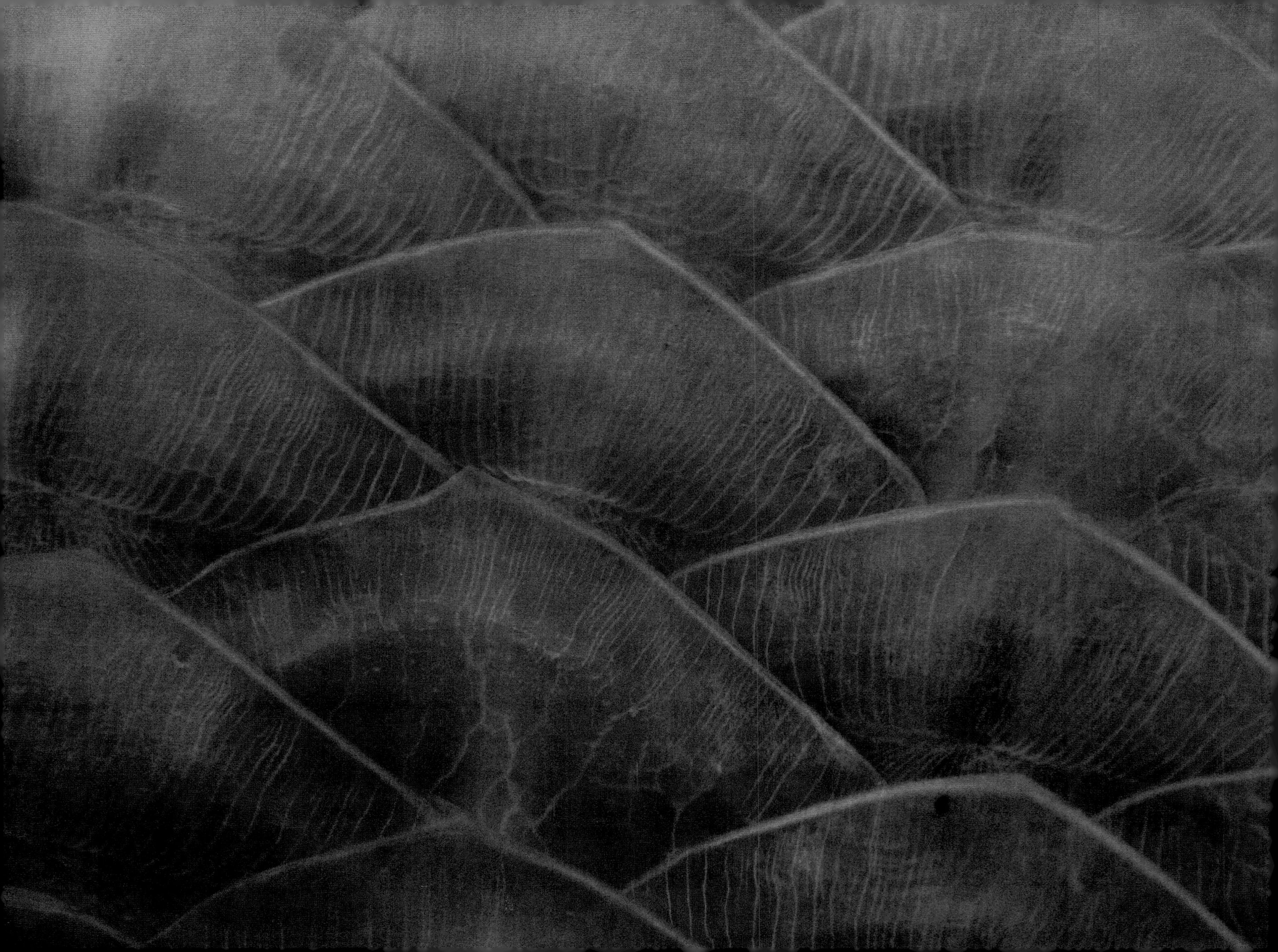

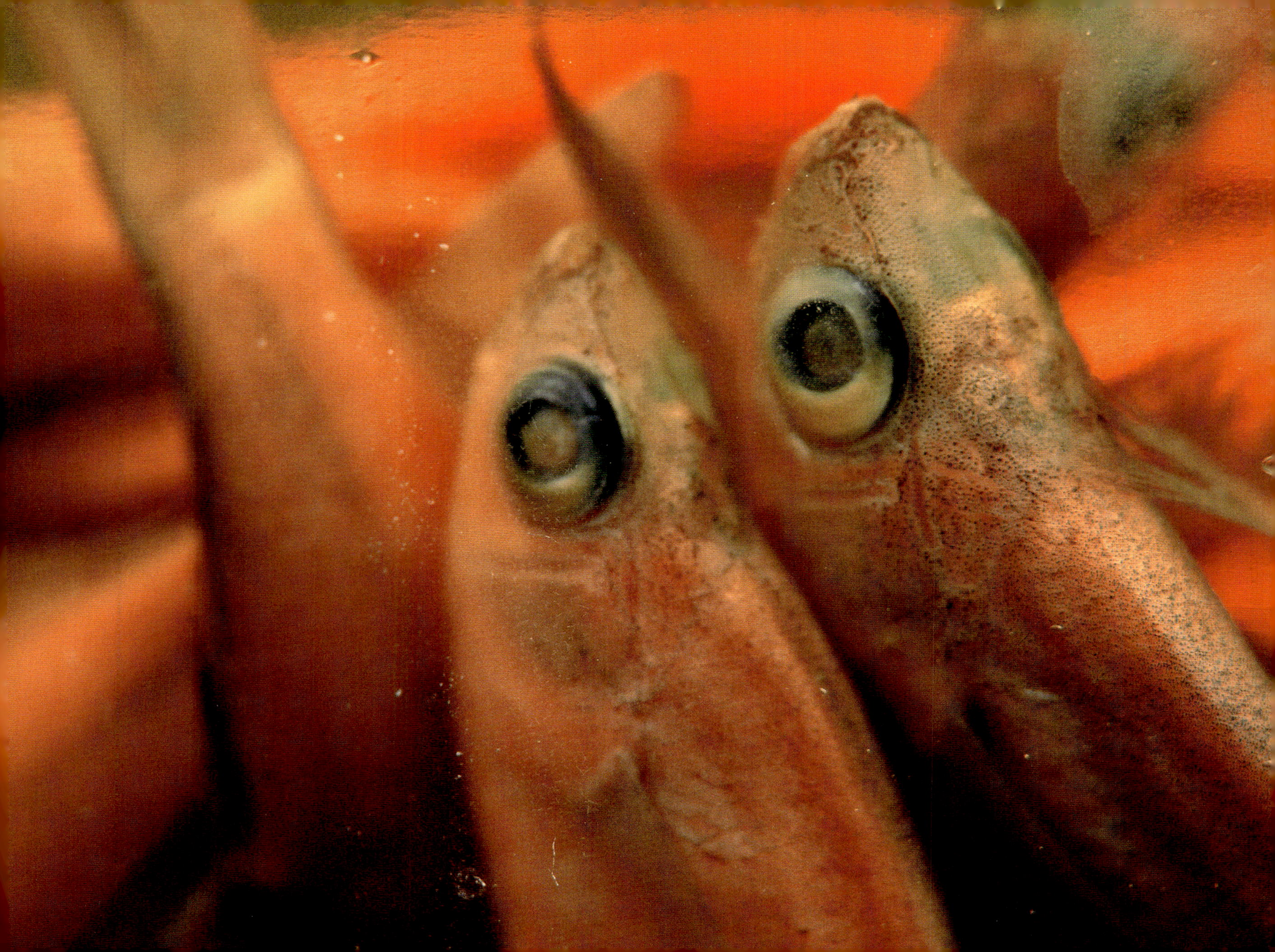

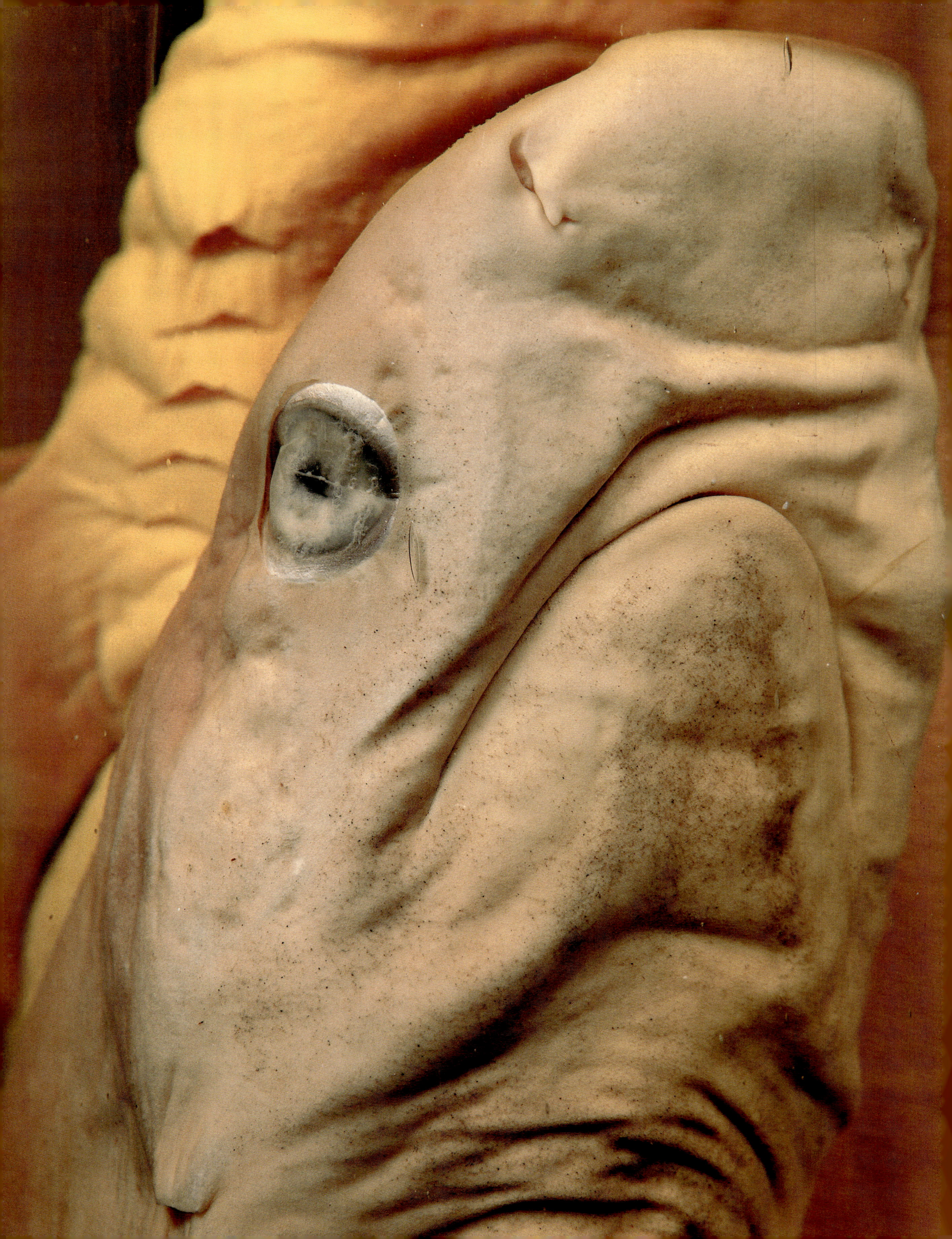

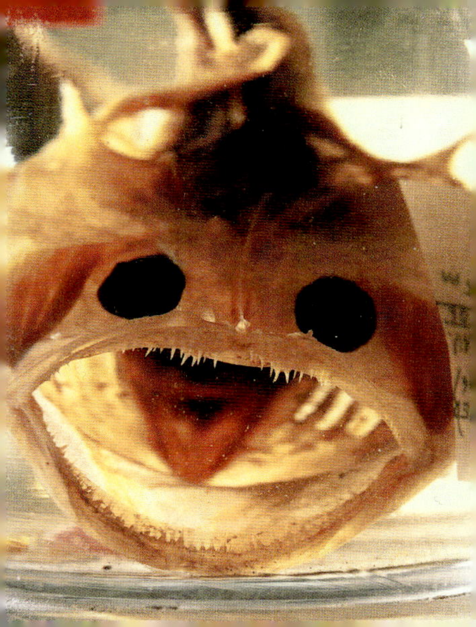

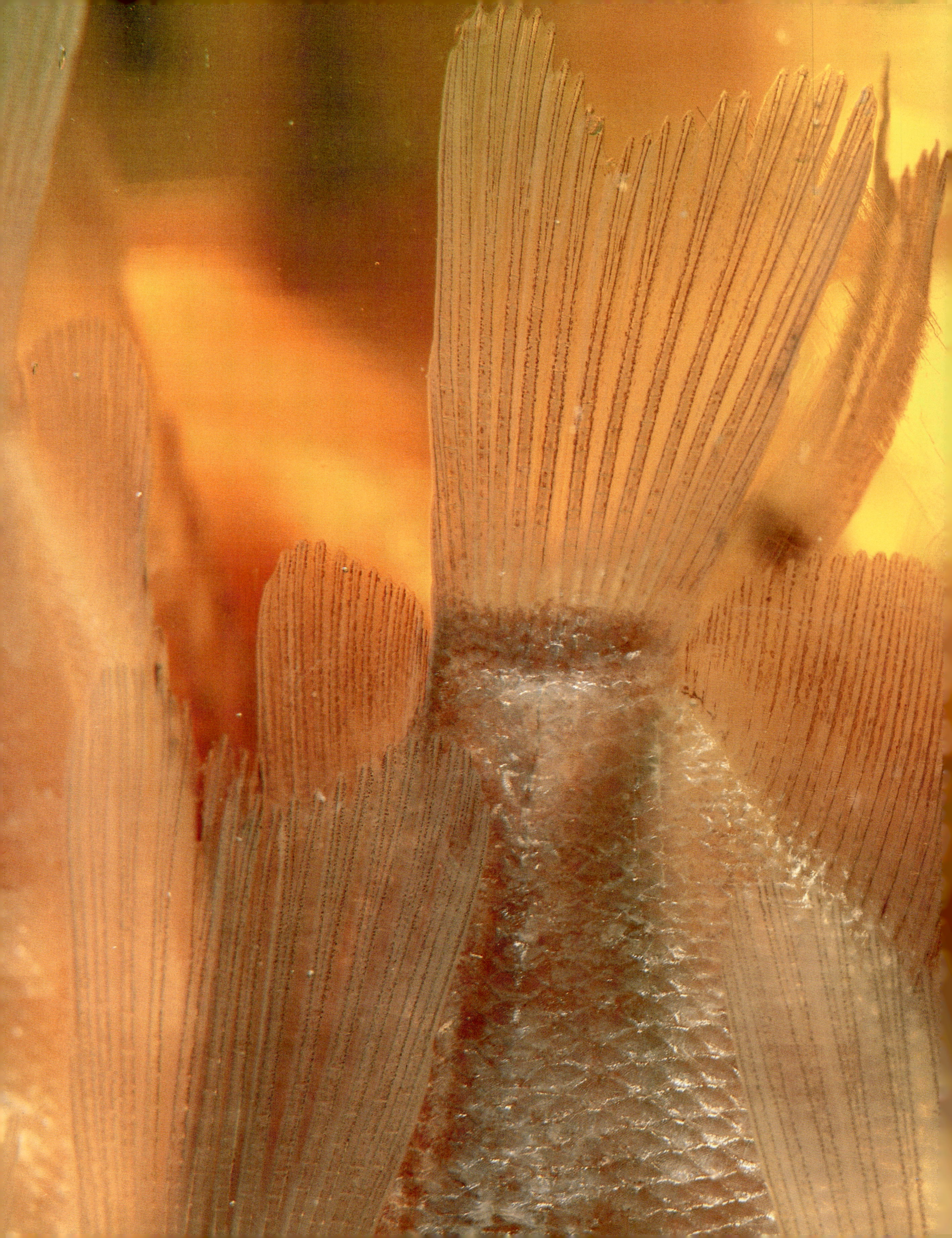

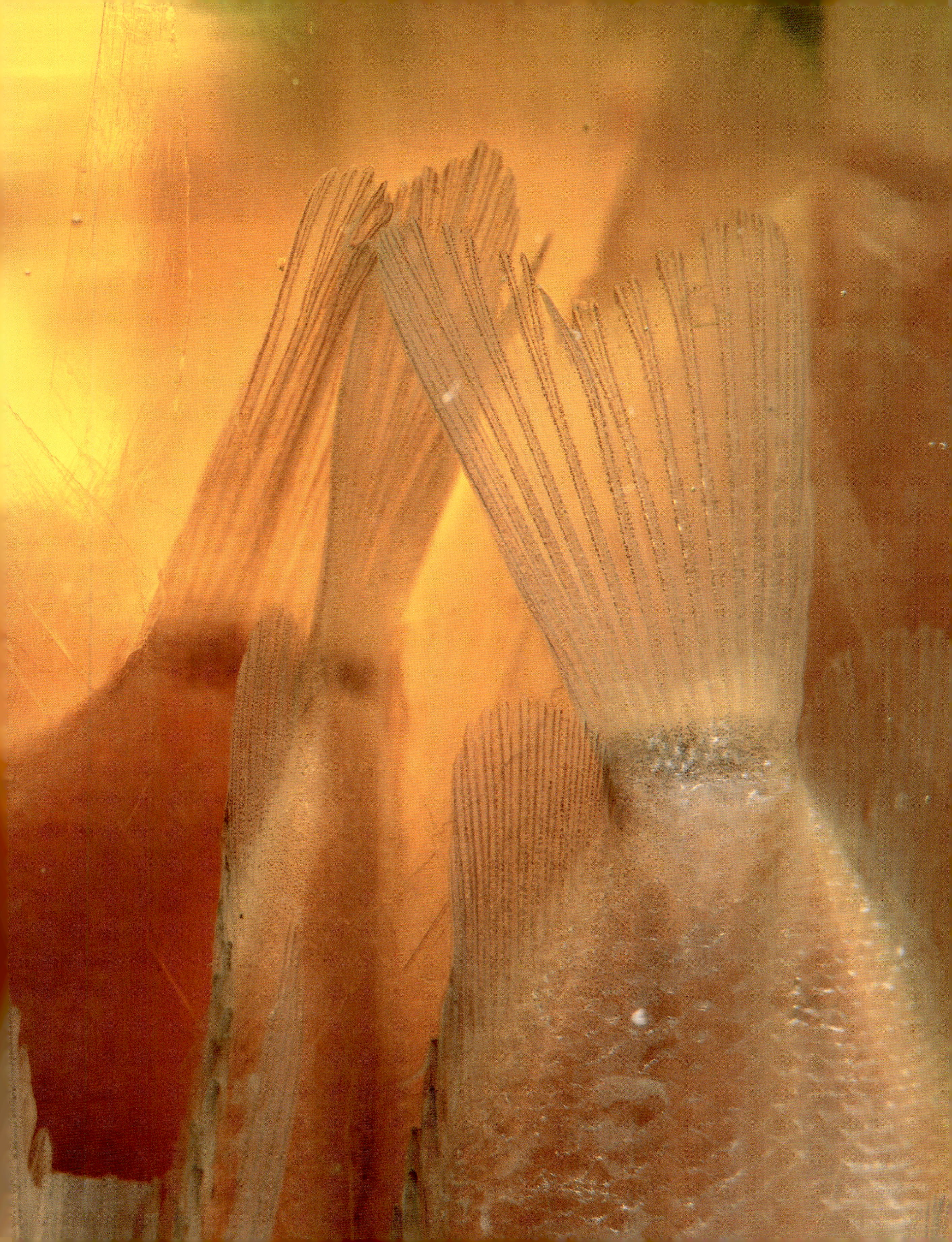

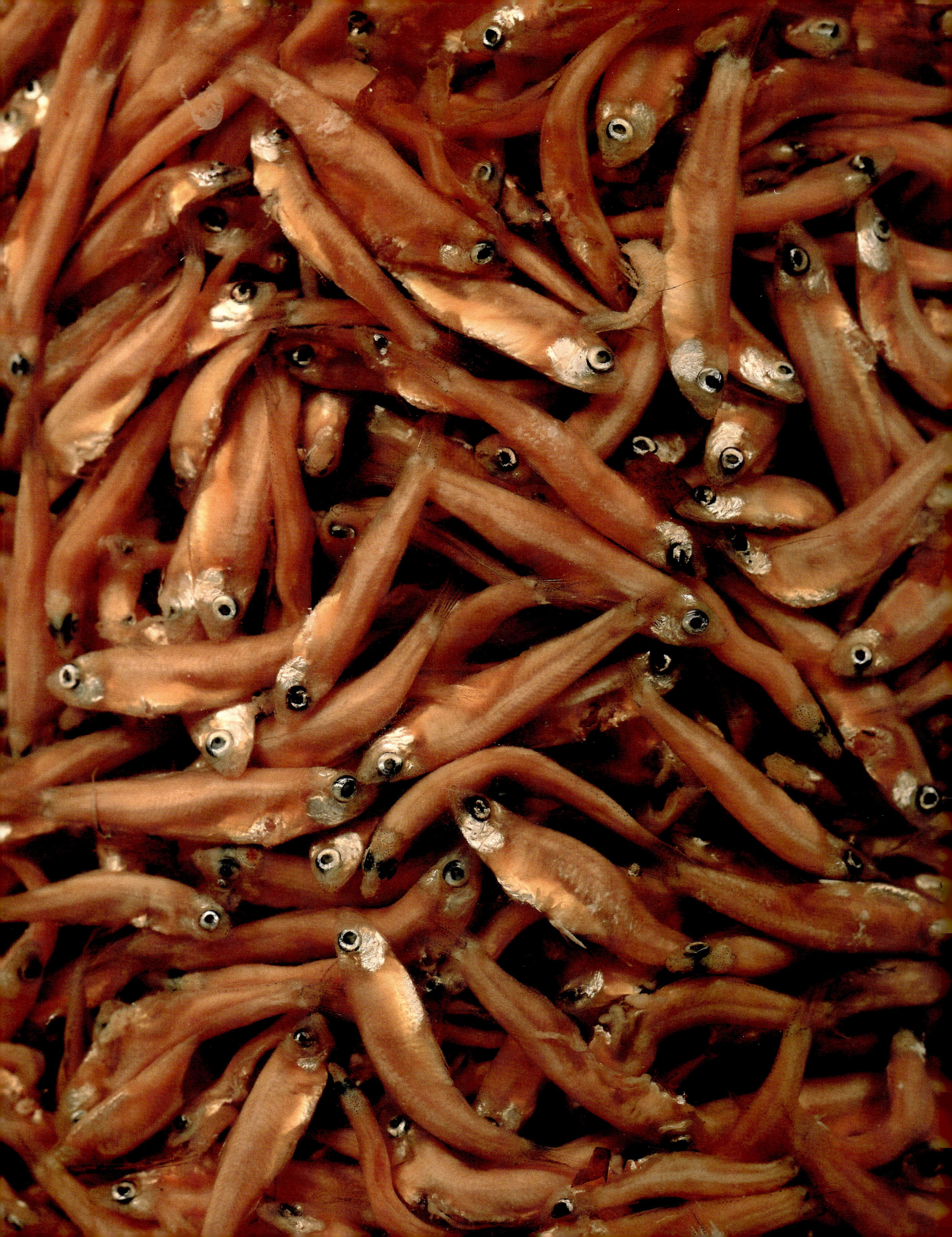

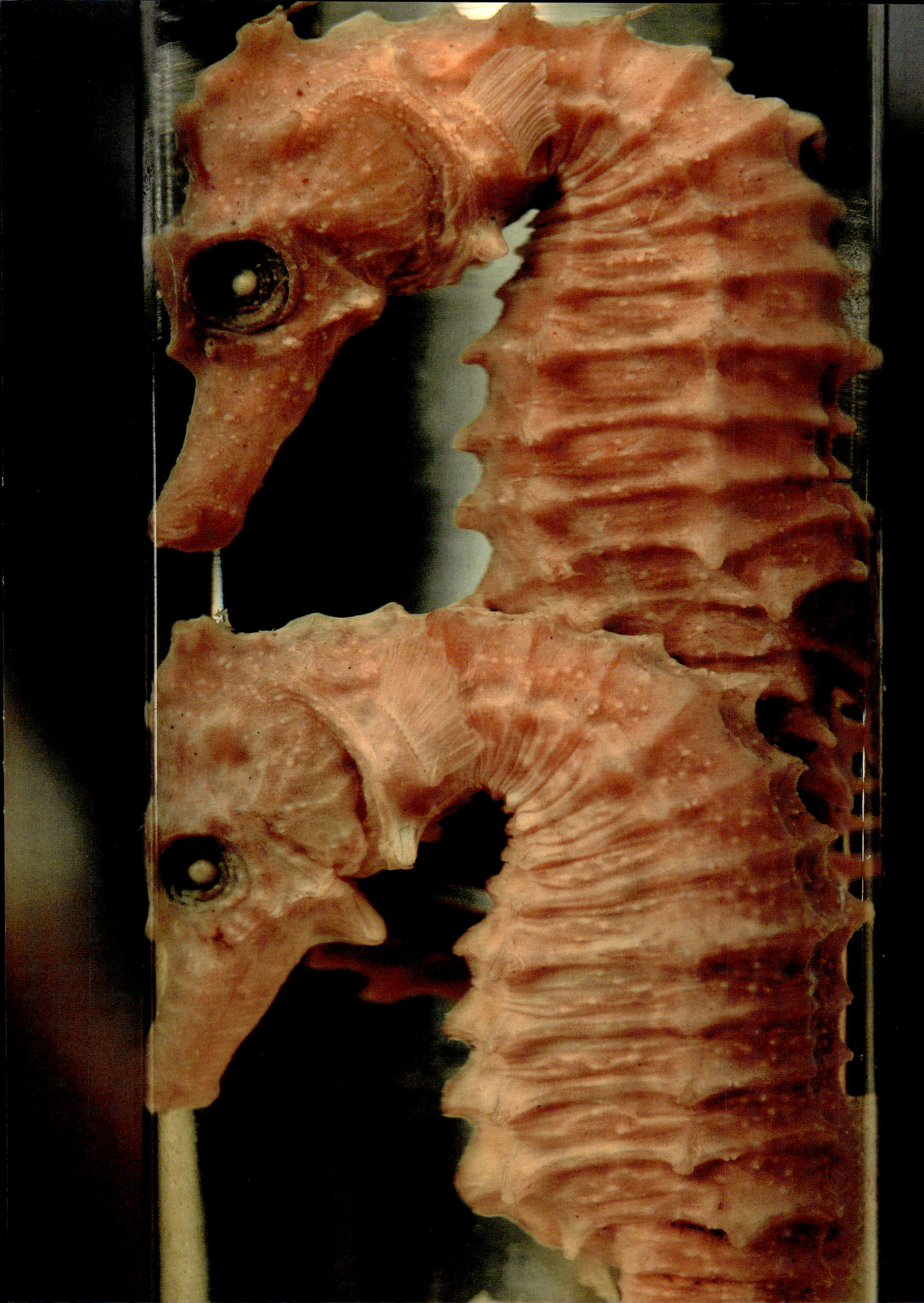

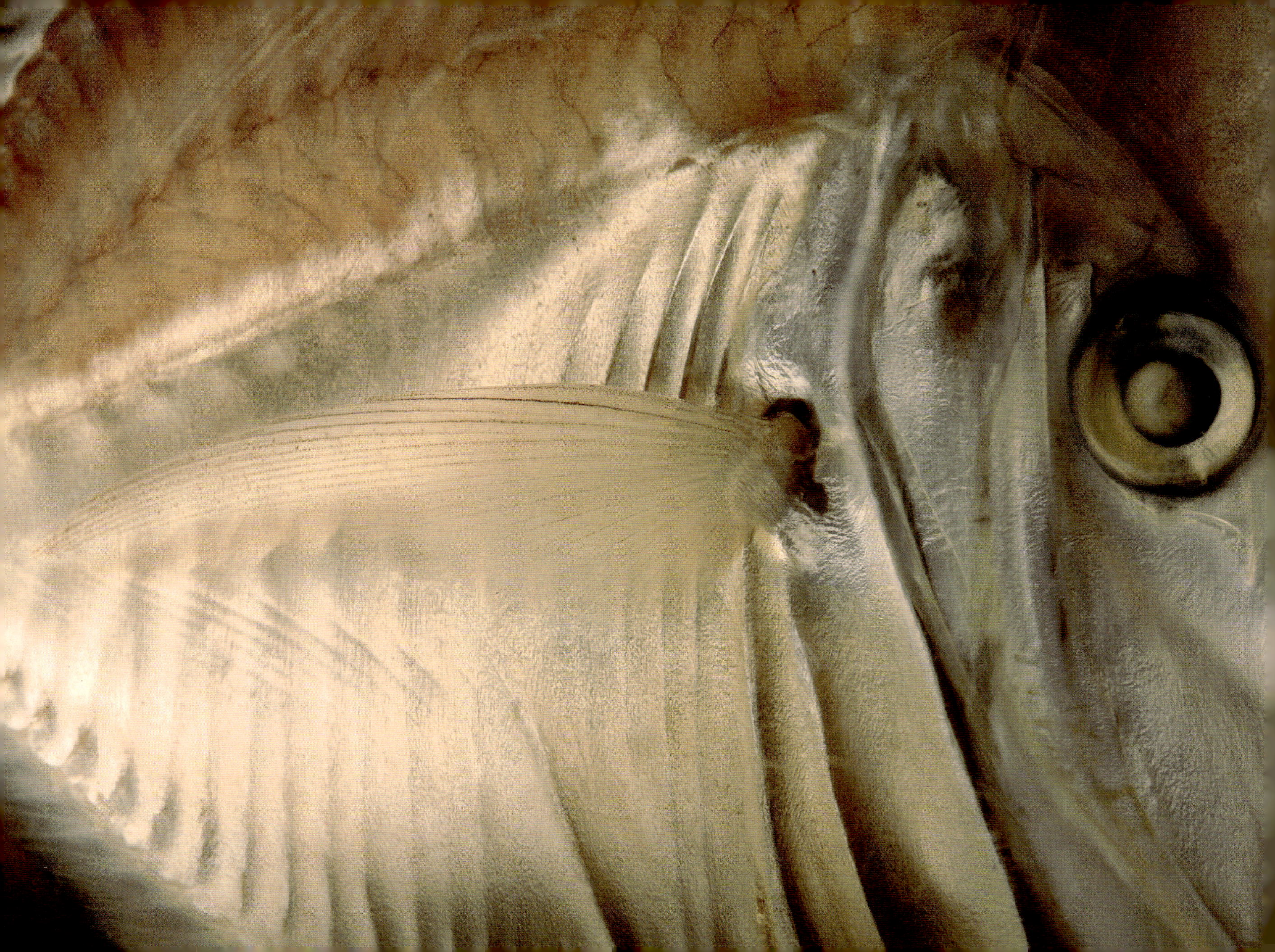

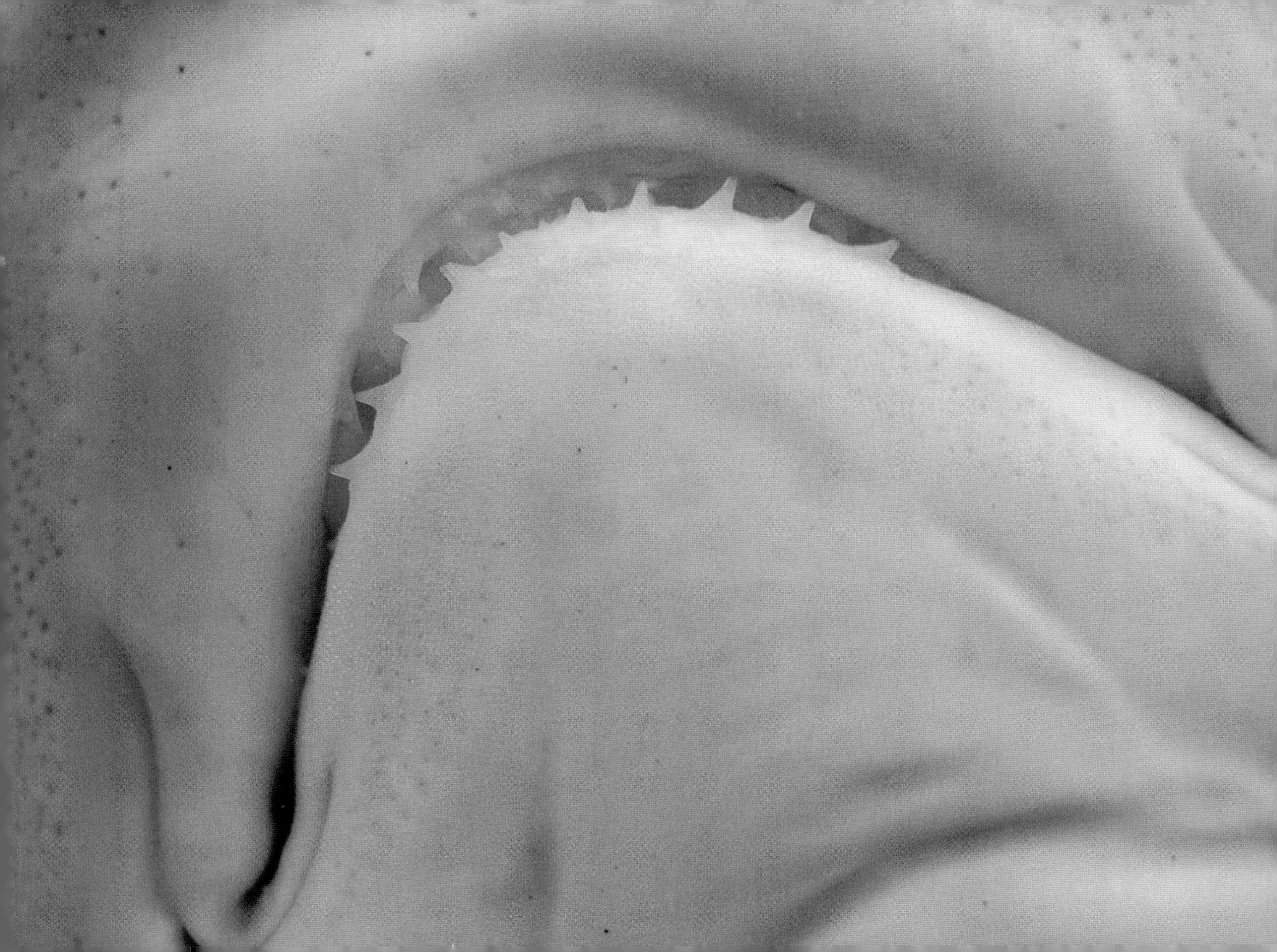

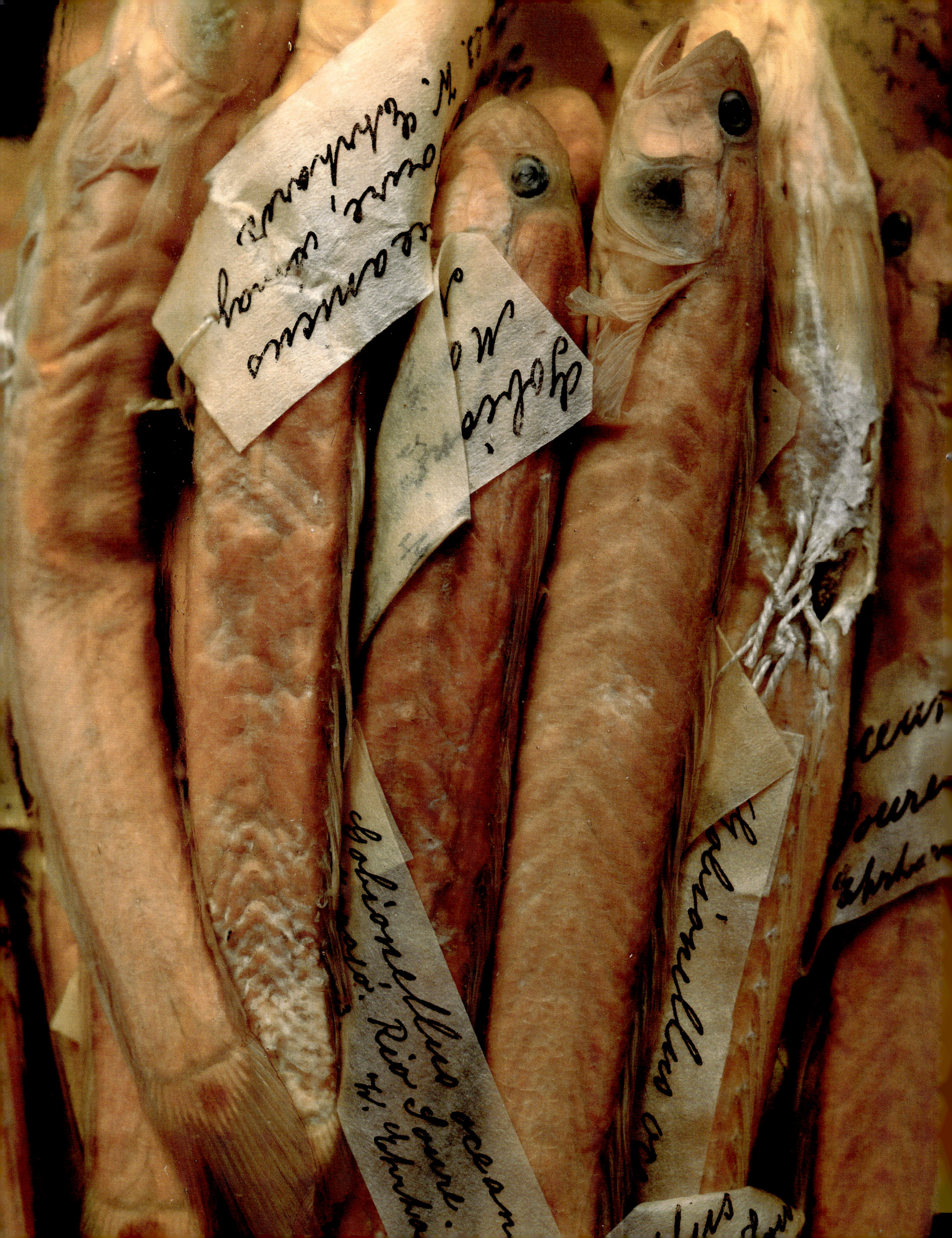

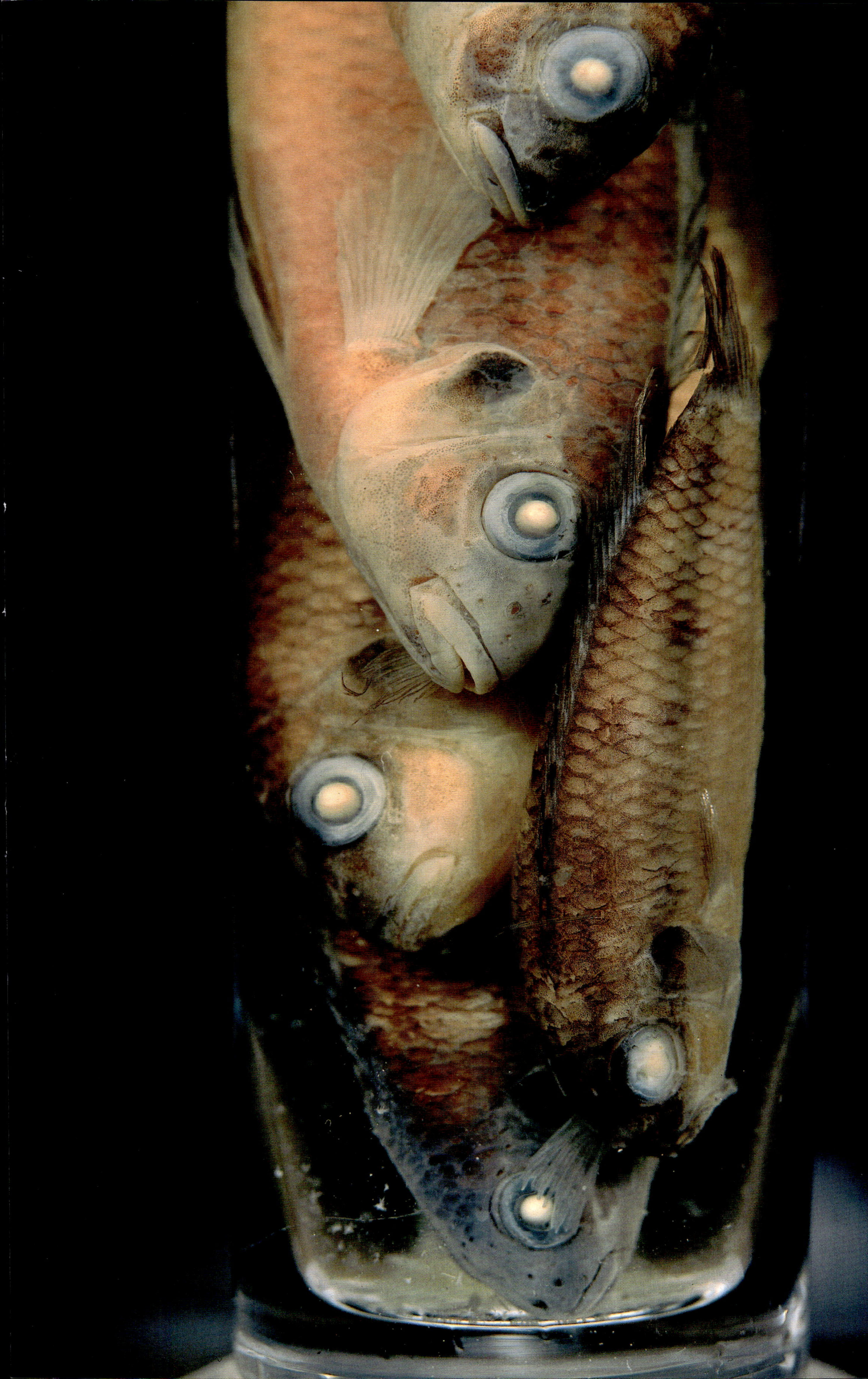

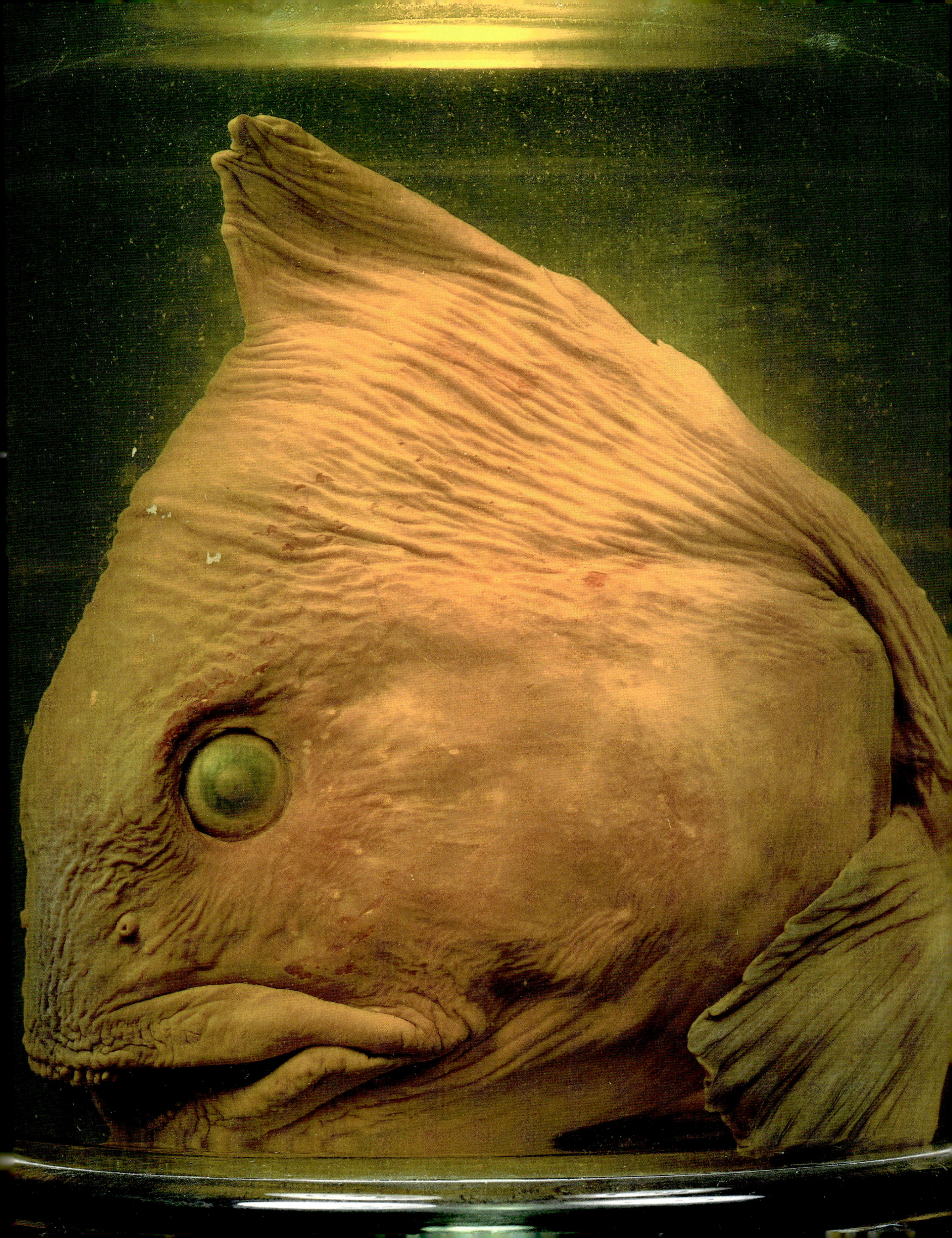

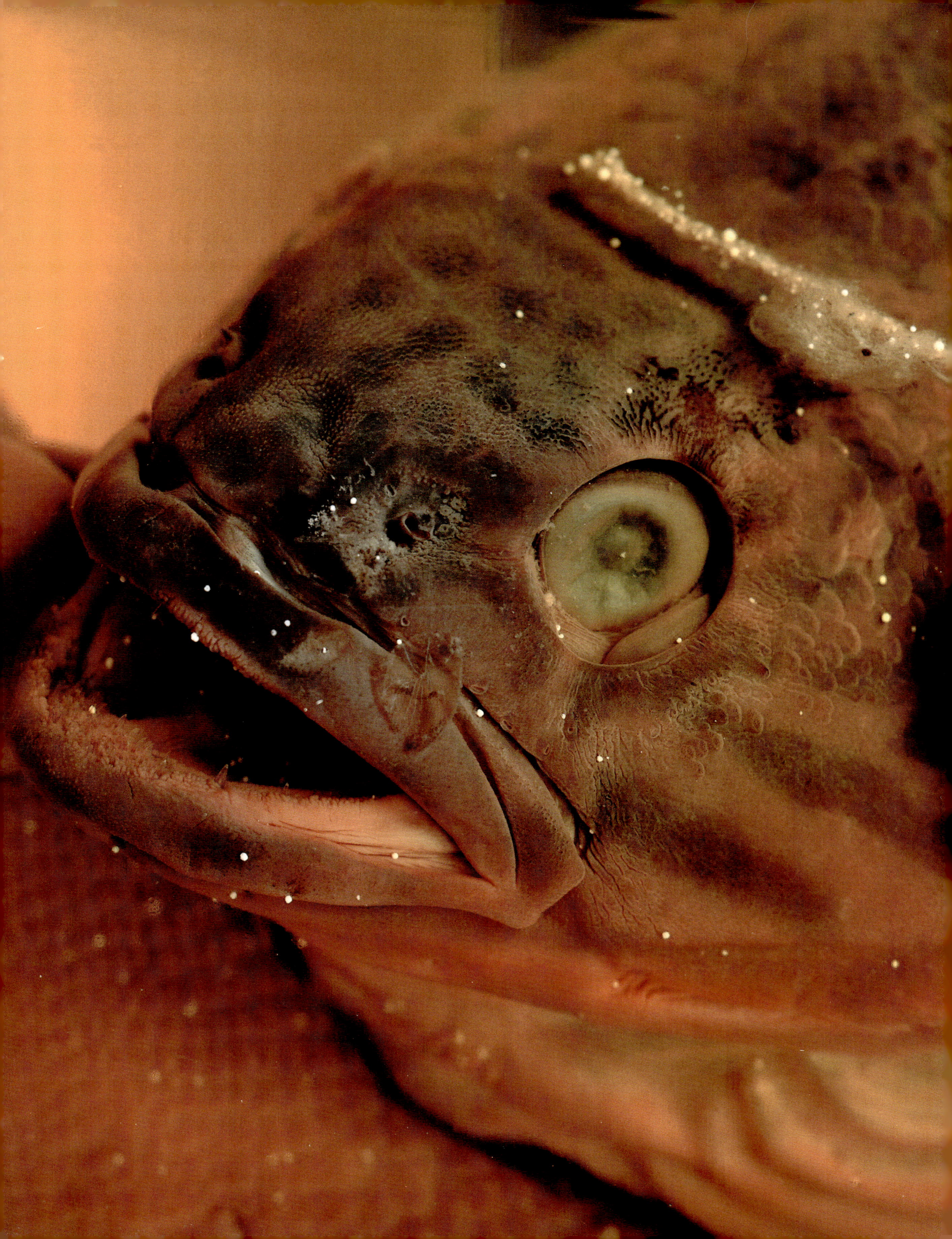

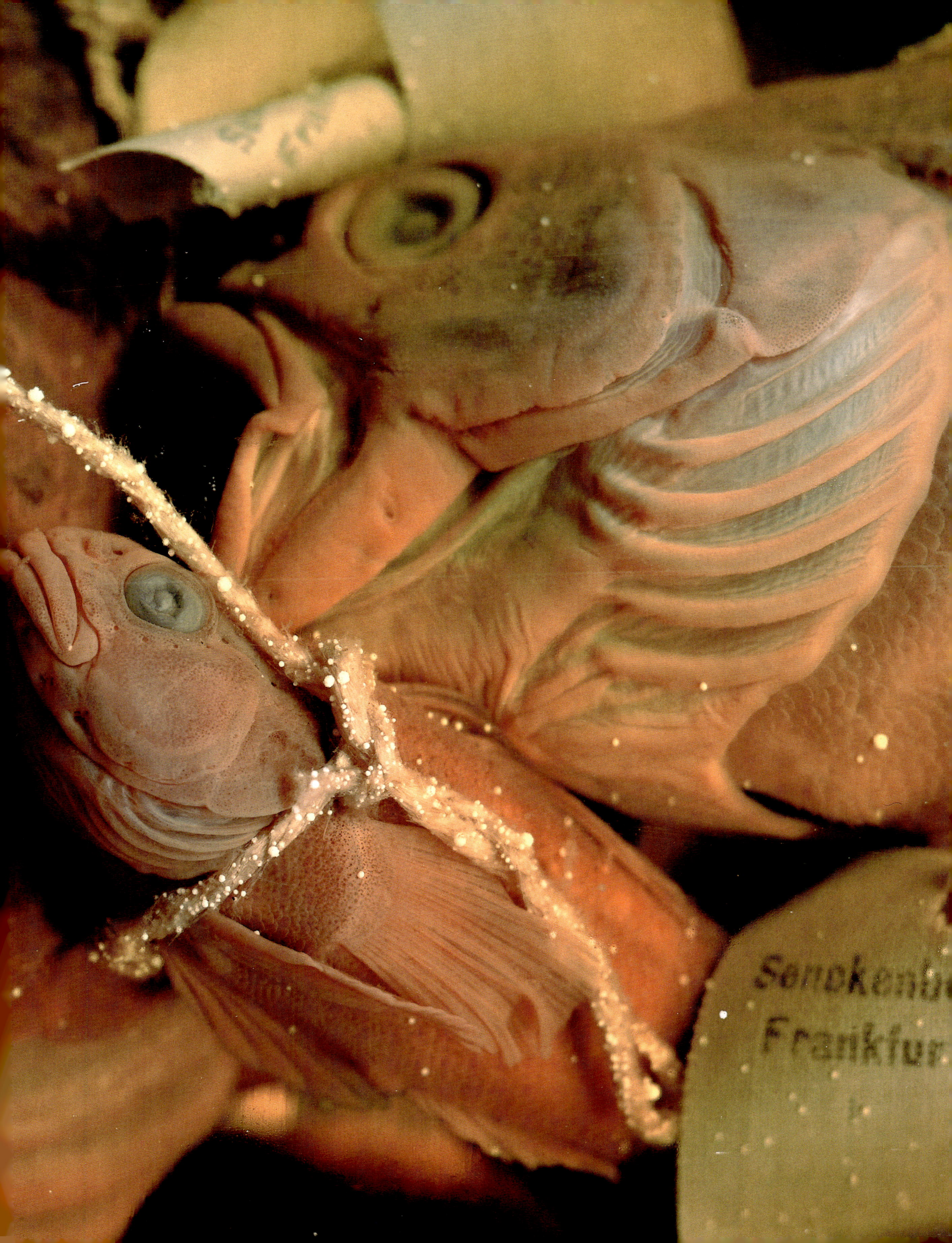

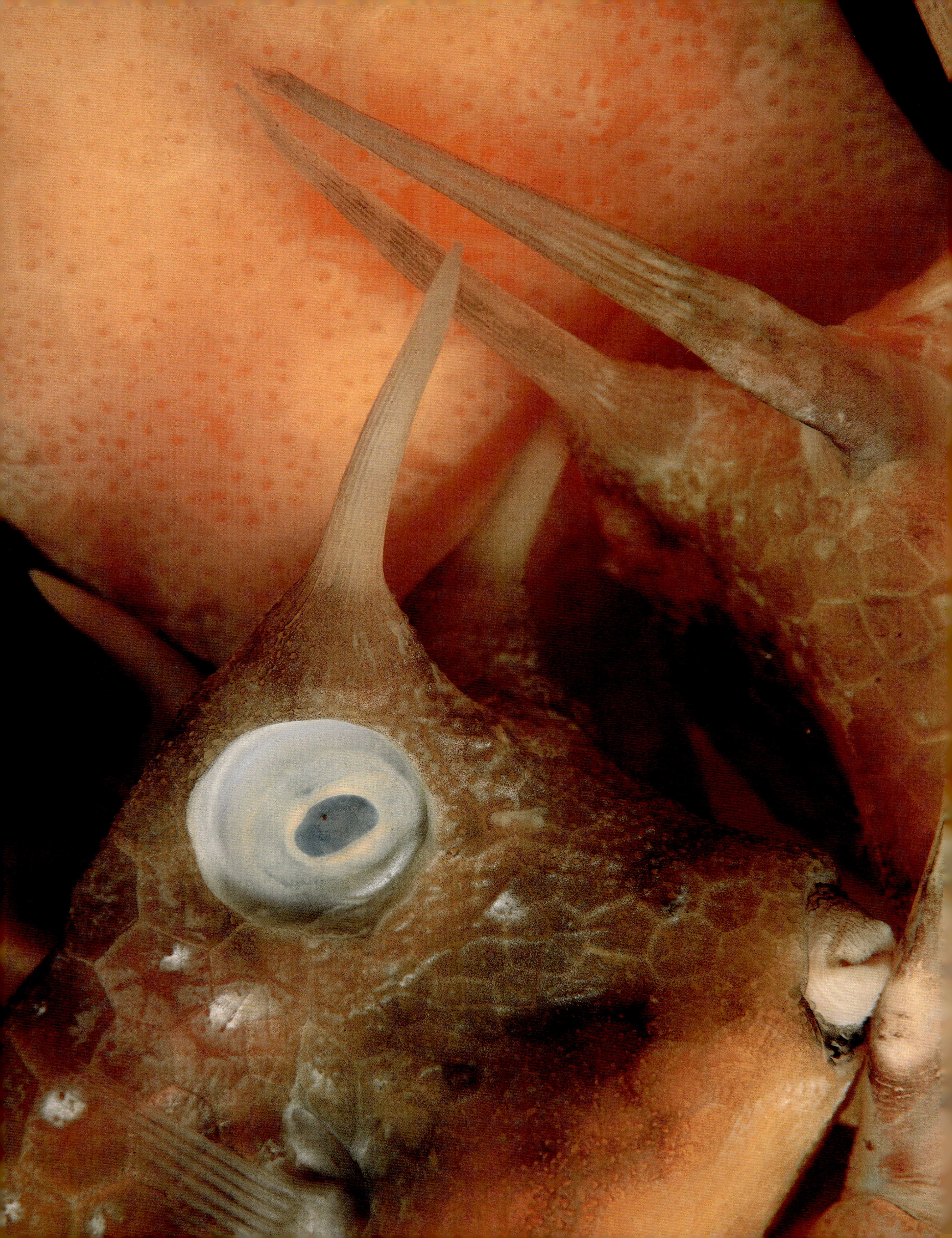

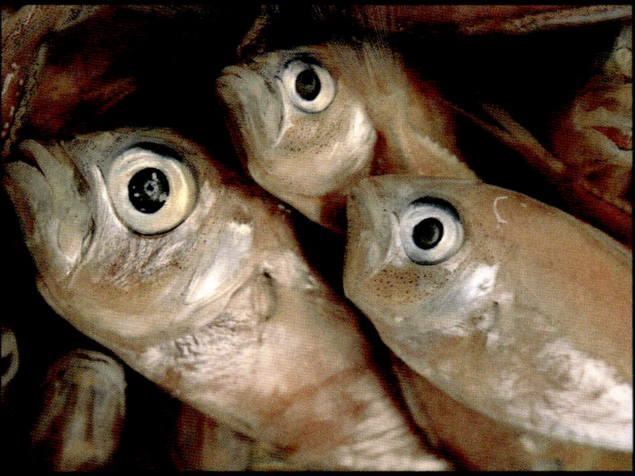

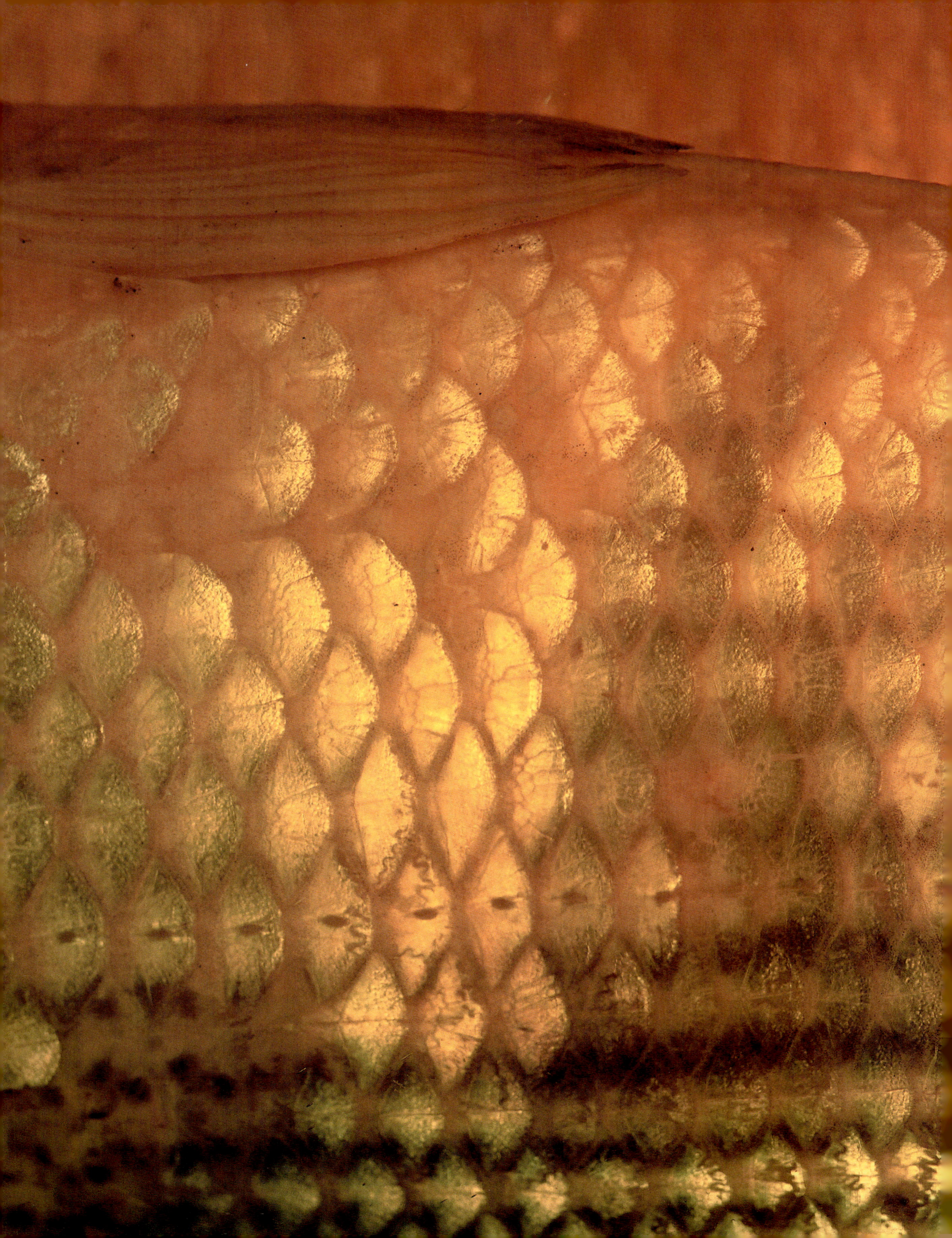

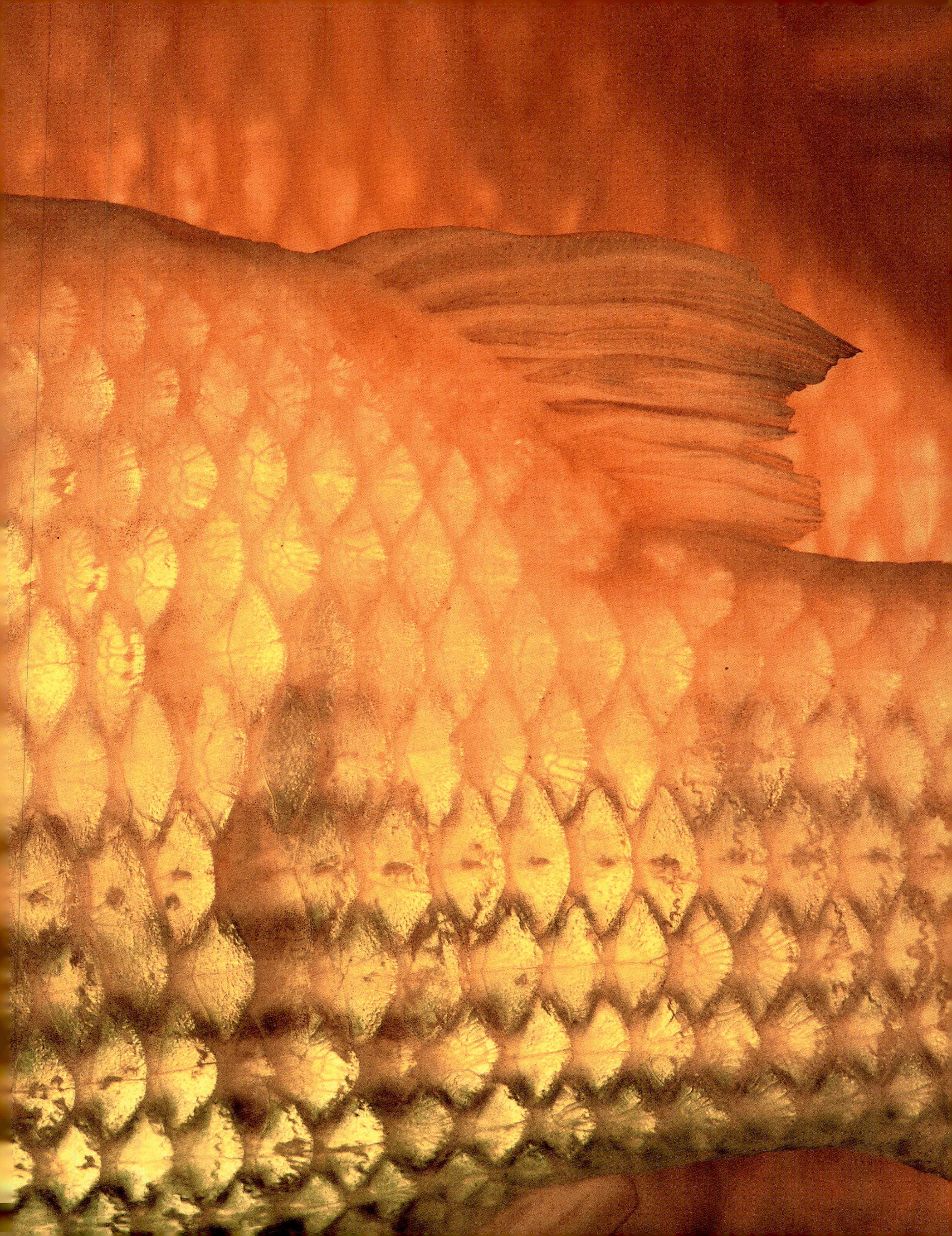

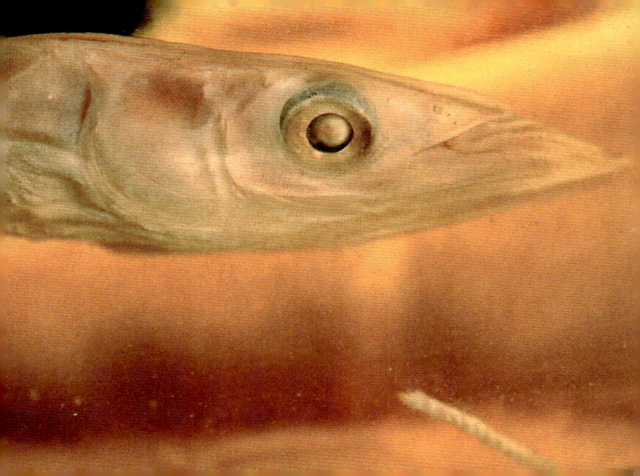

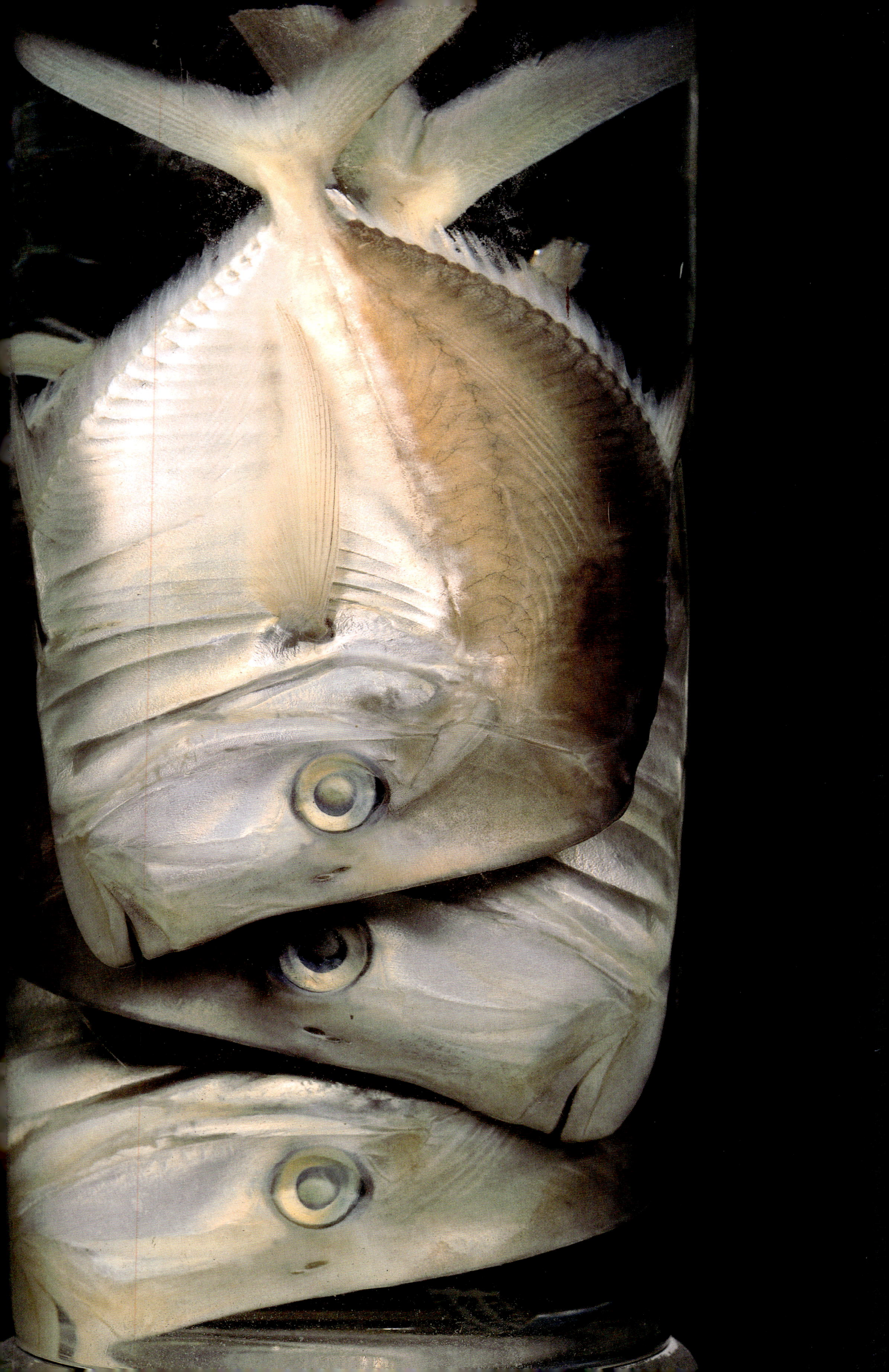

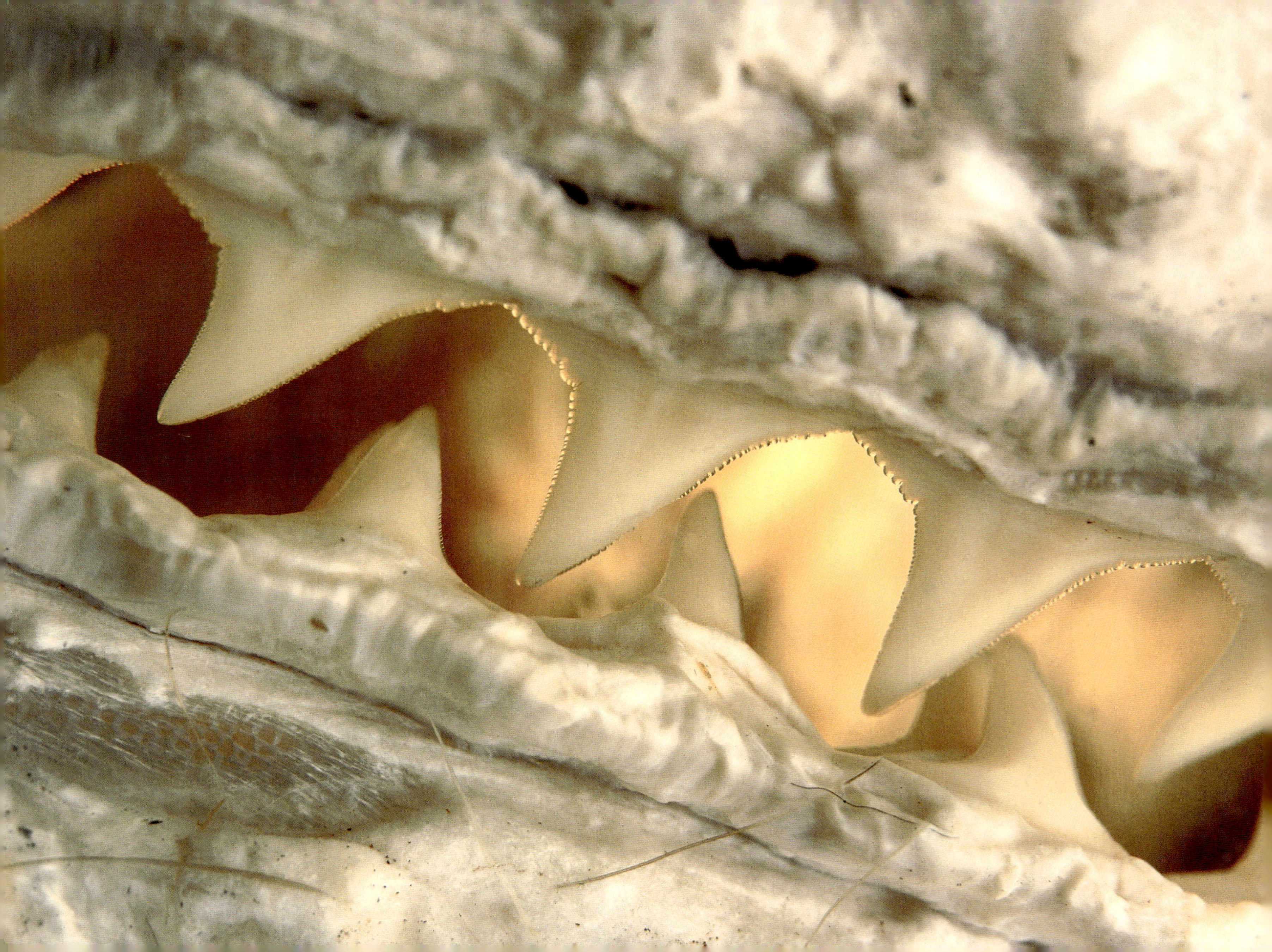

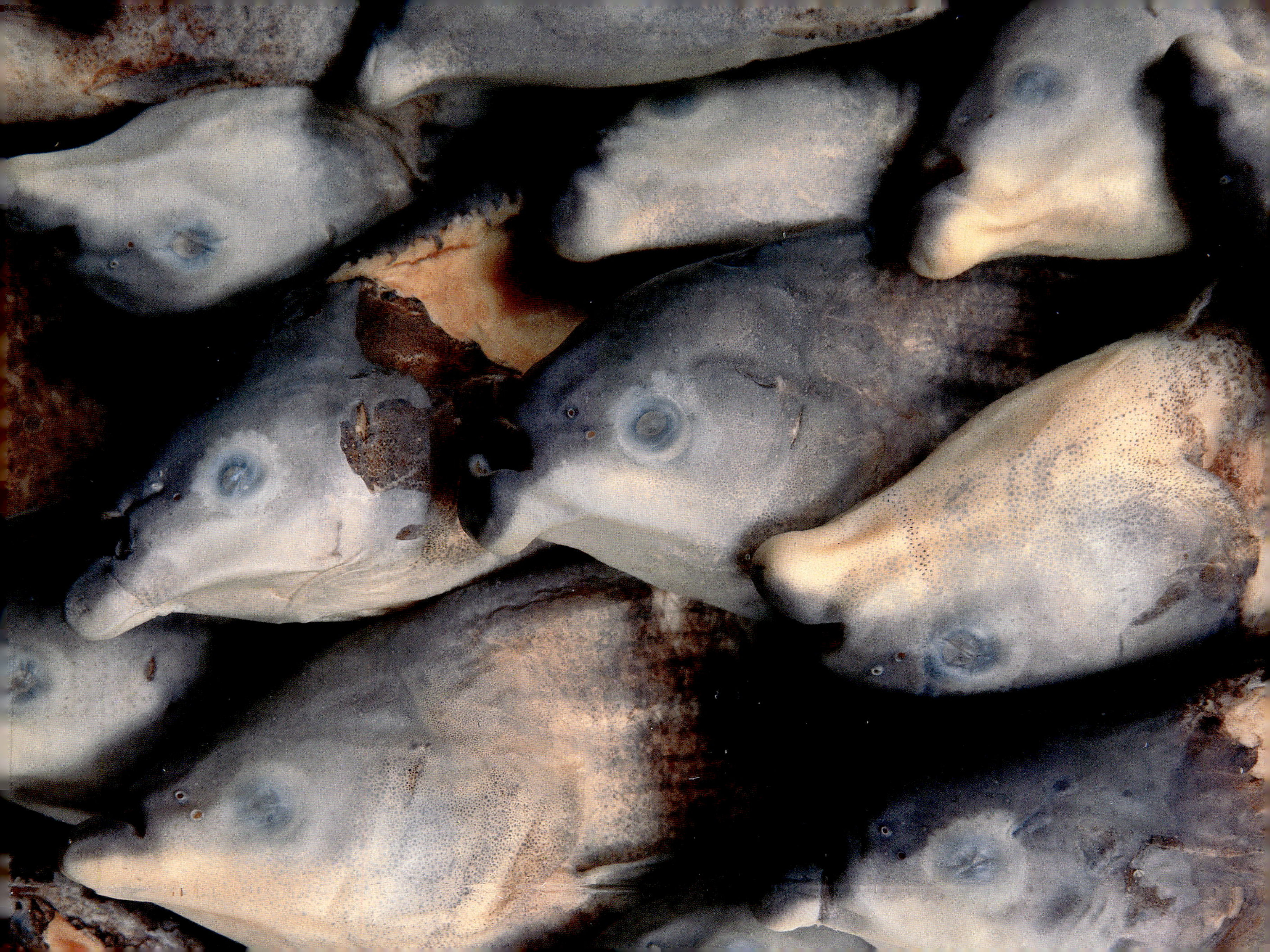

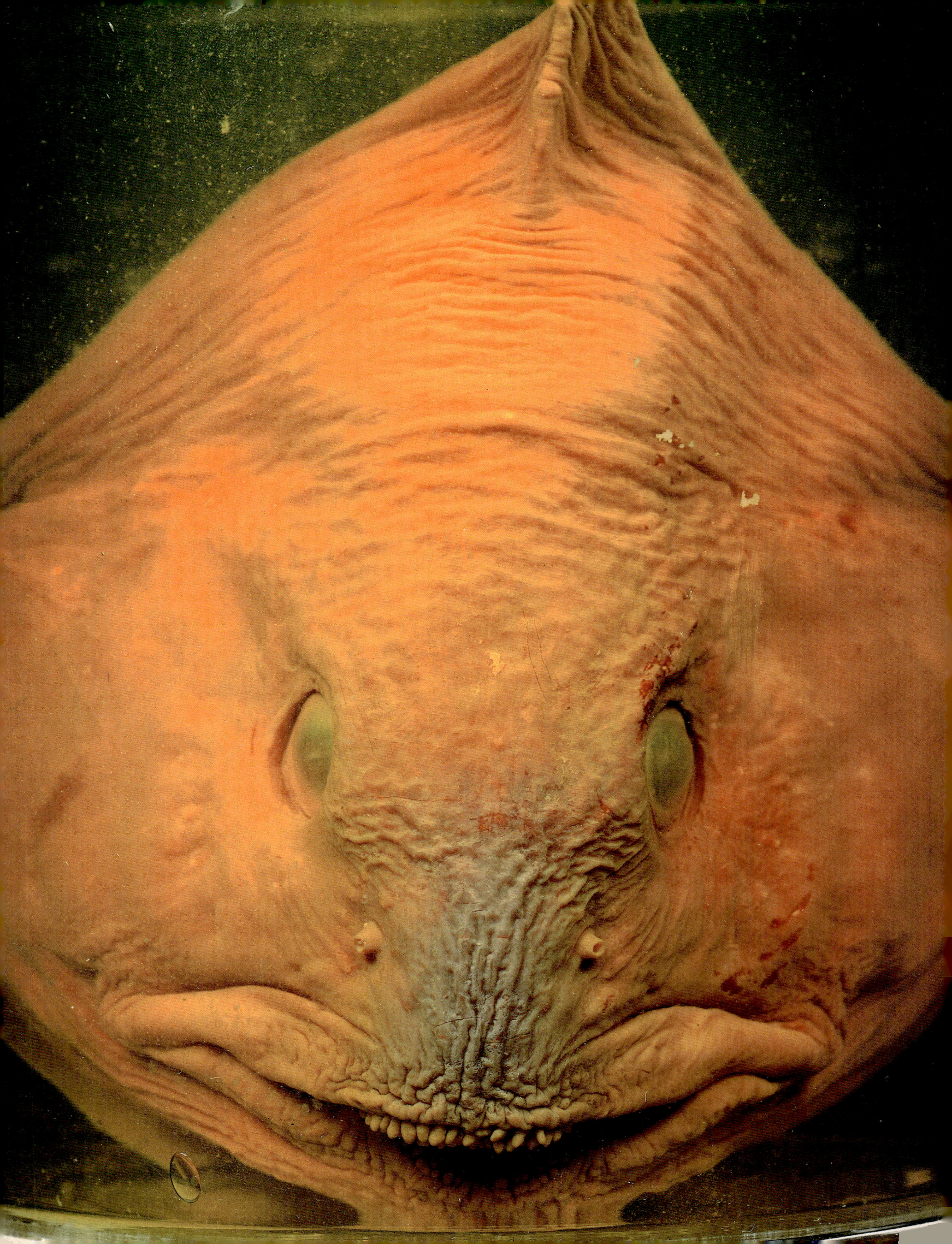

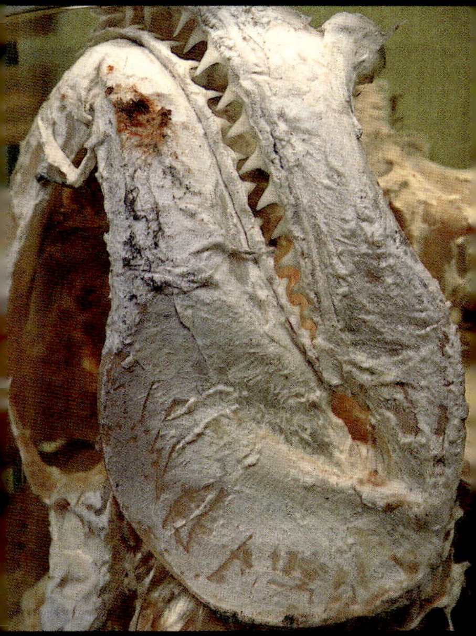

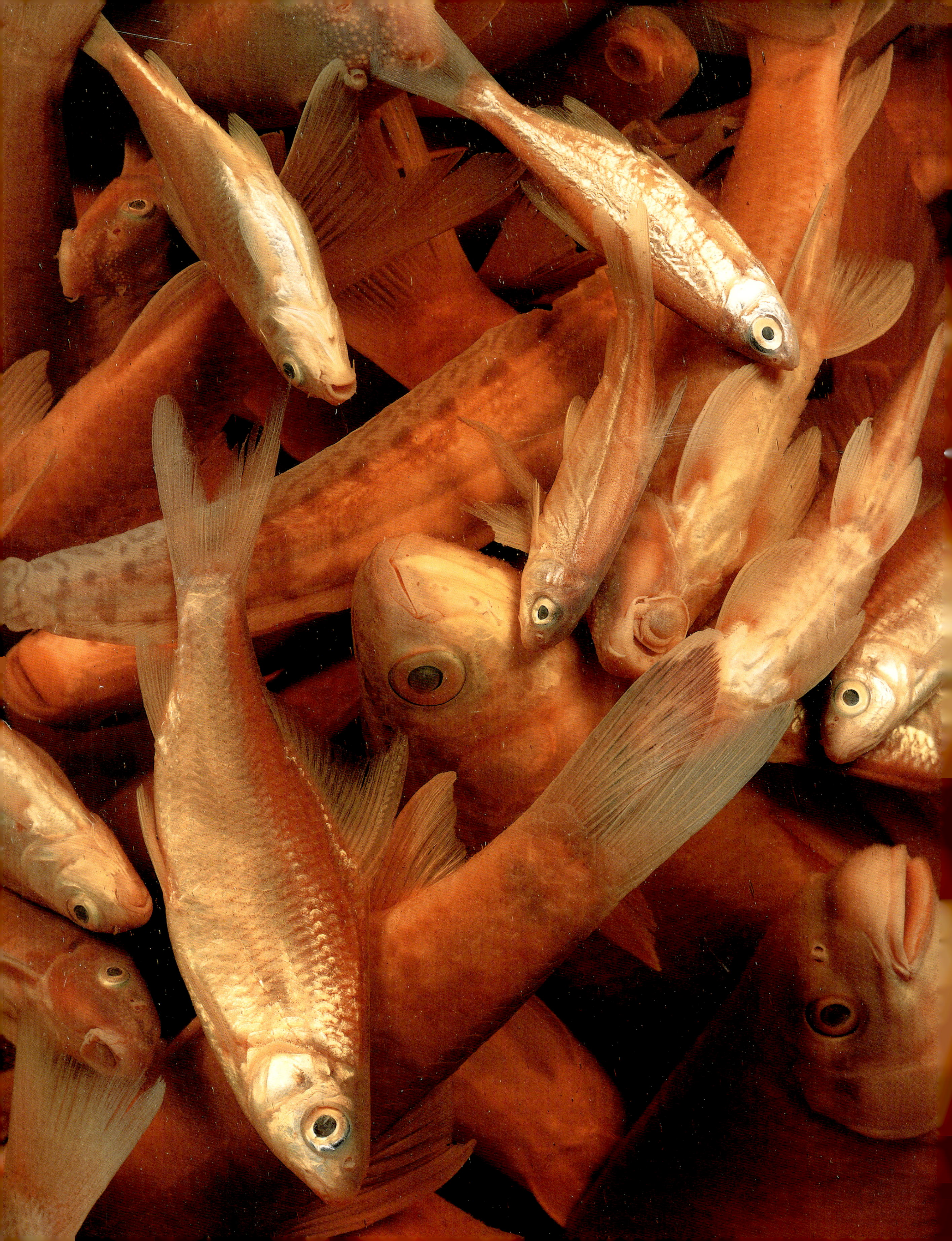

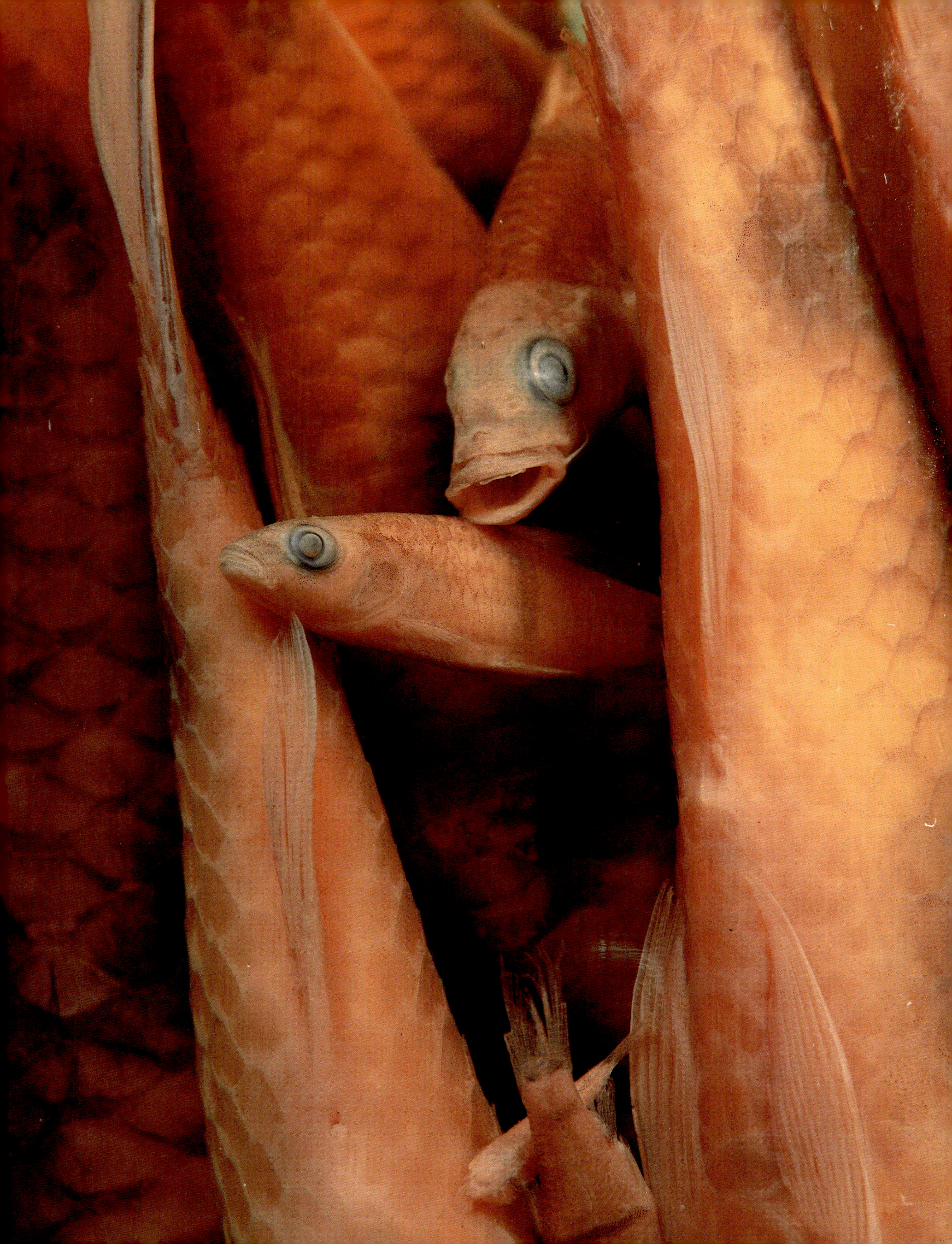

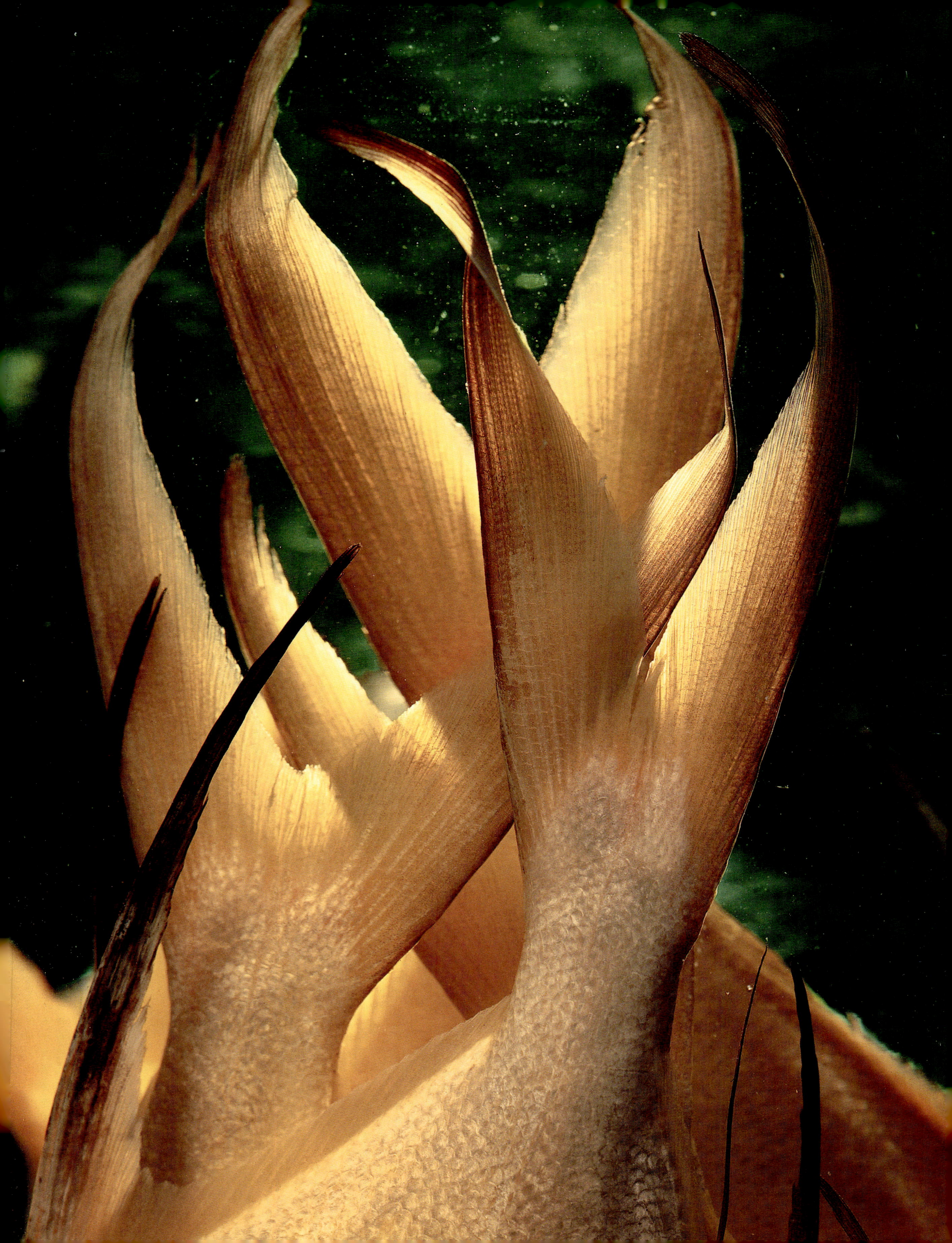

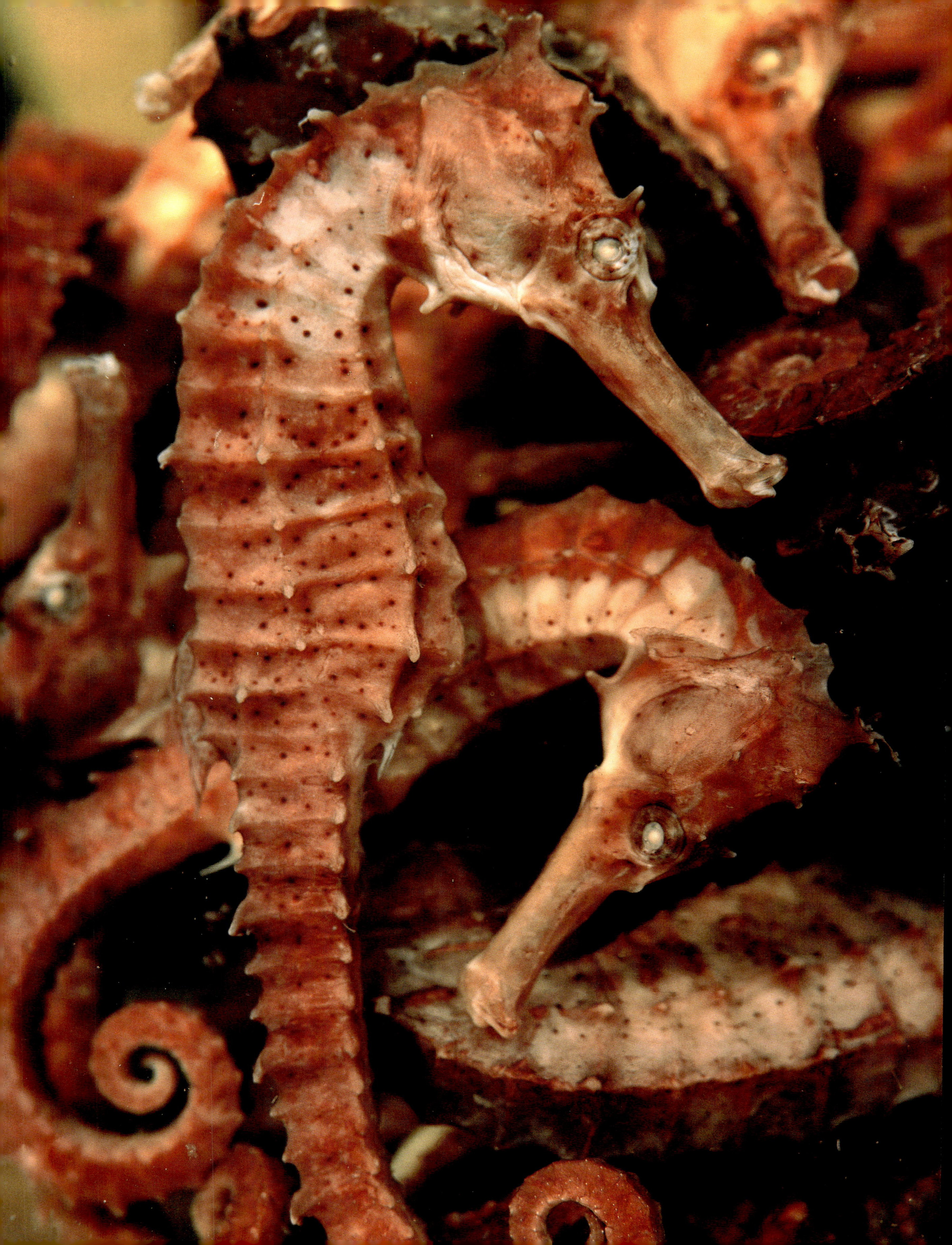

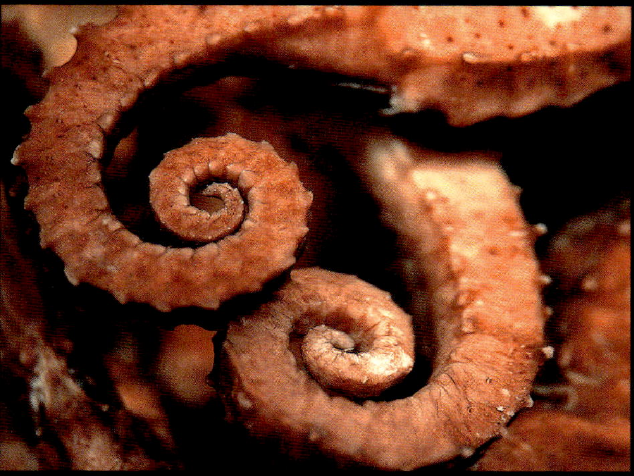

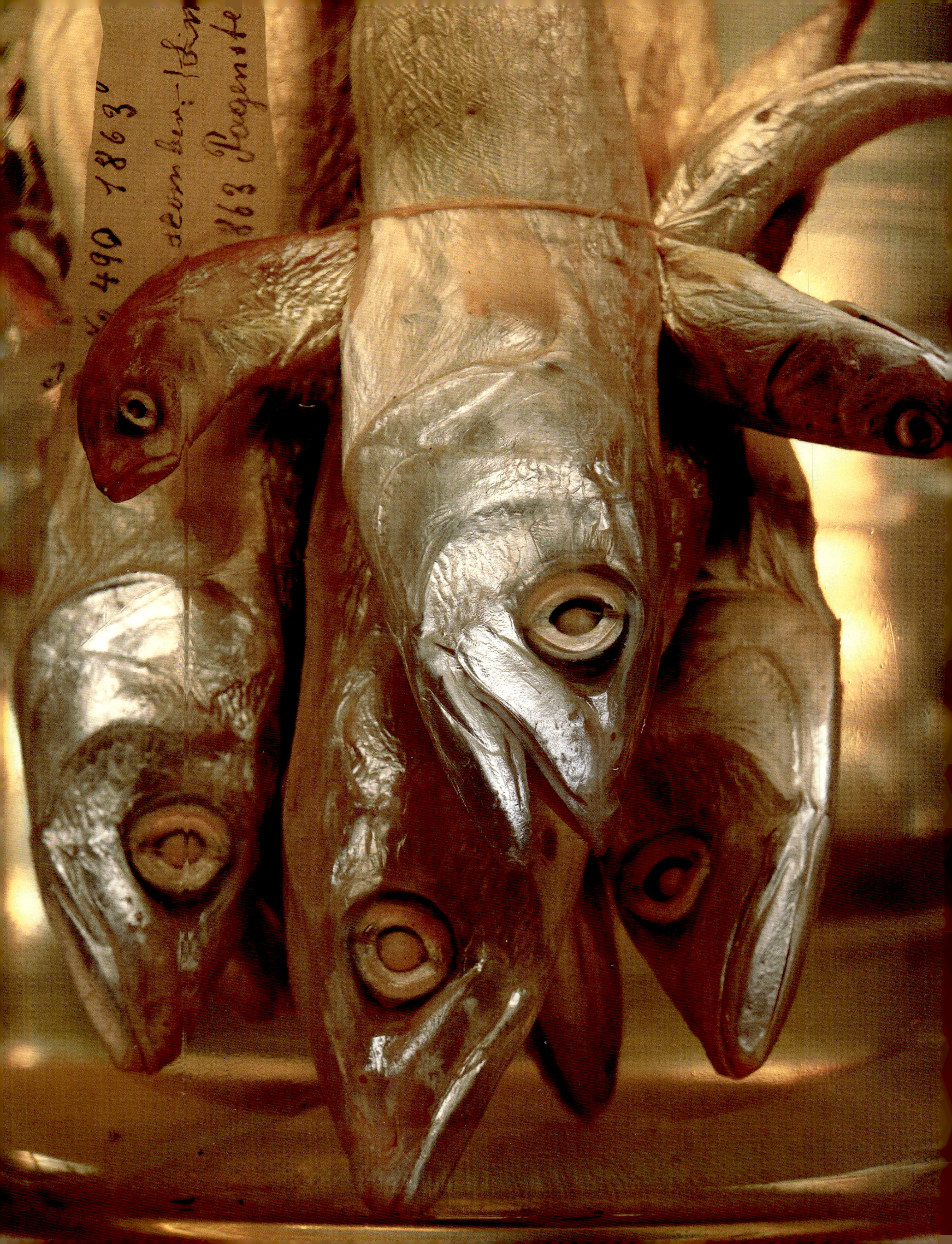
490 1863

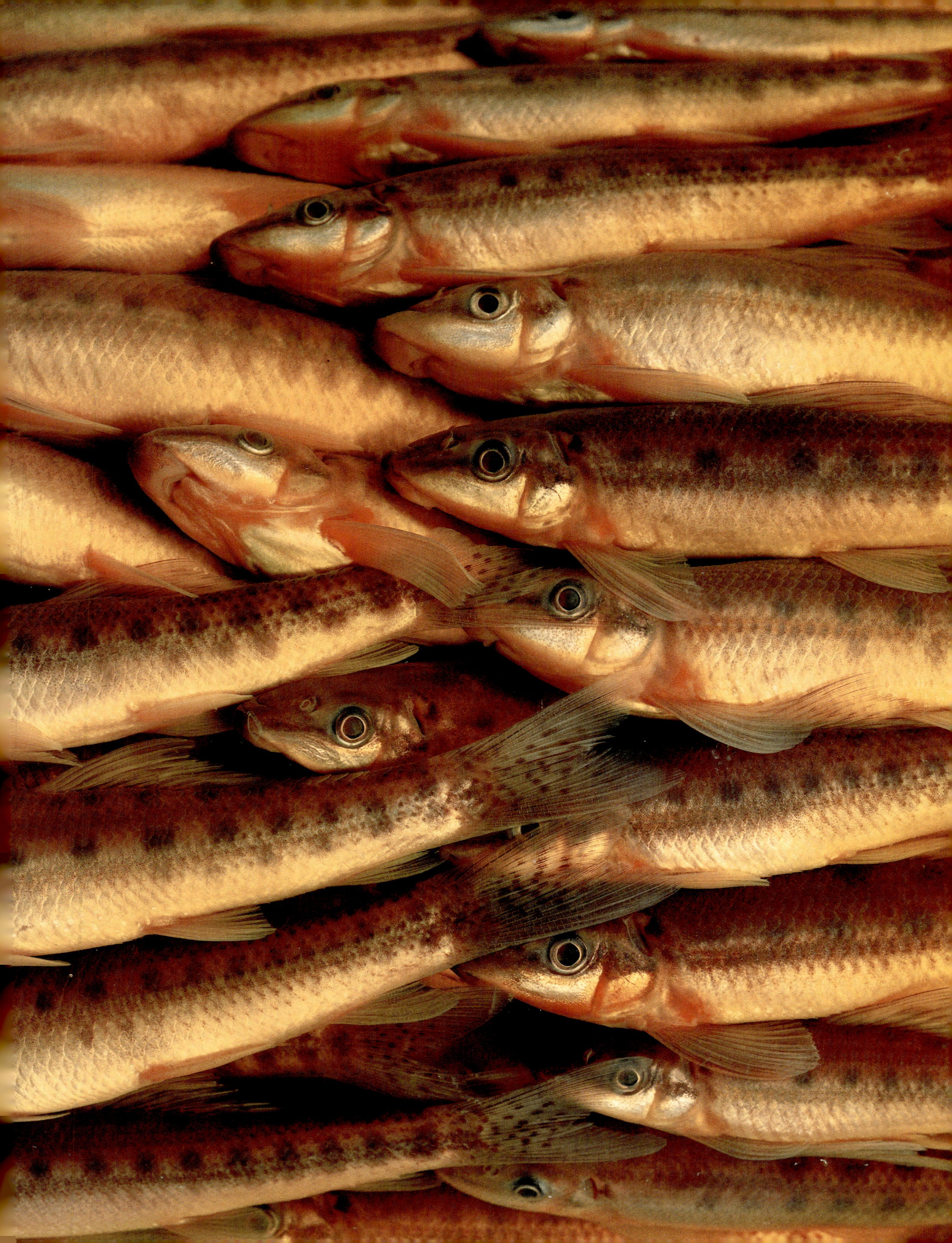

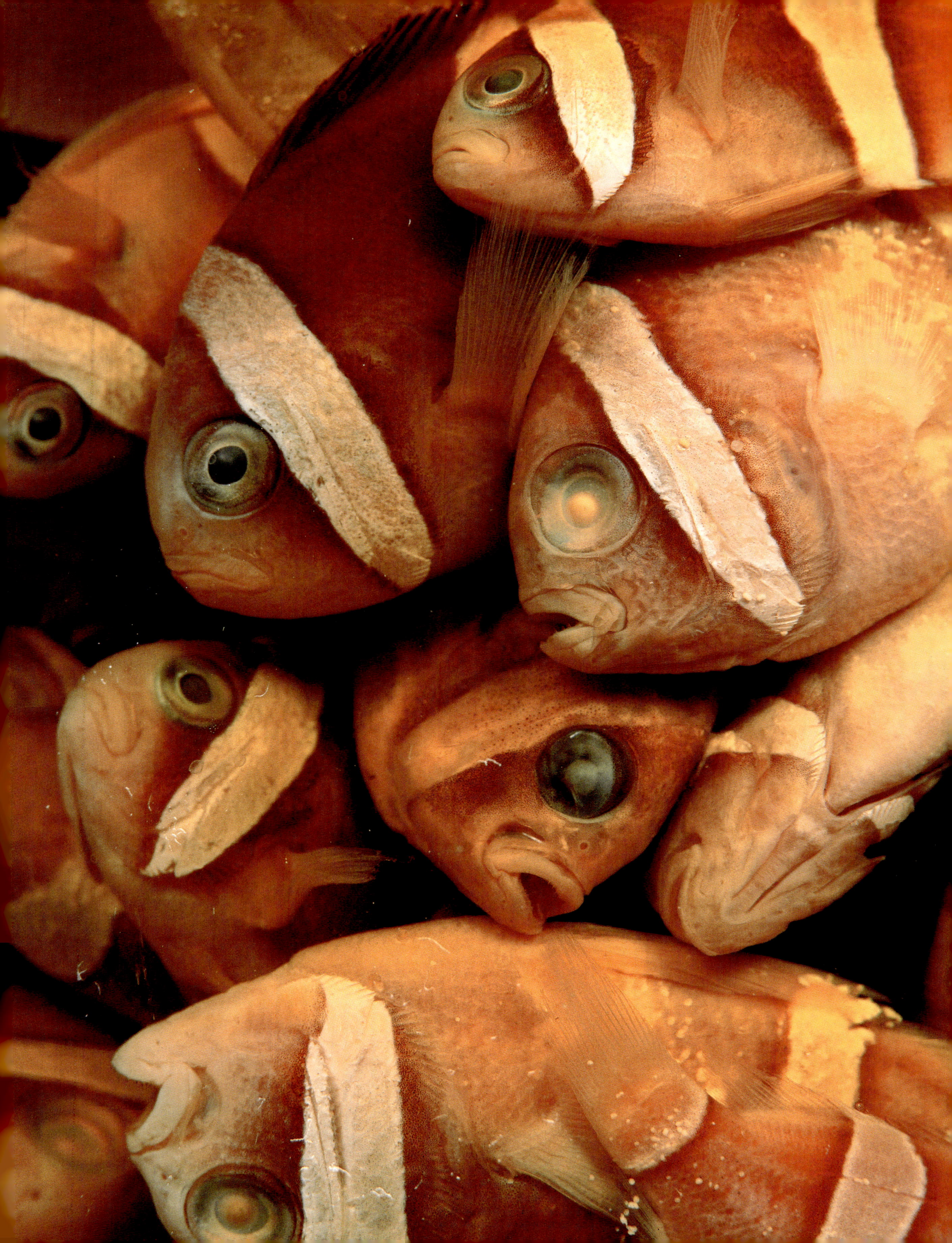

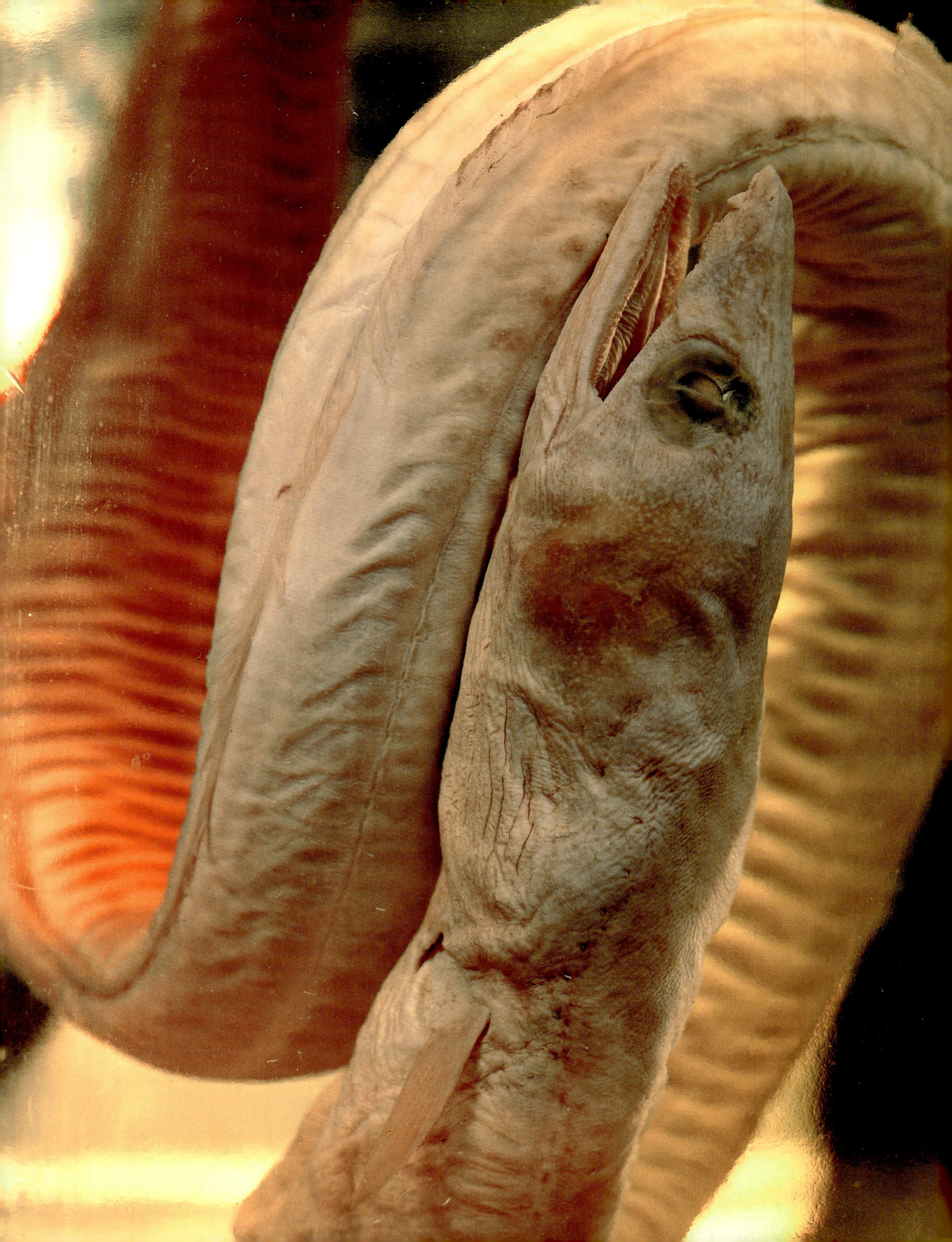

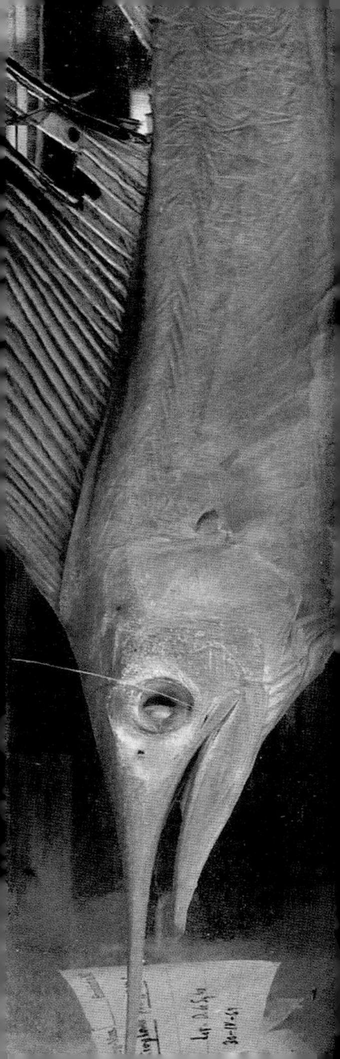

CONSERVING ANIMALS

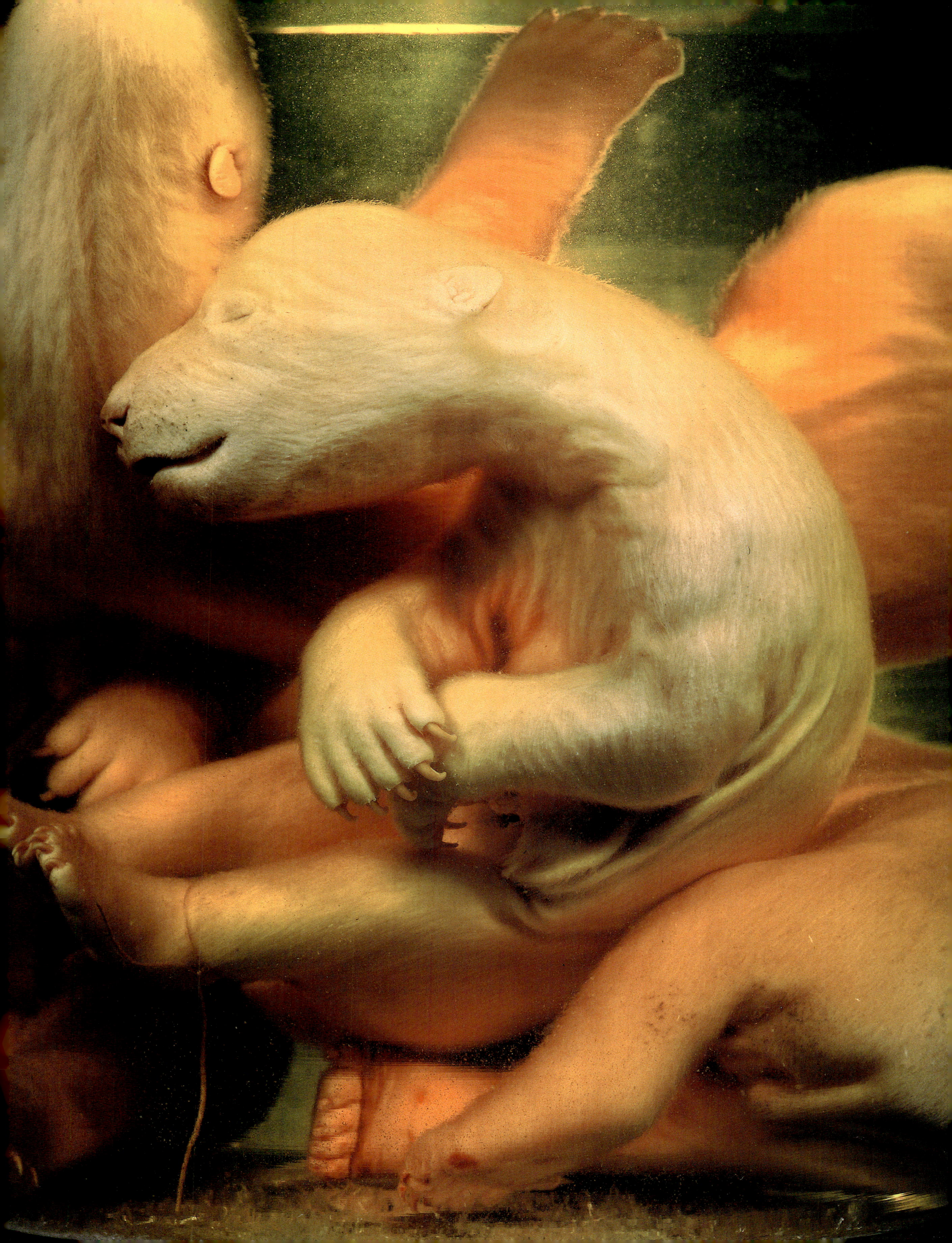

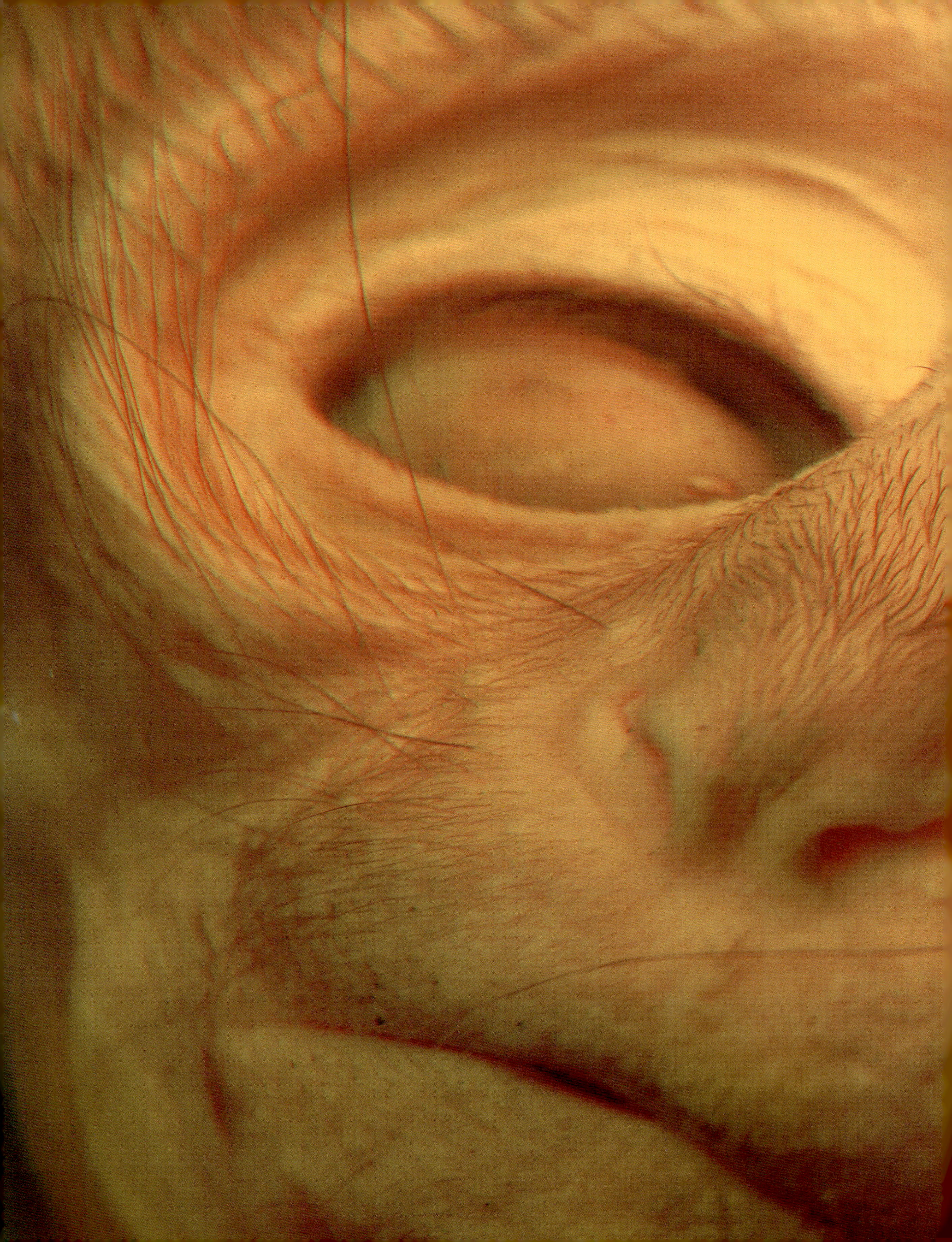

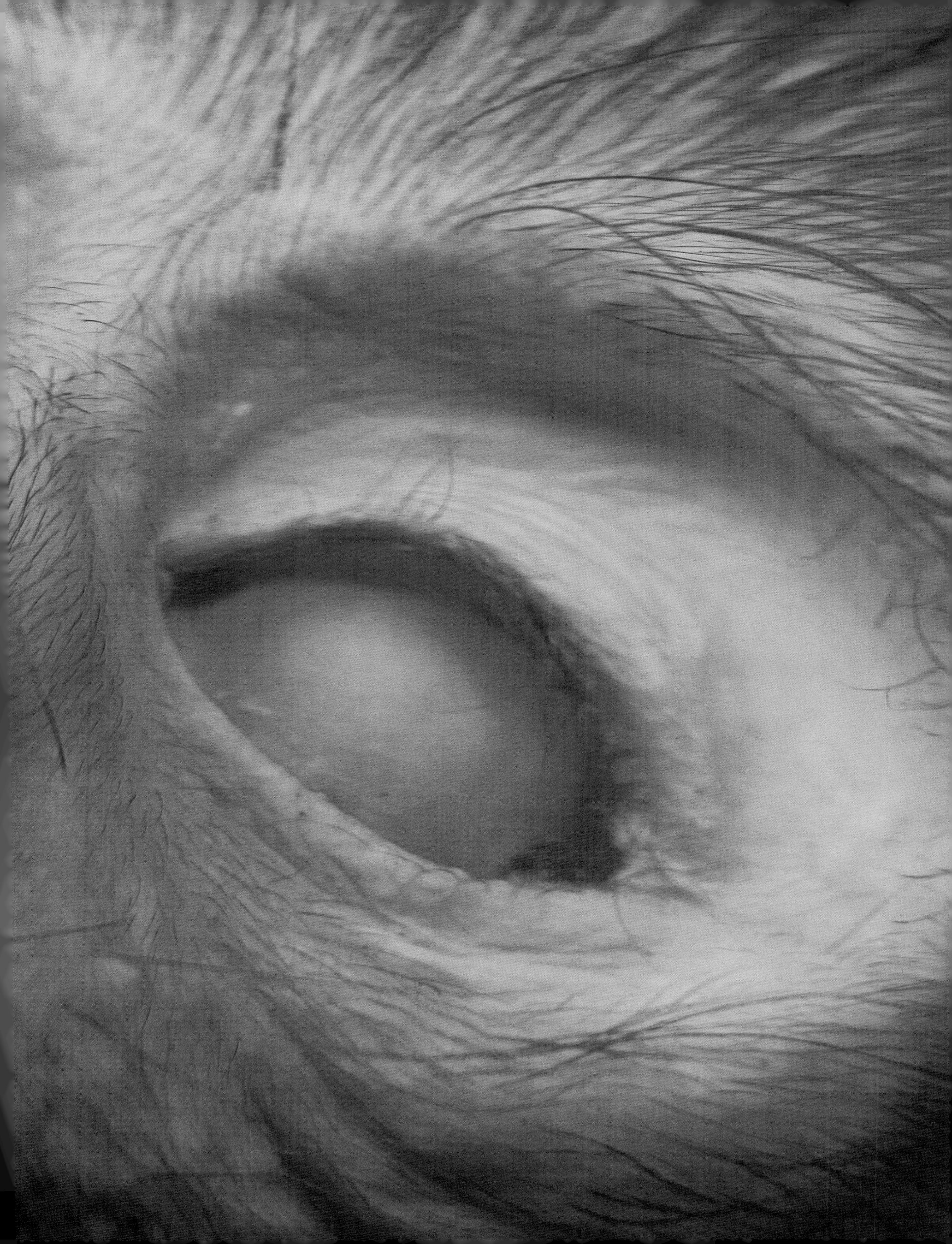

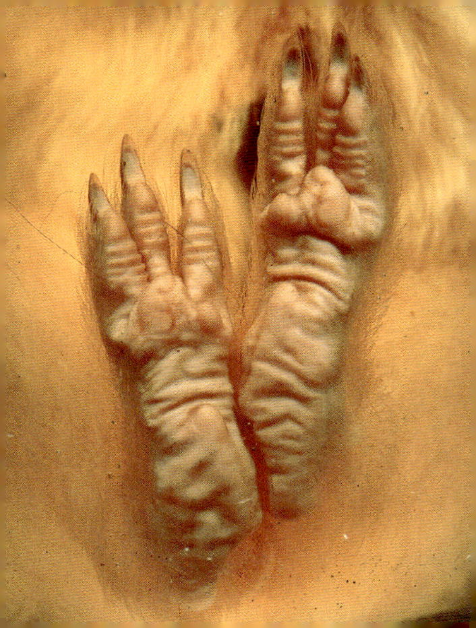

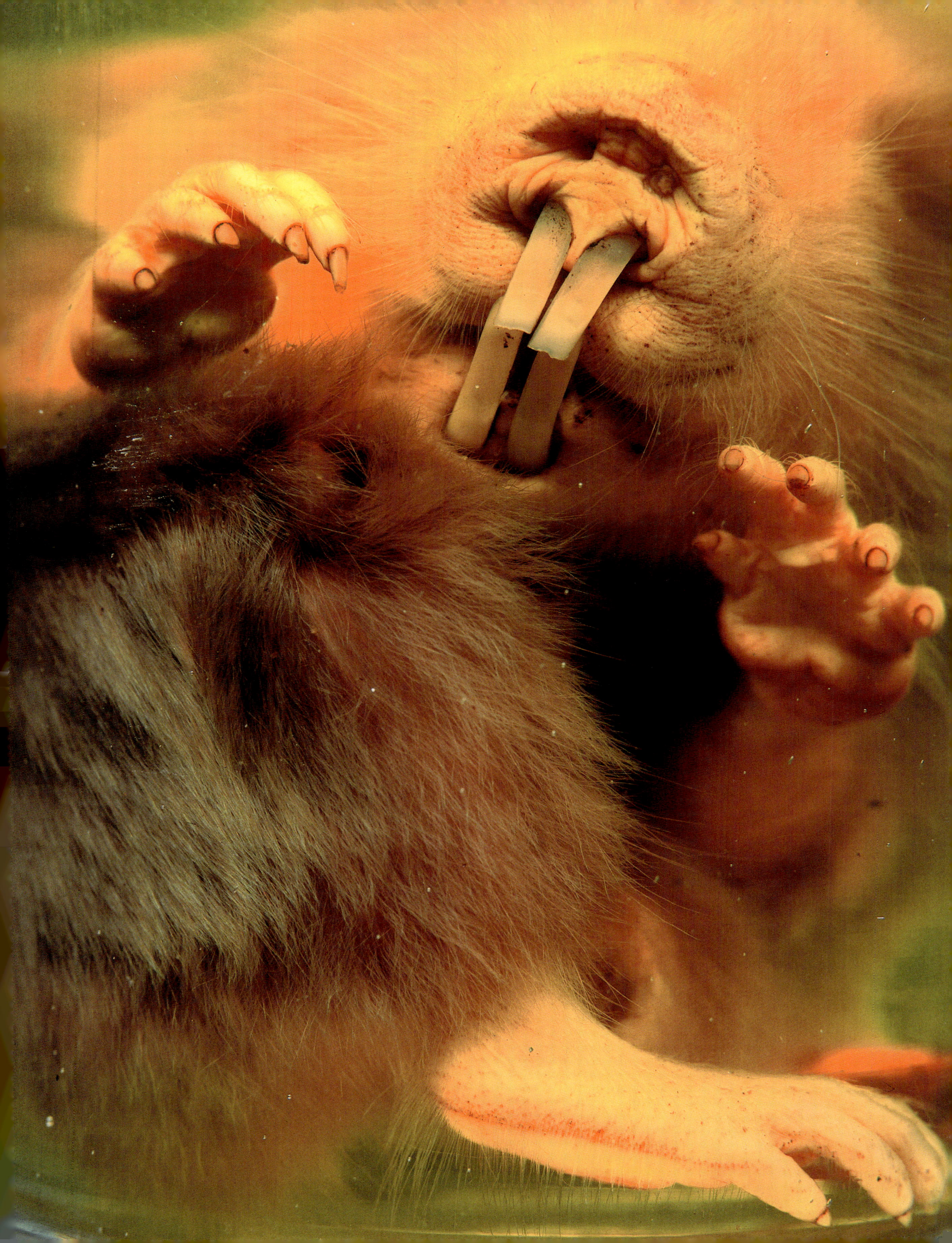

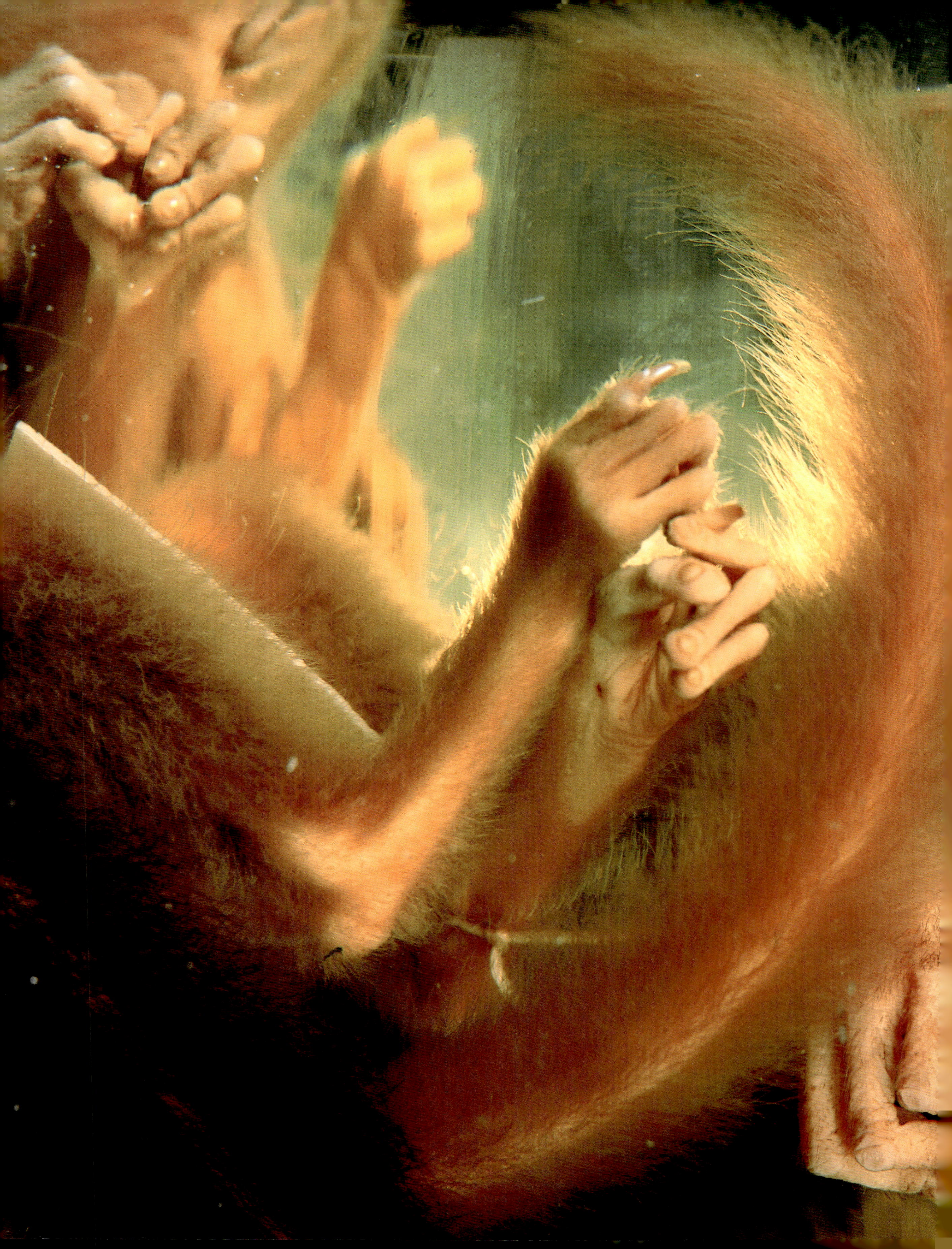

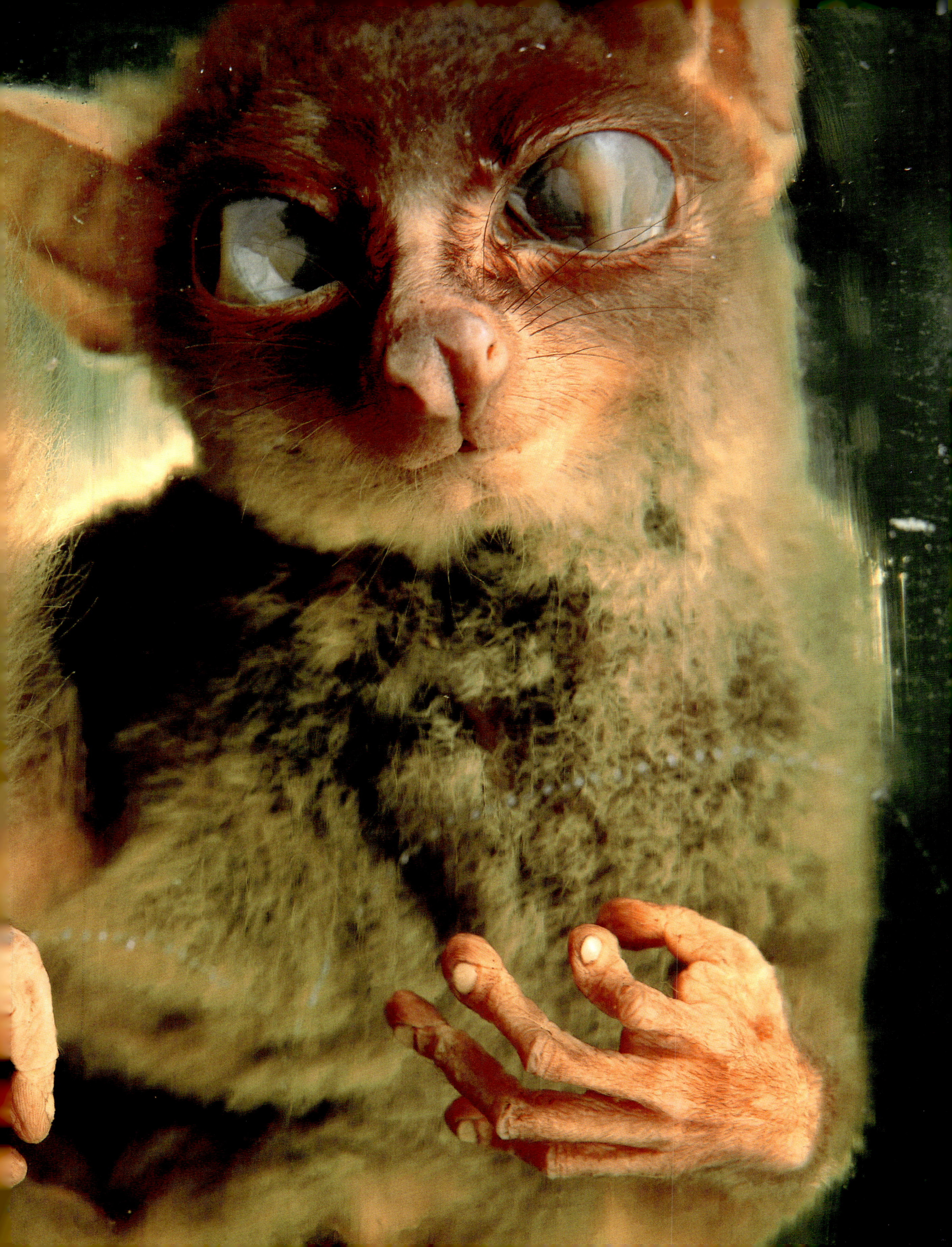

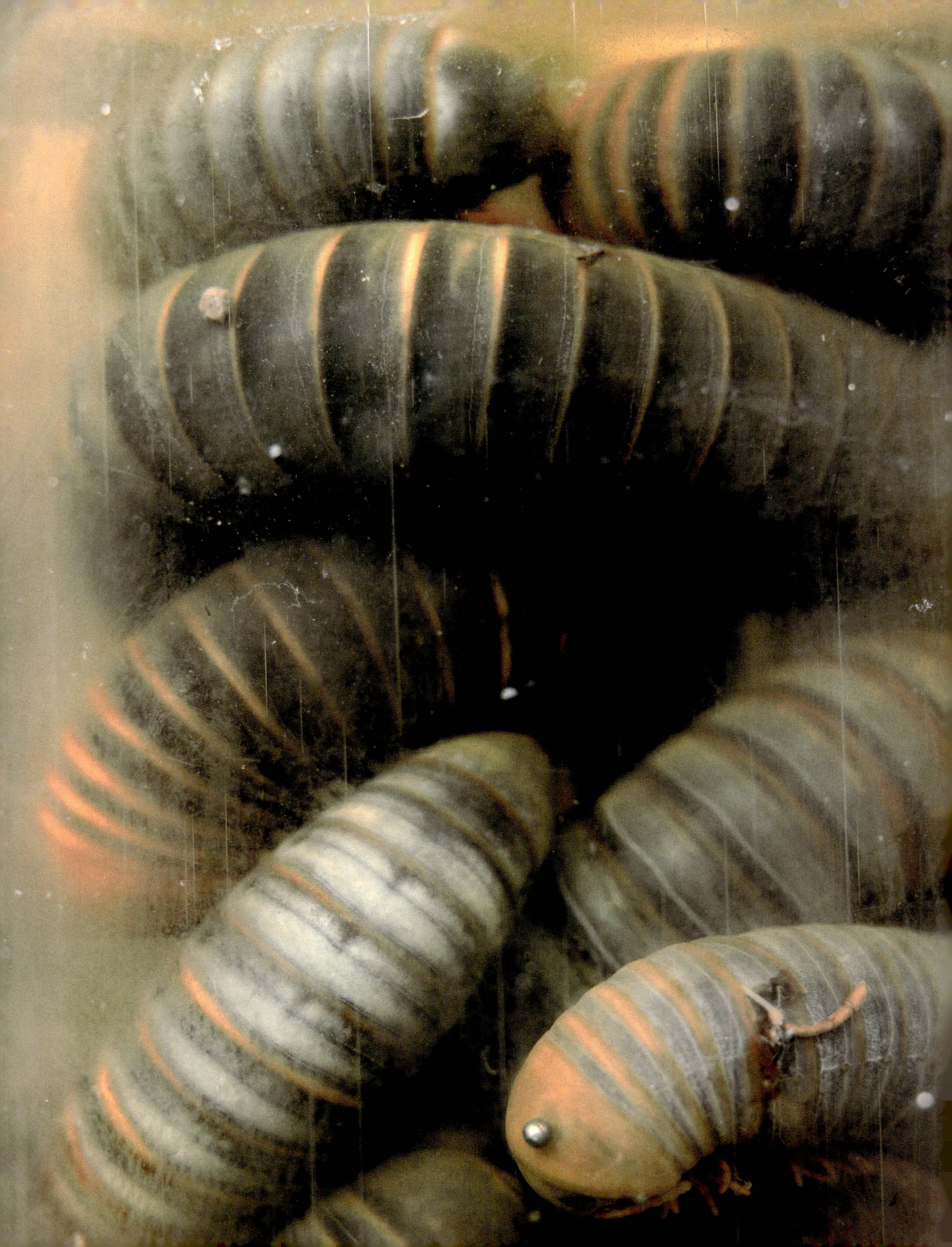

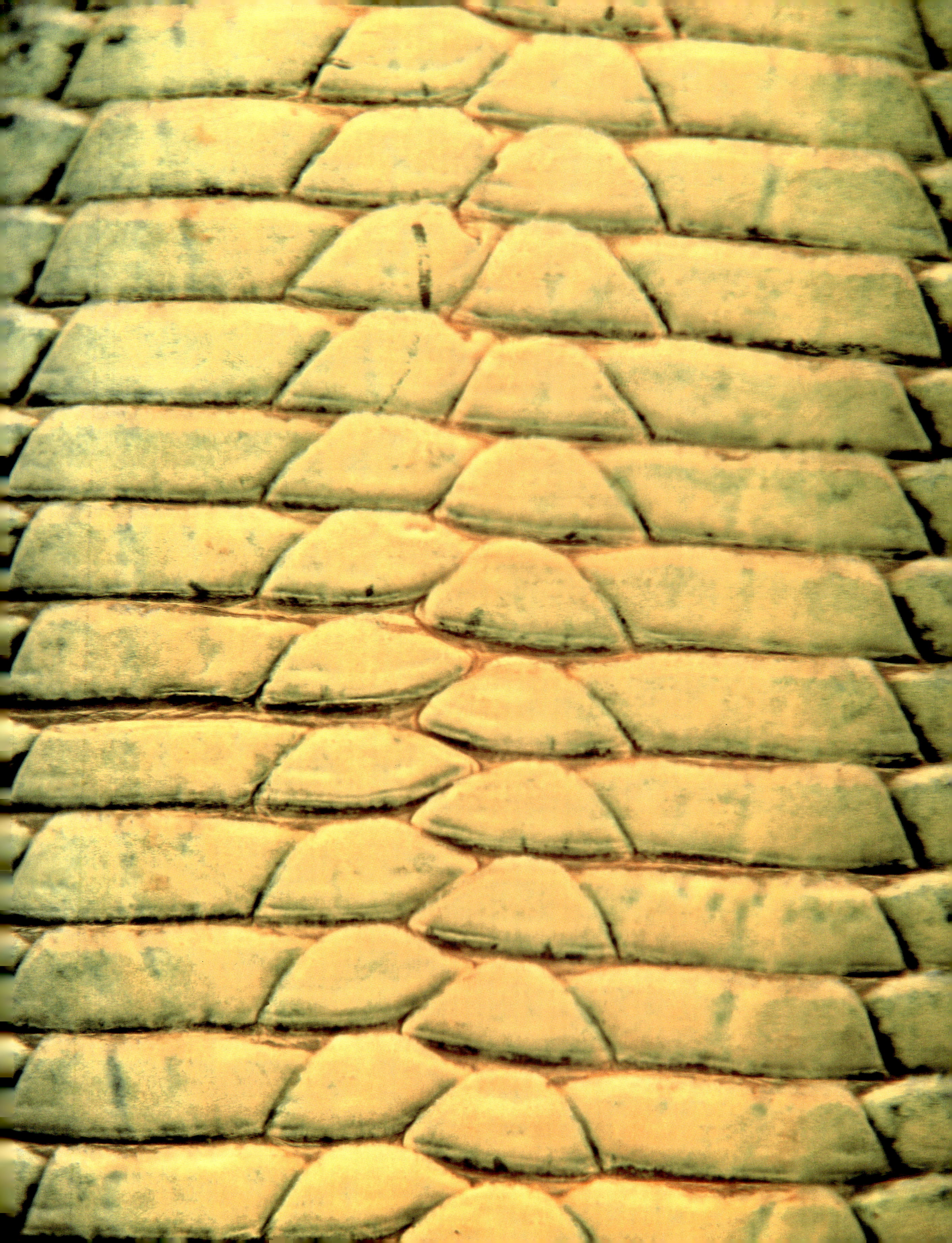

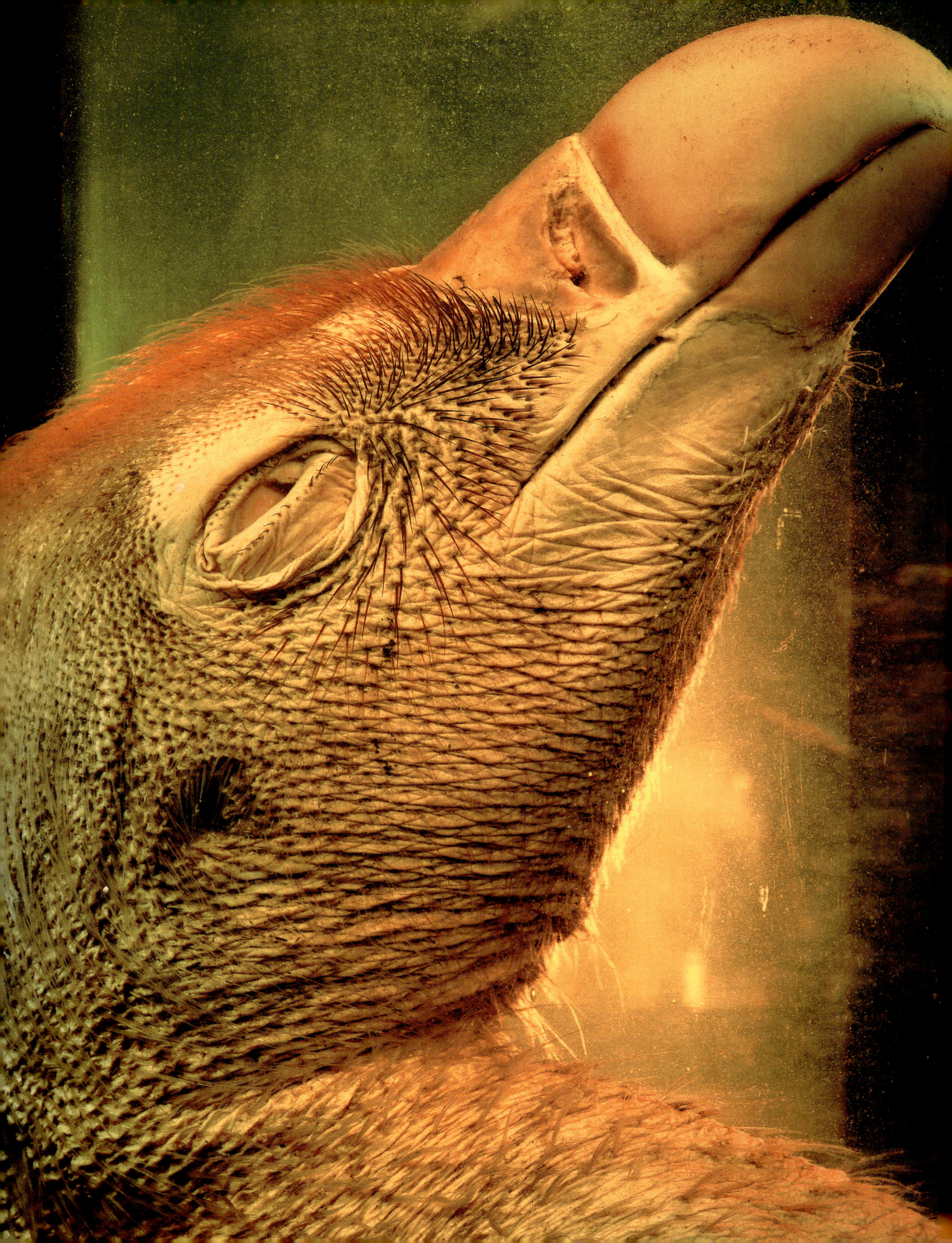

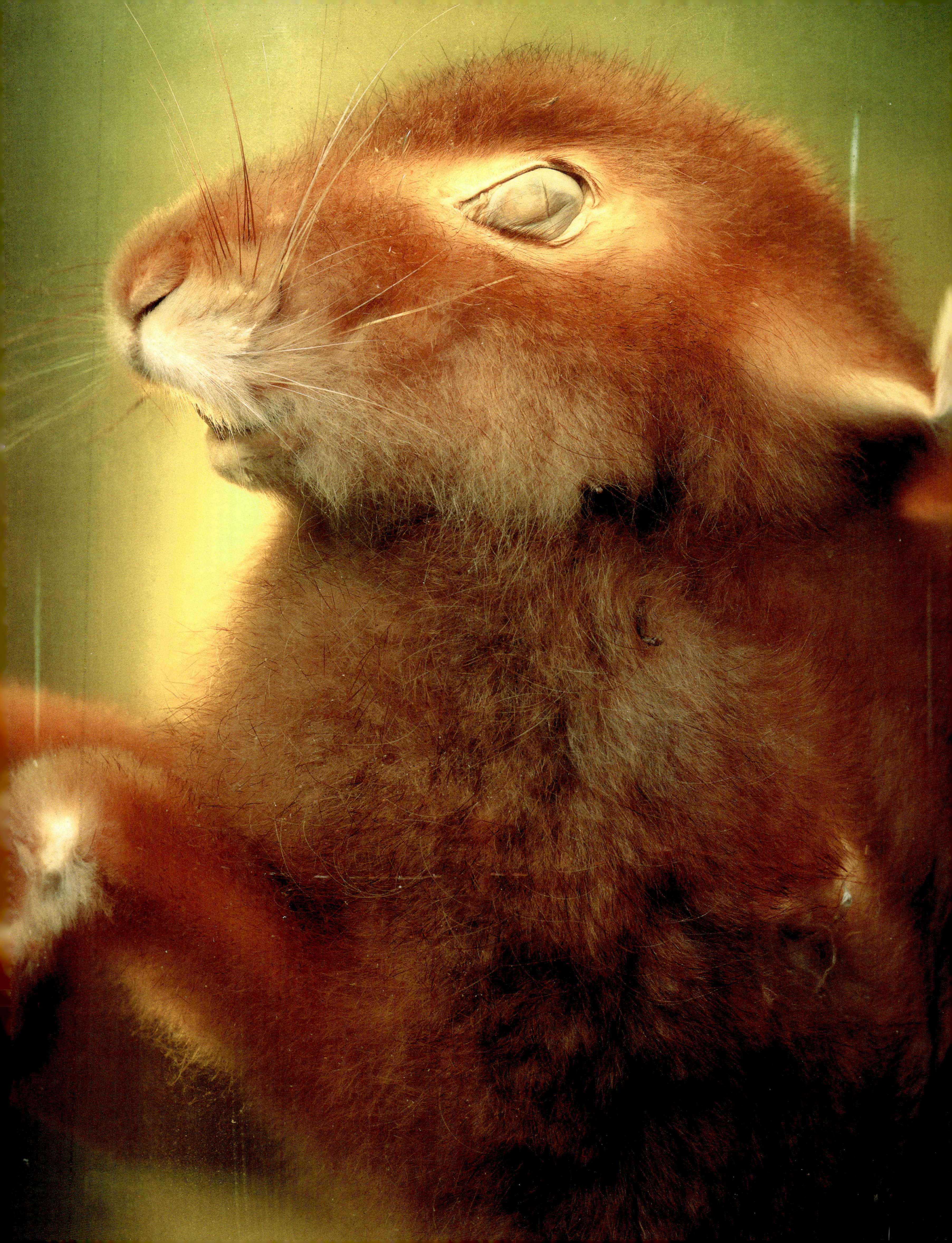

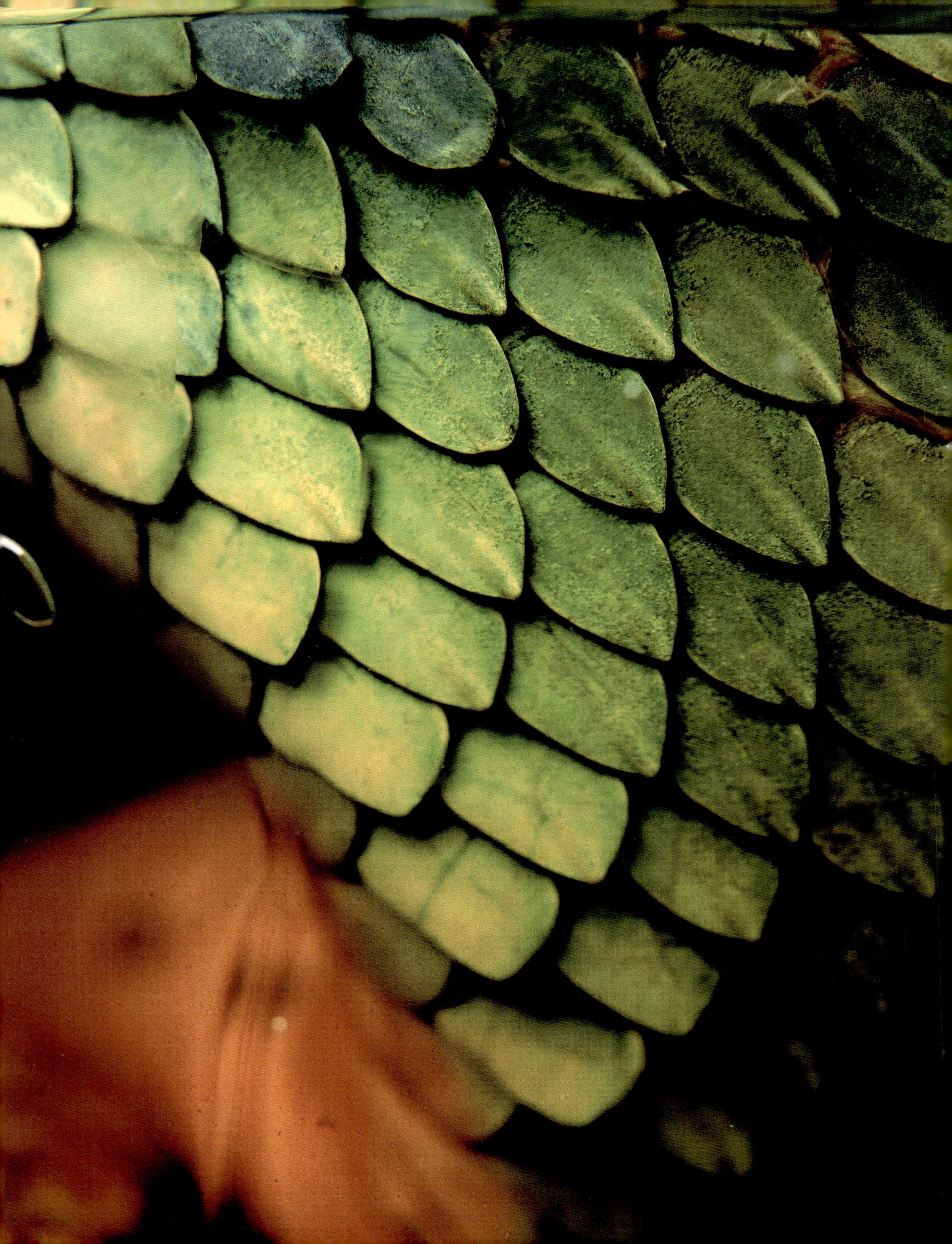

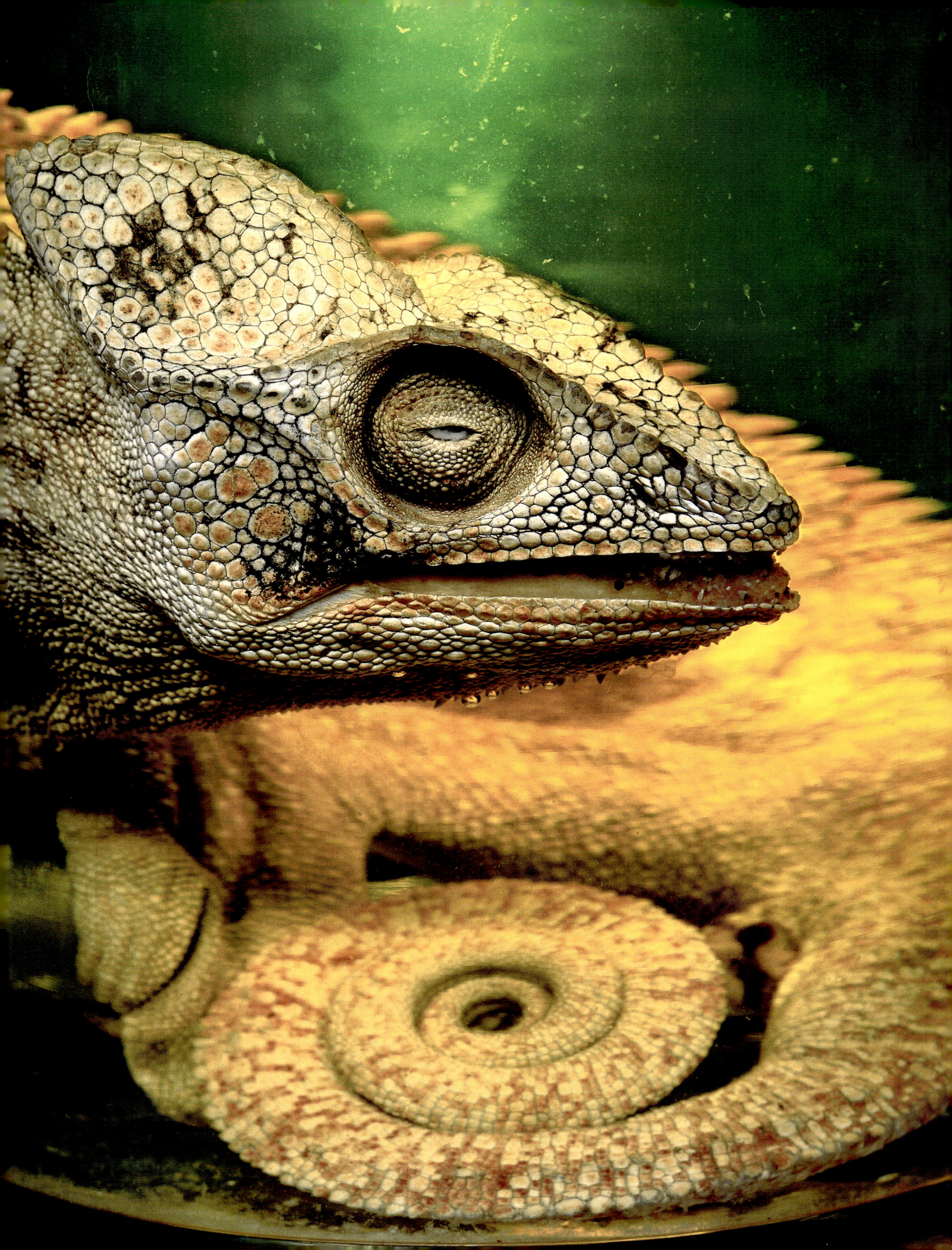

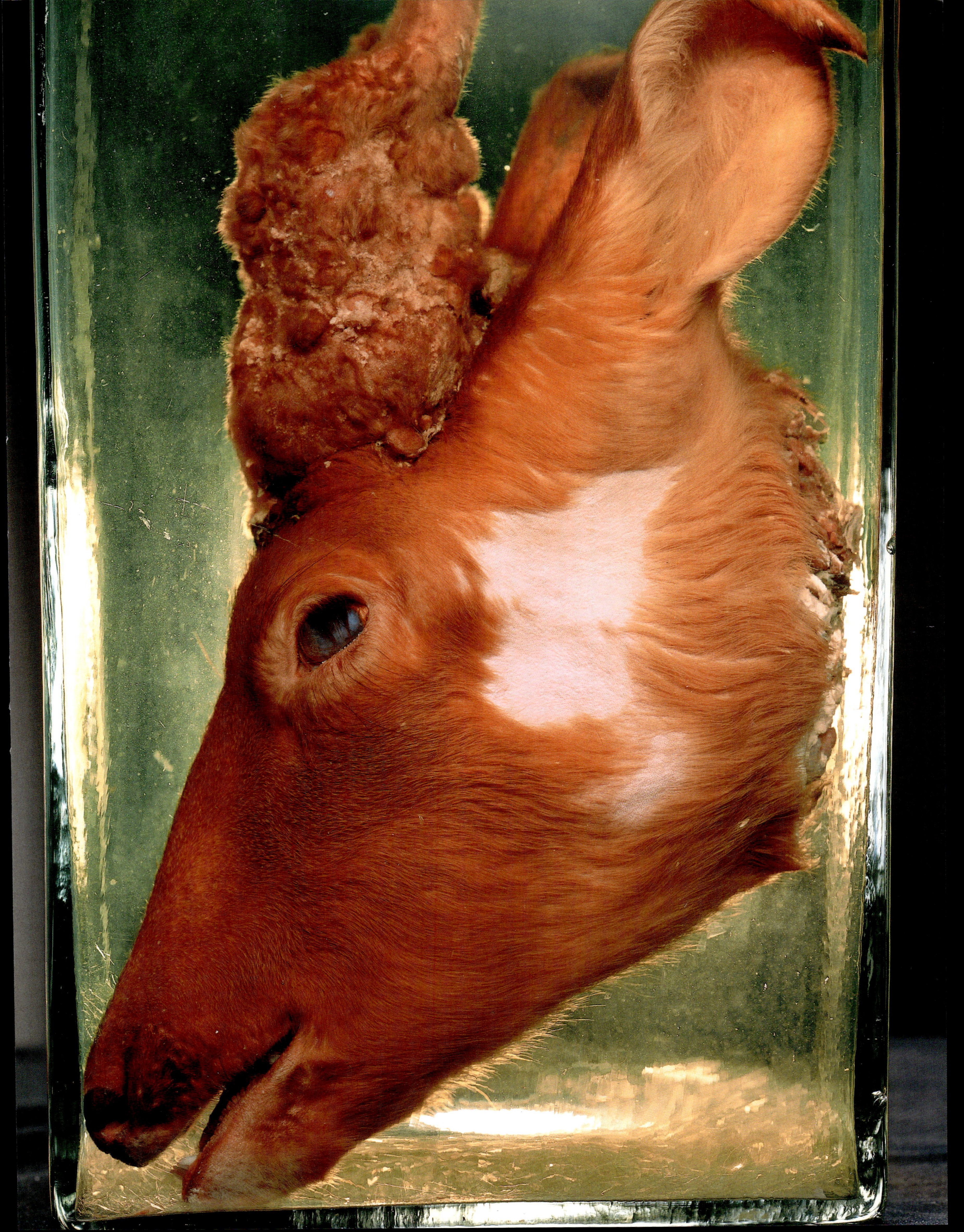

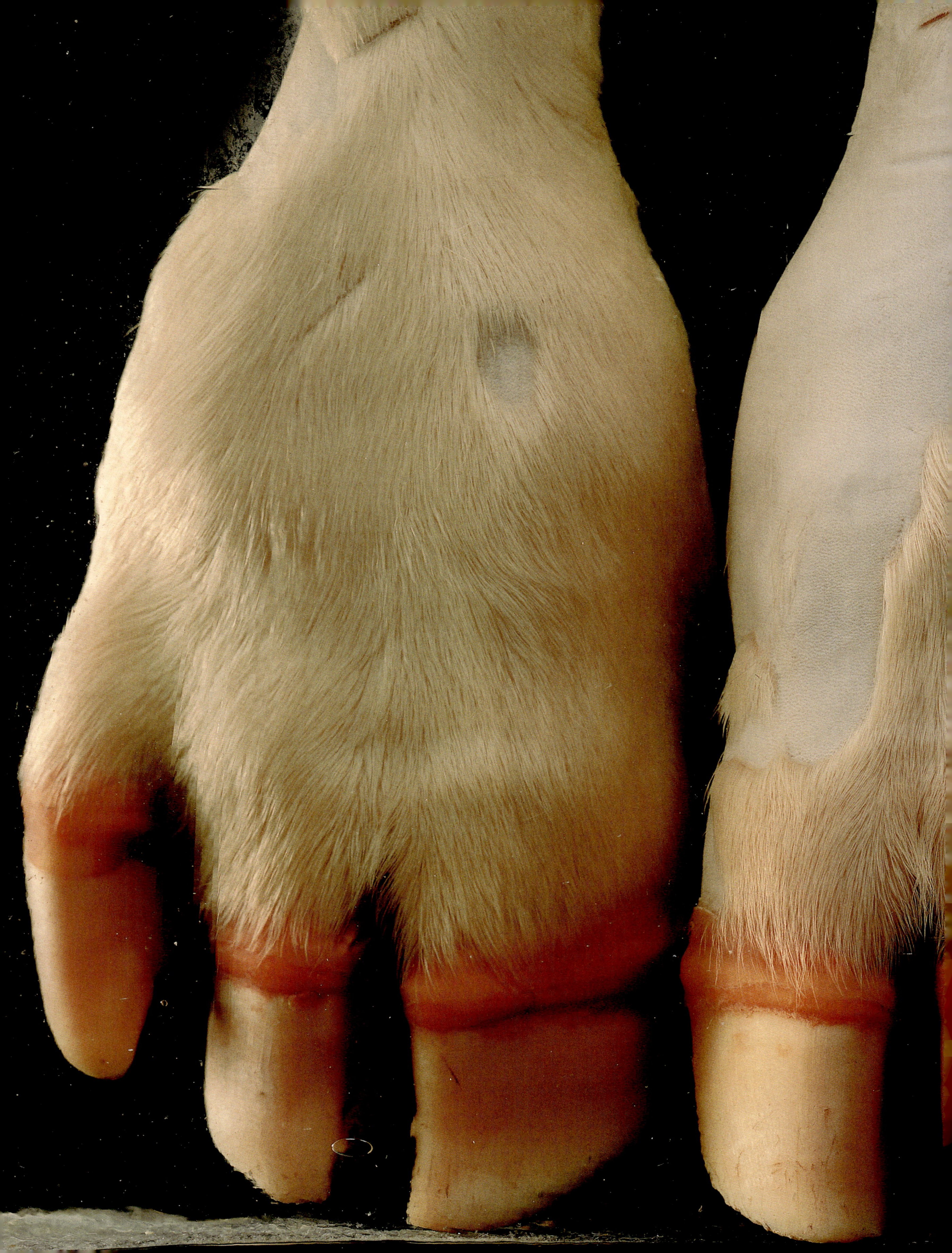

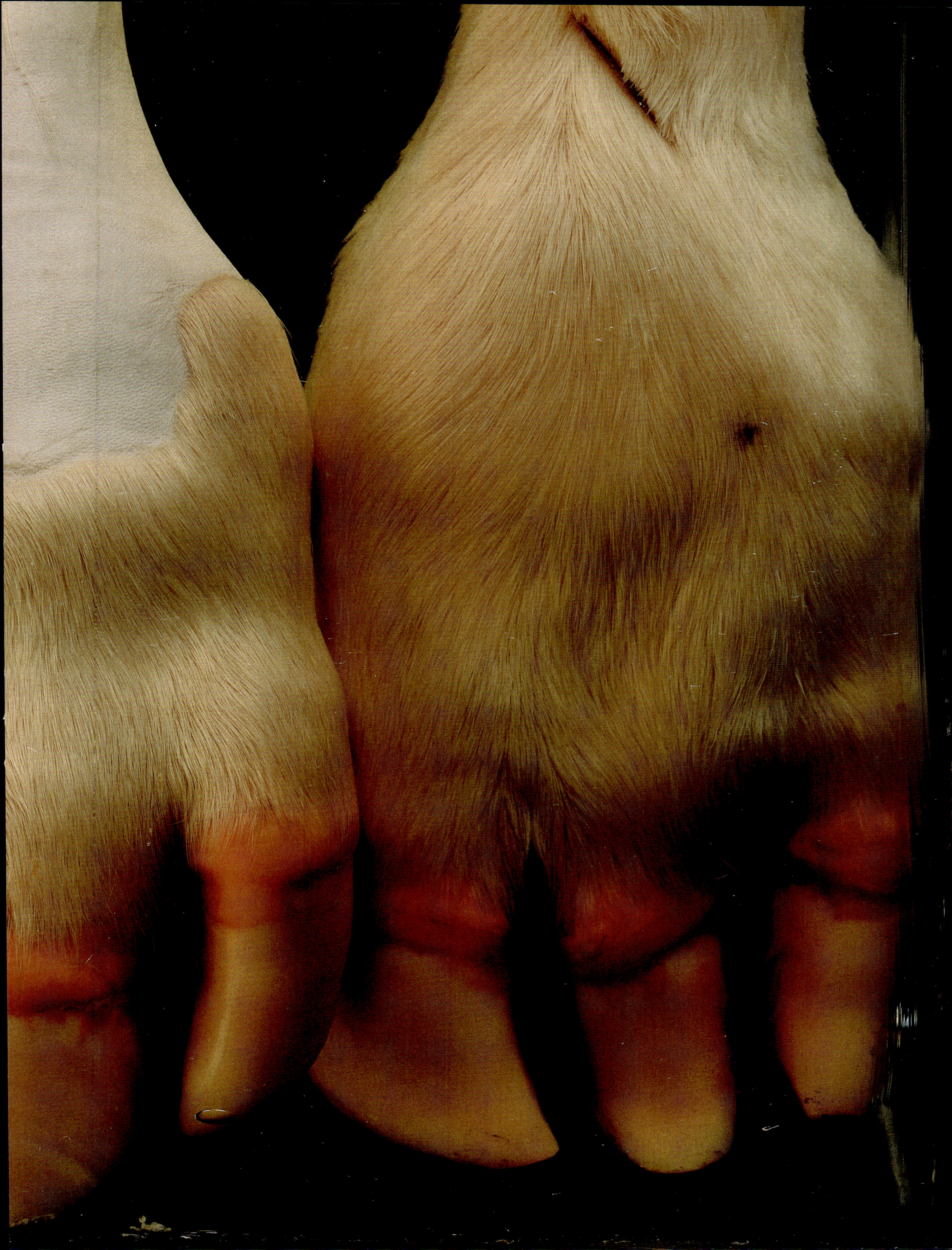

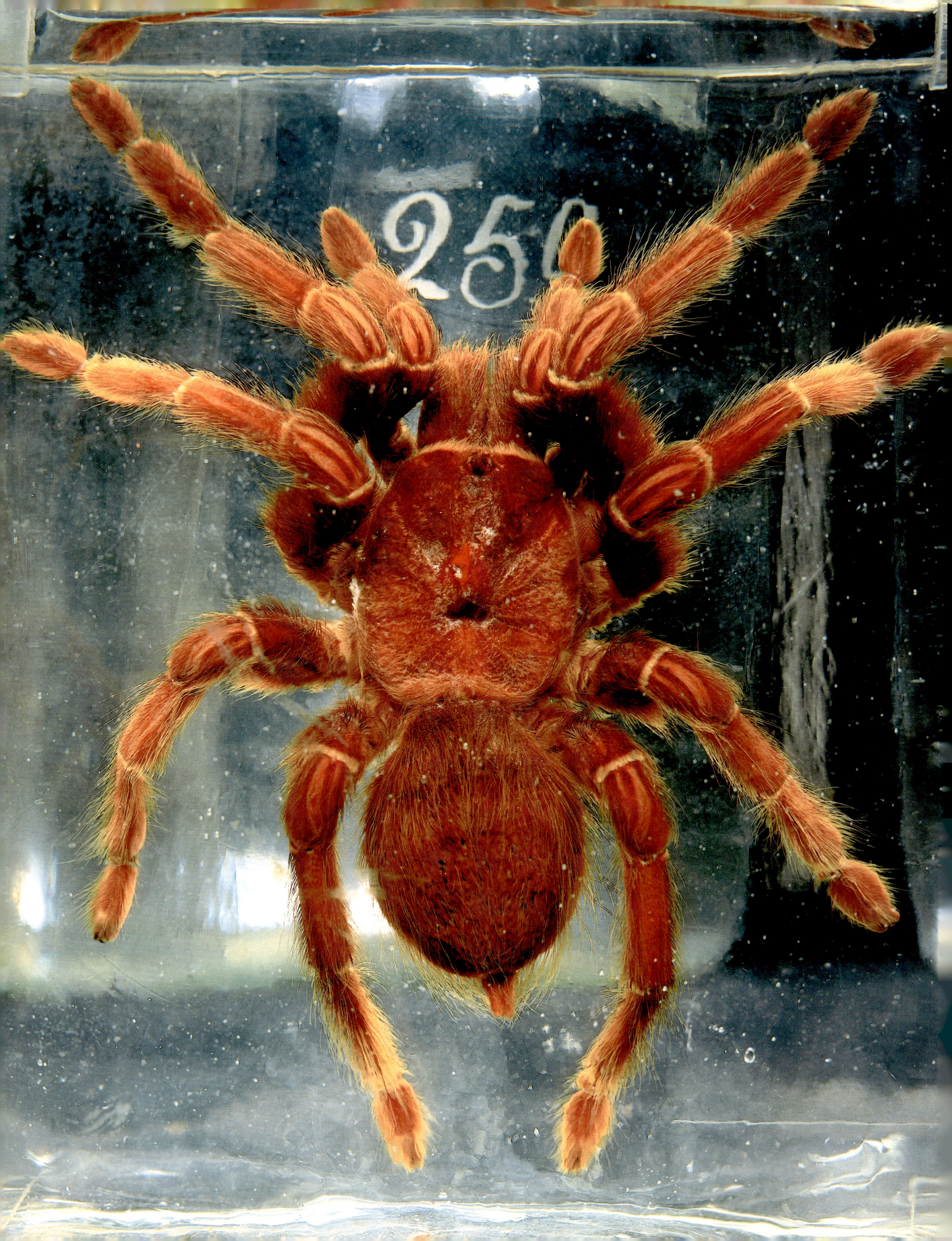
25

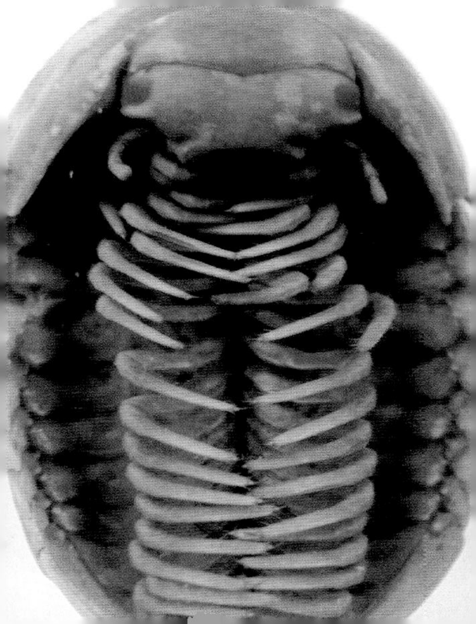

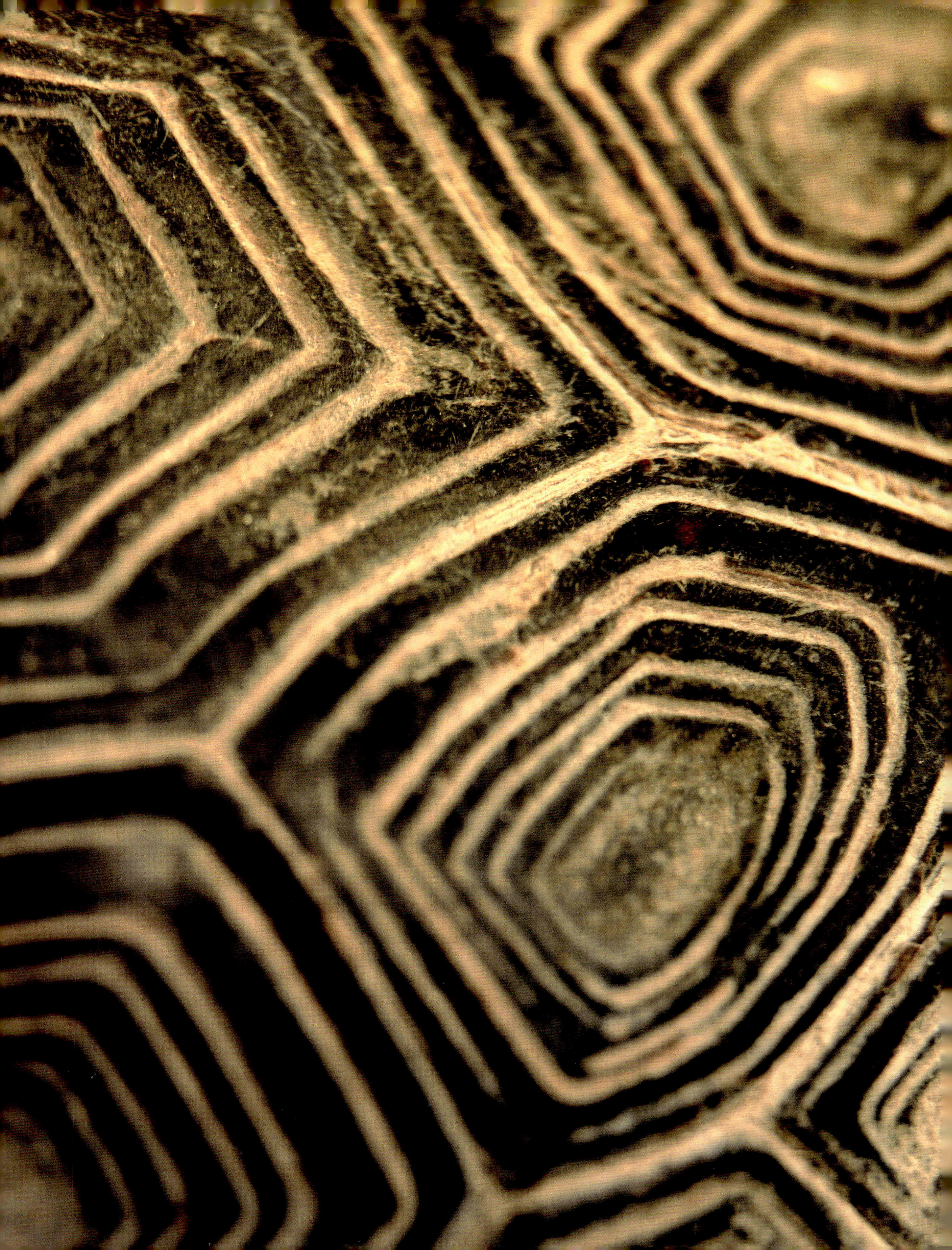

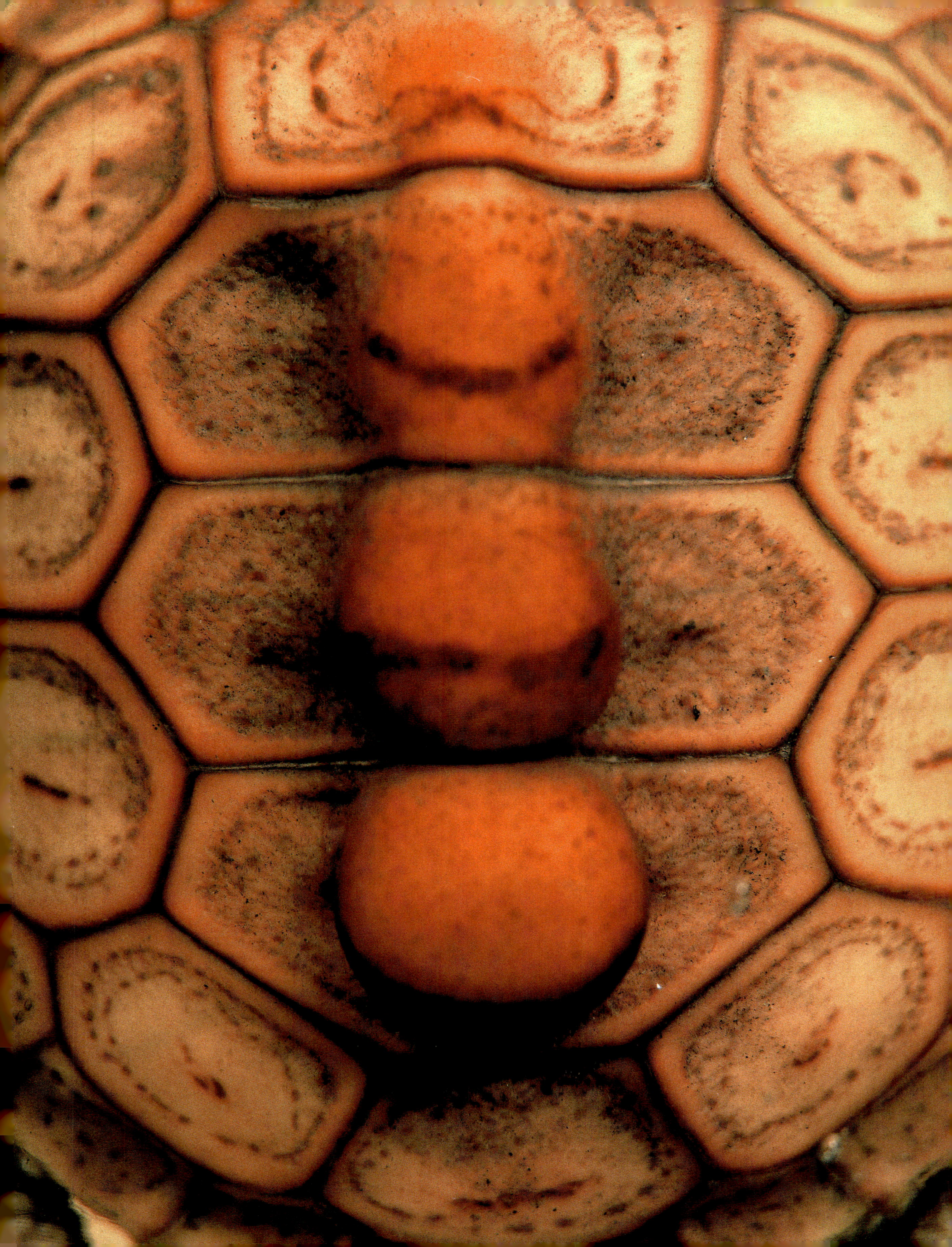

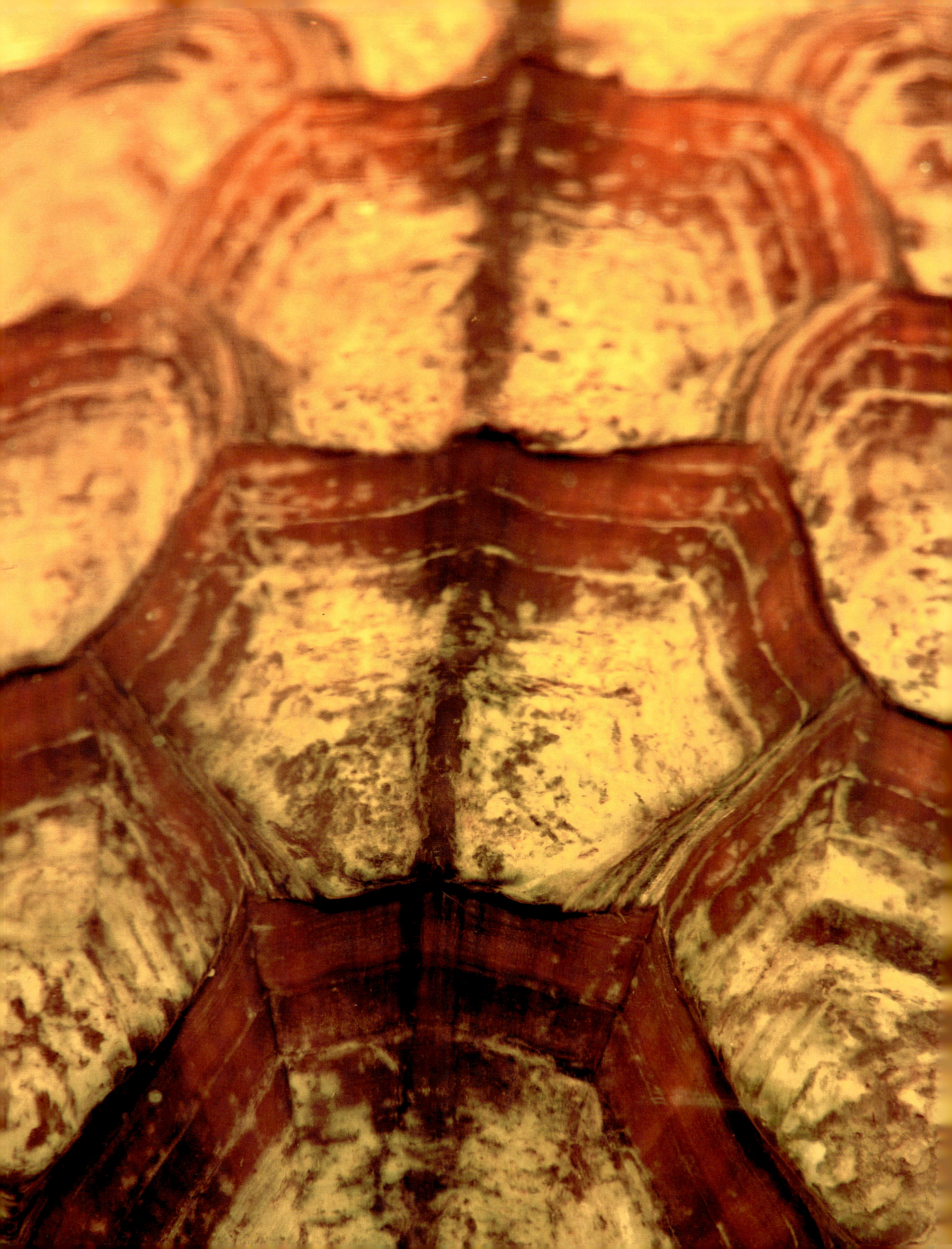

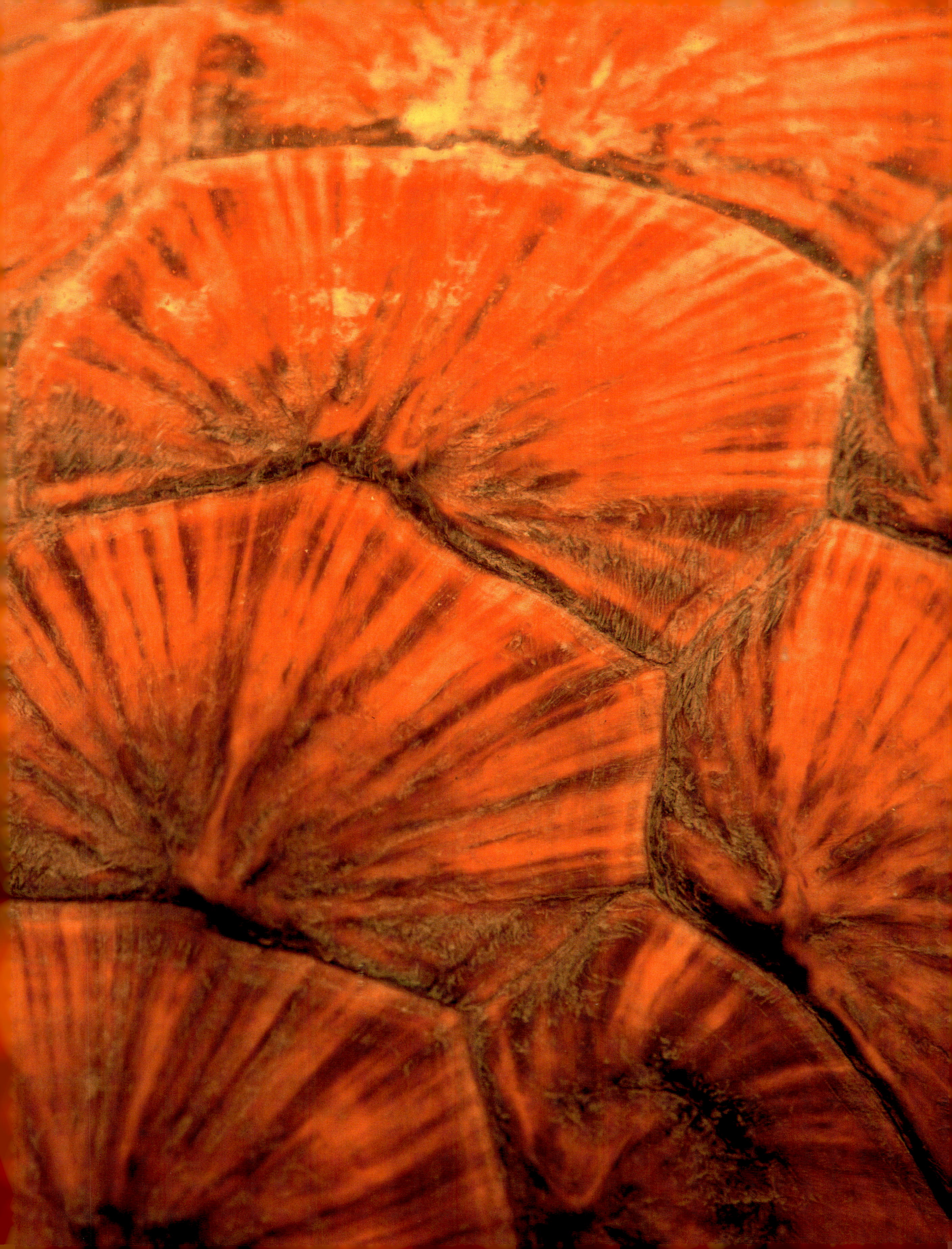

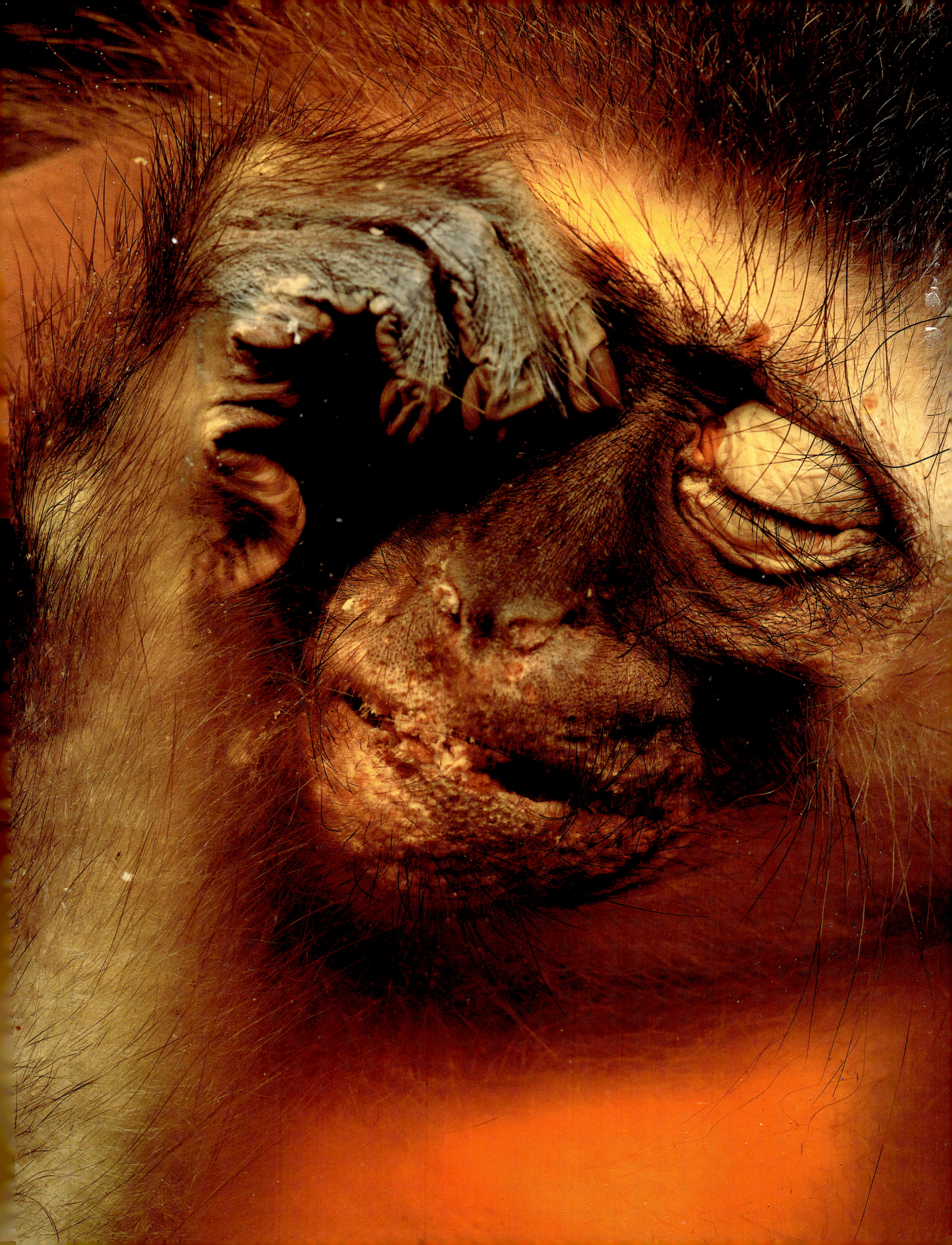

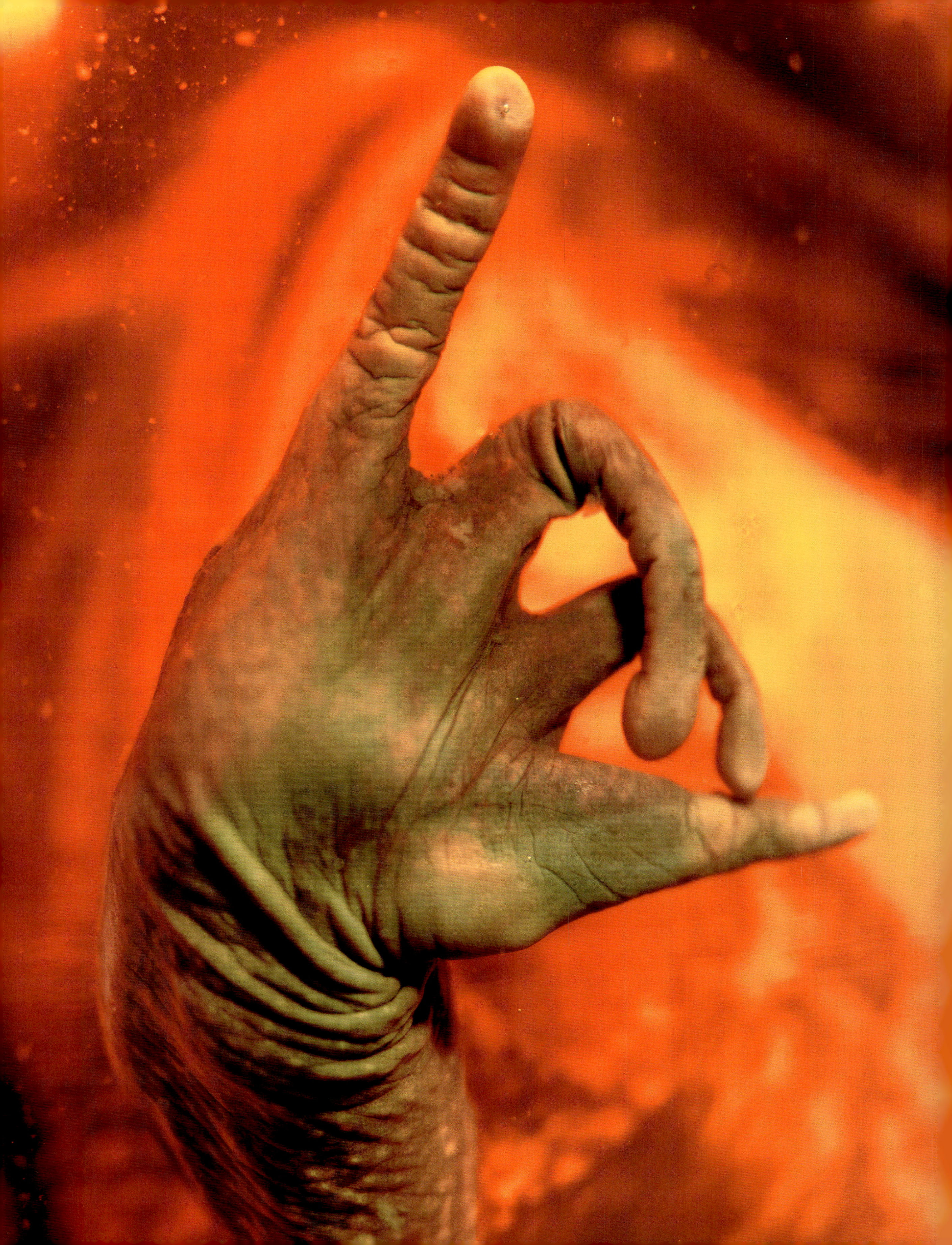

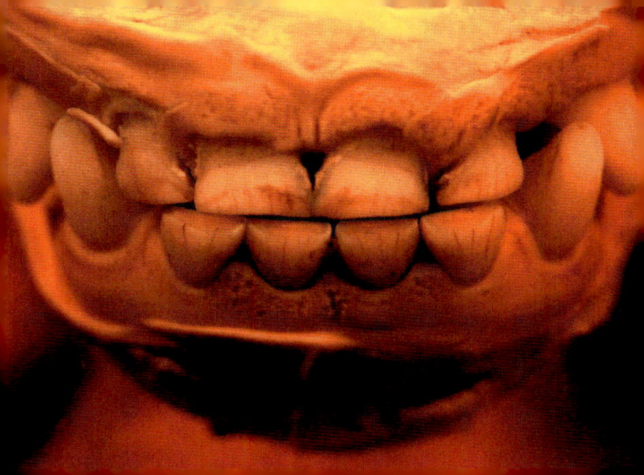

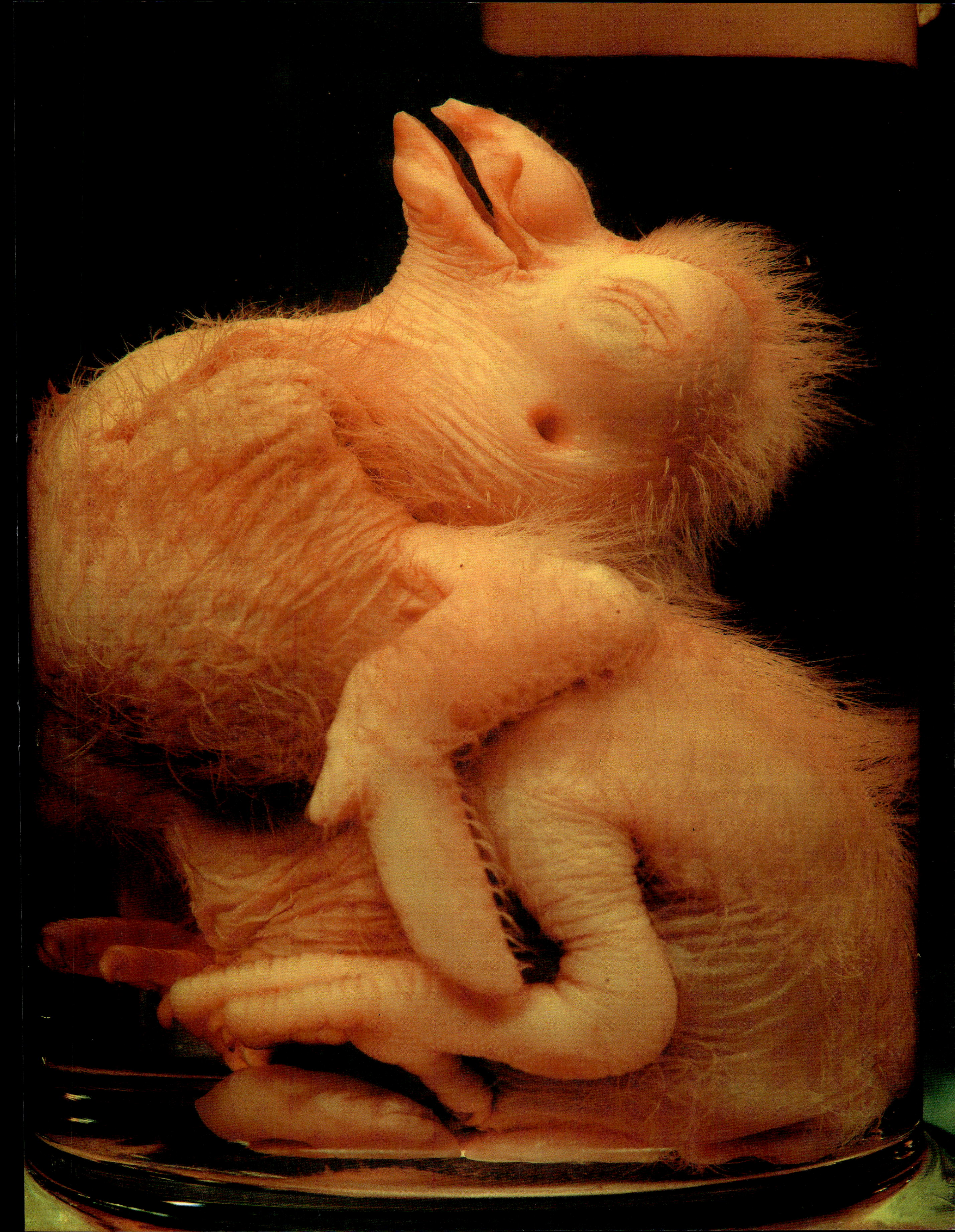

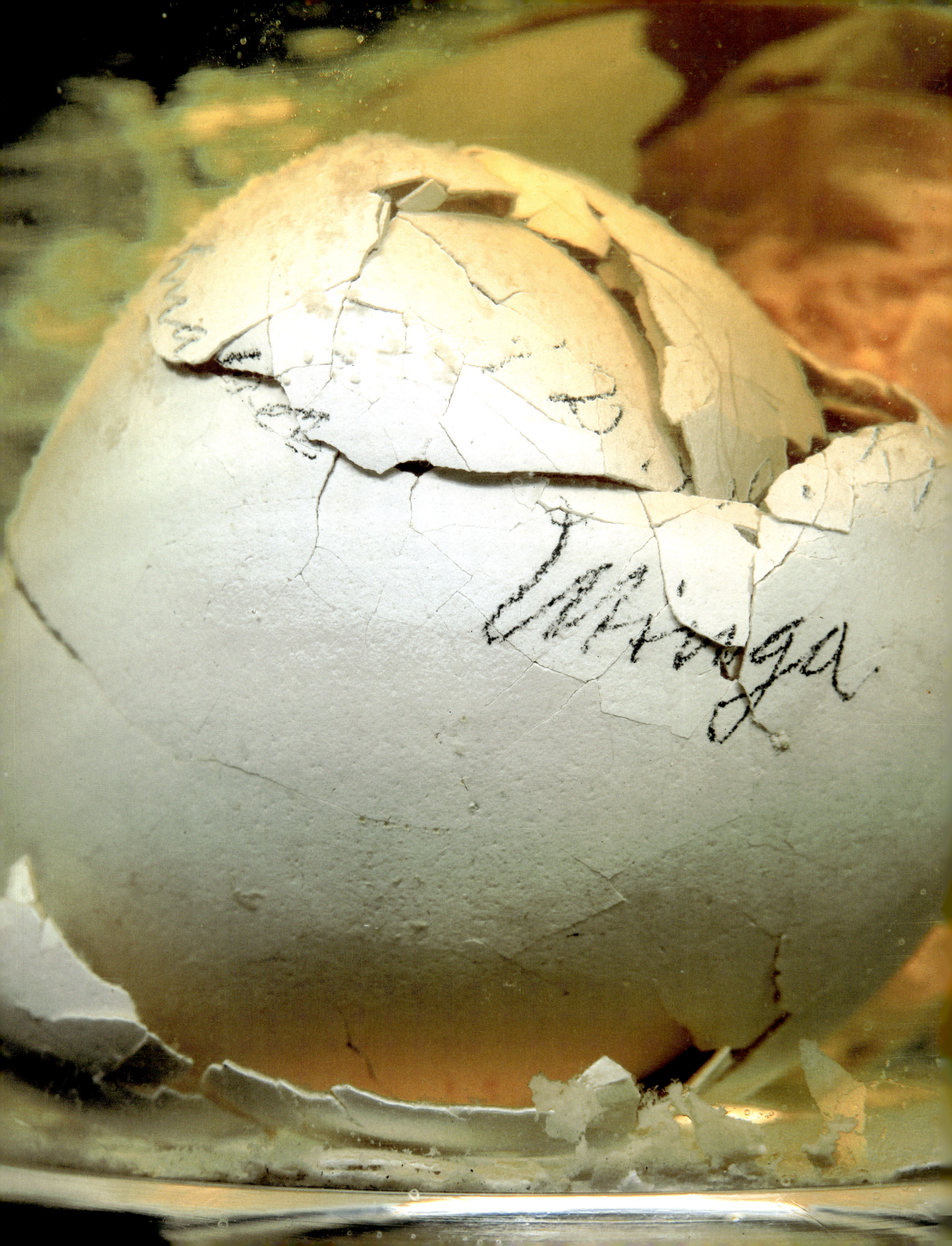

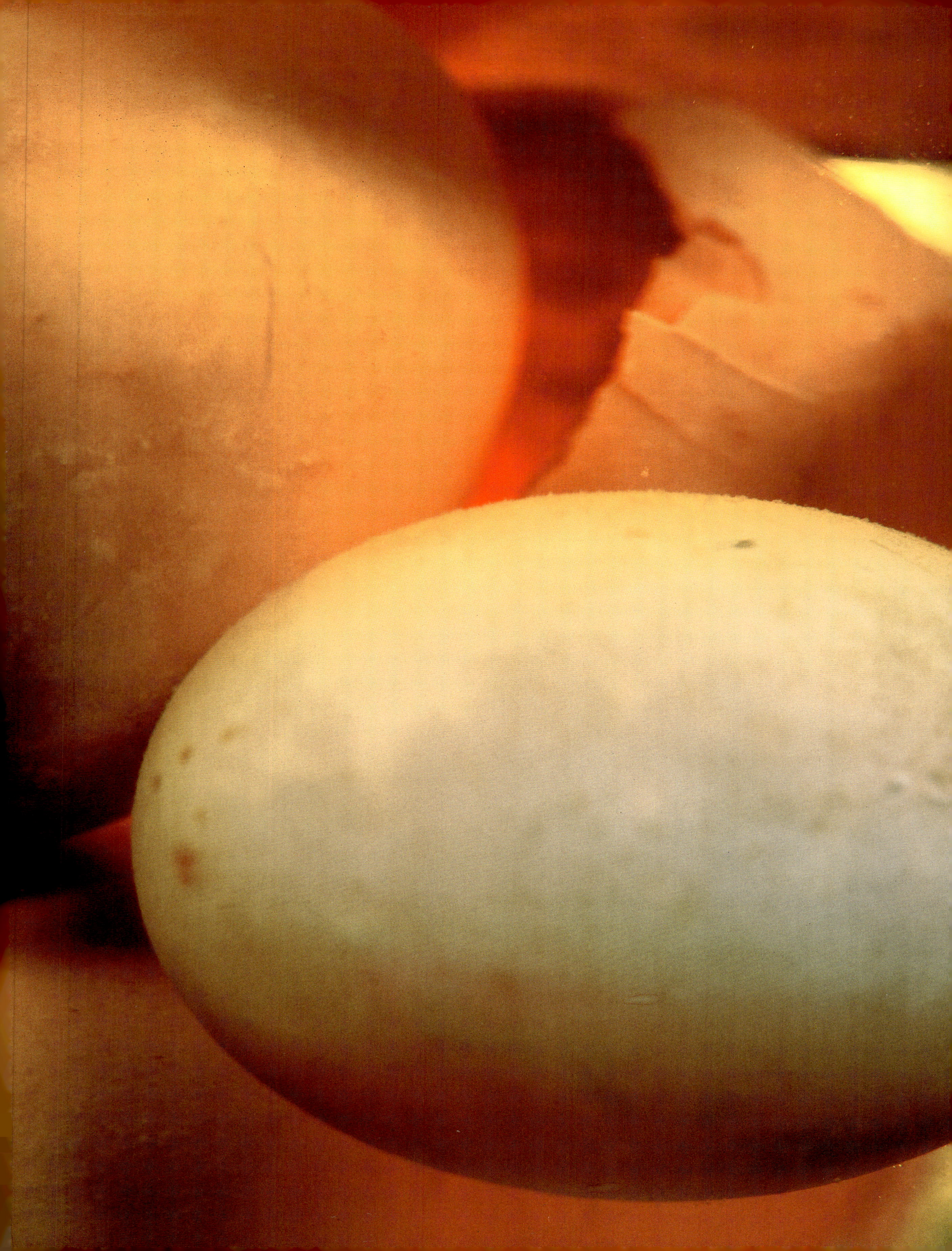

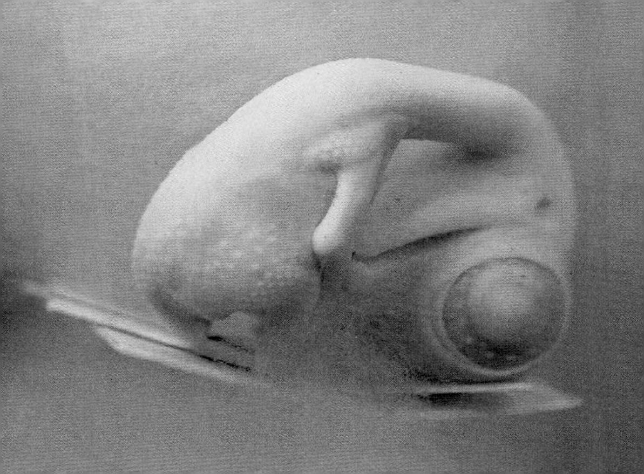

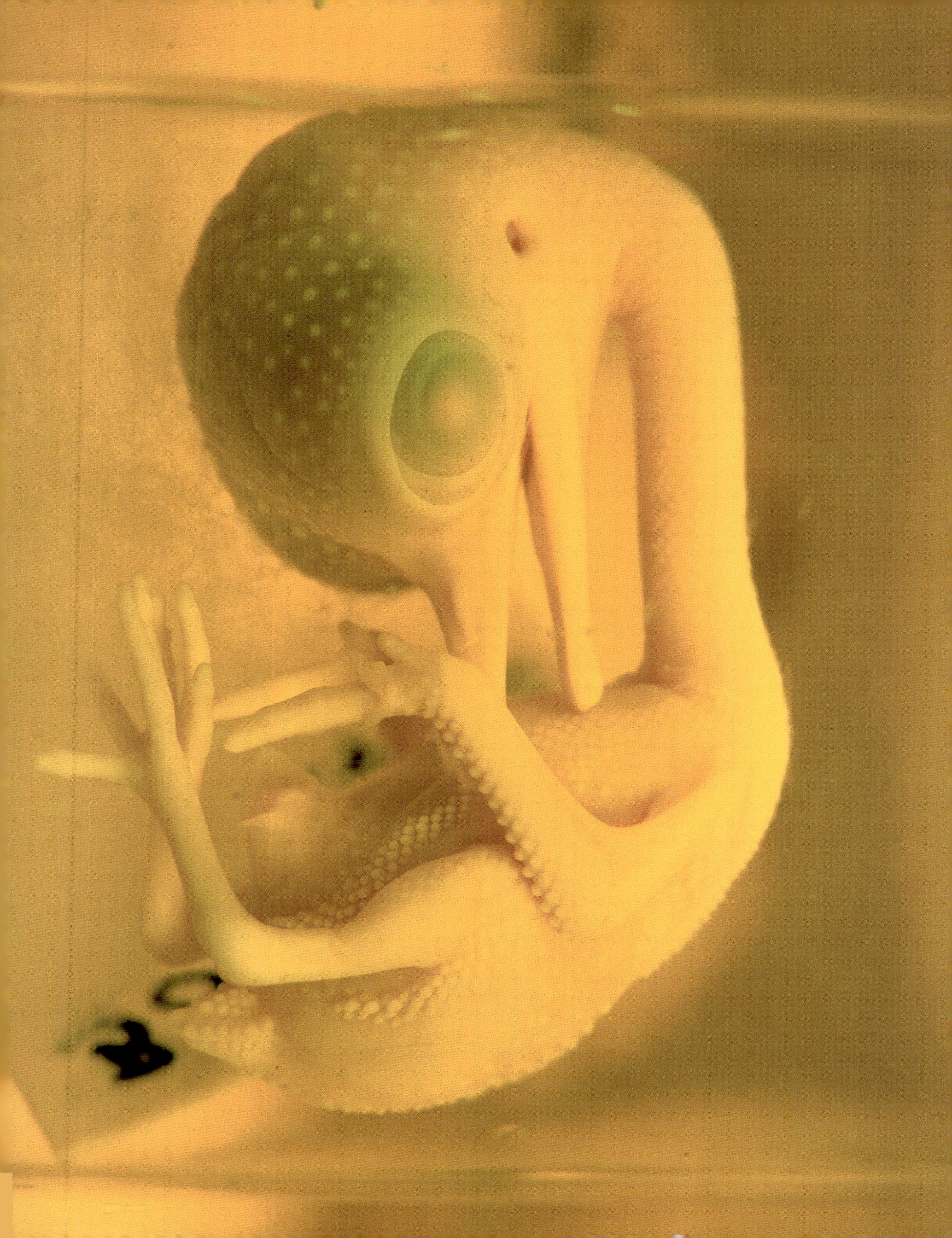

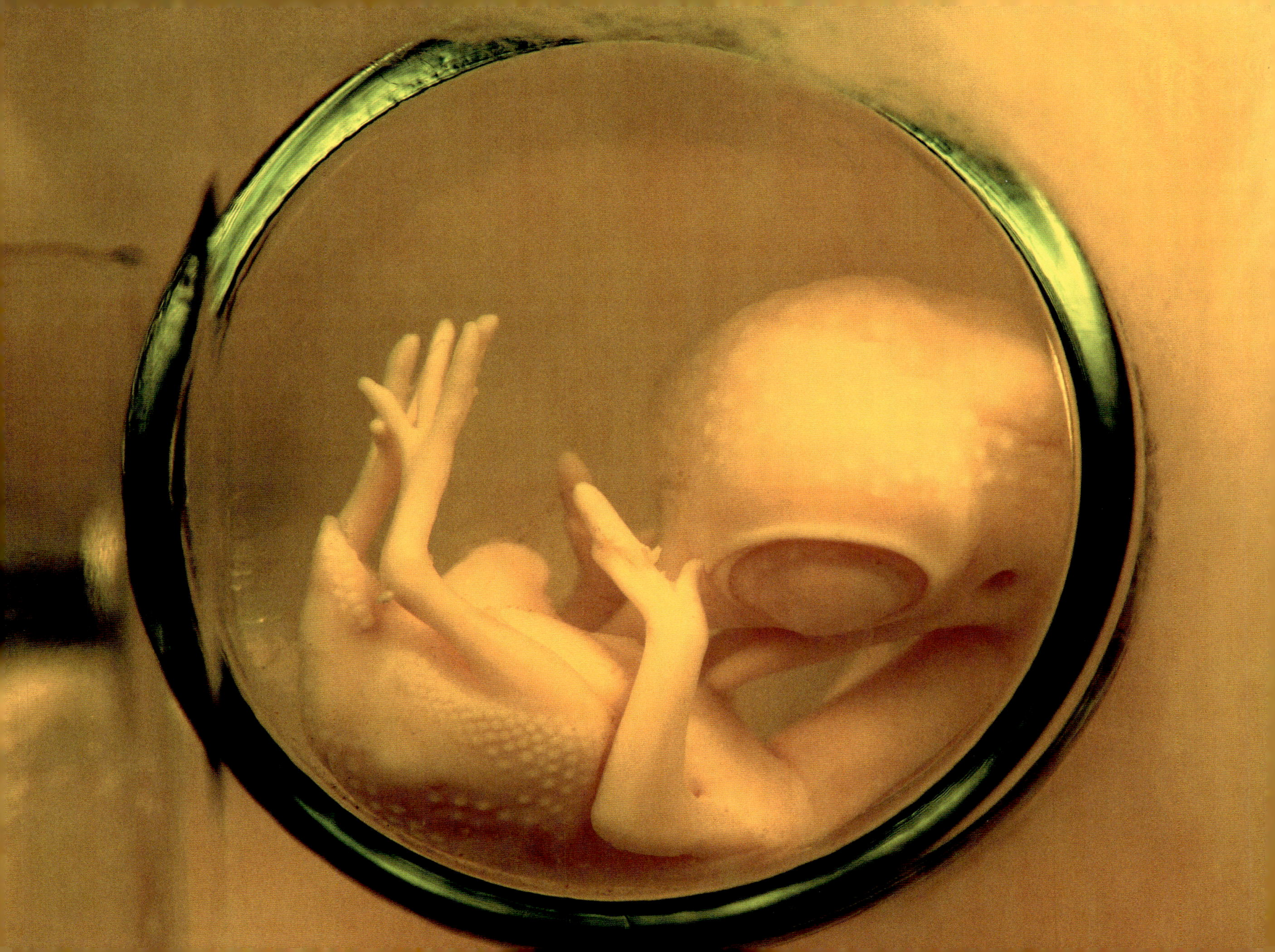

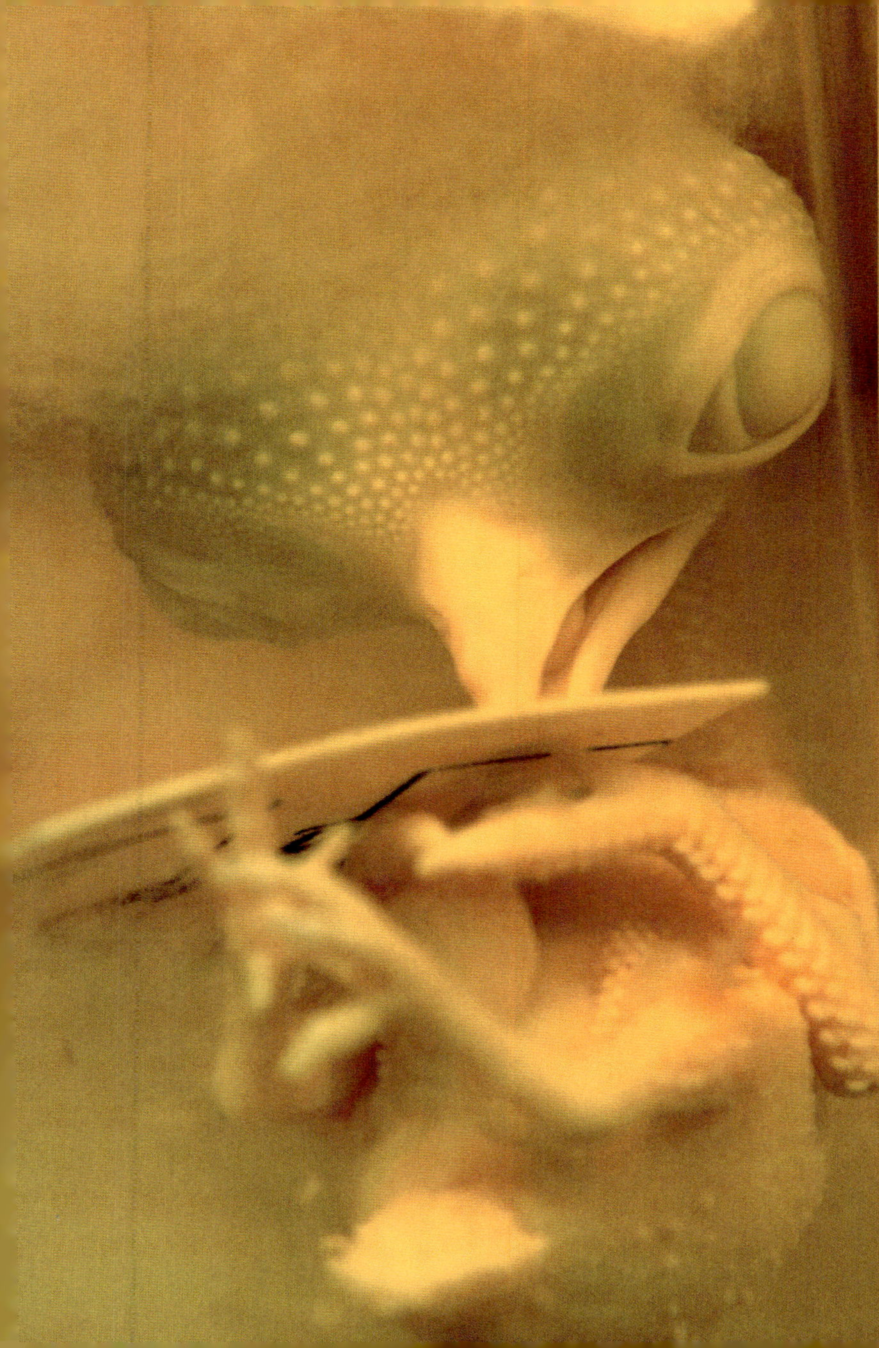
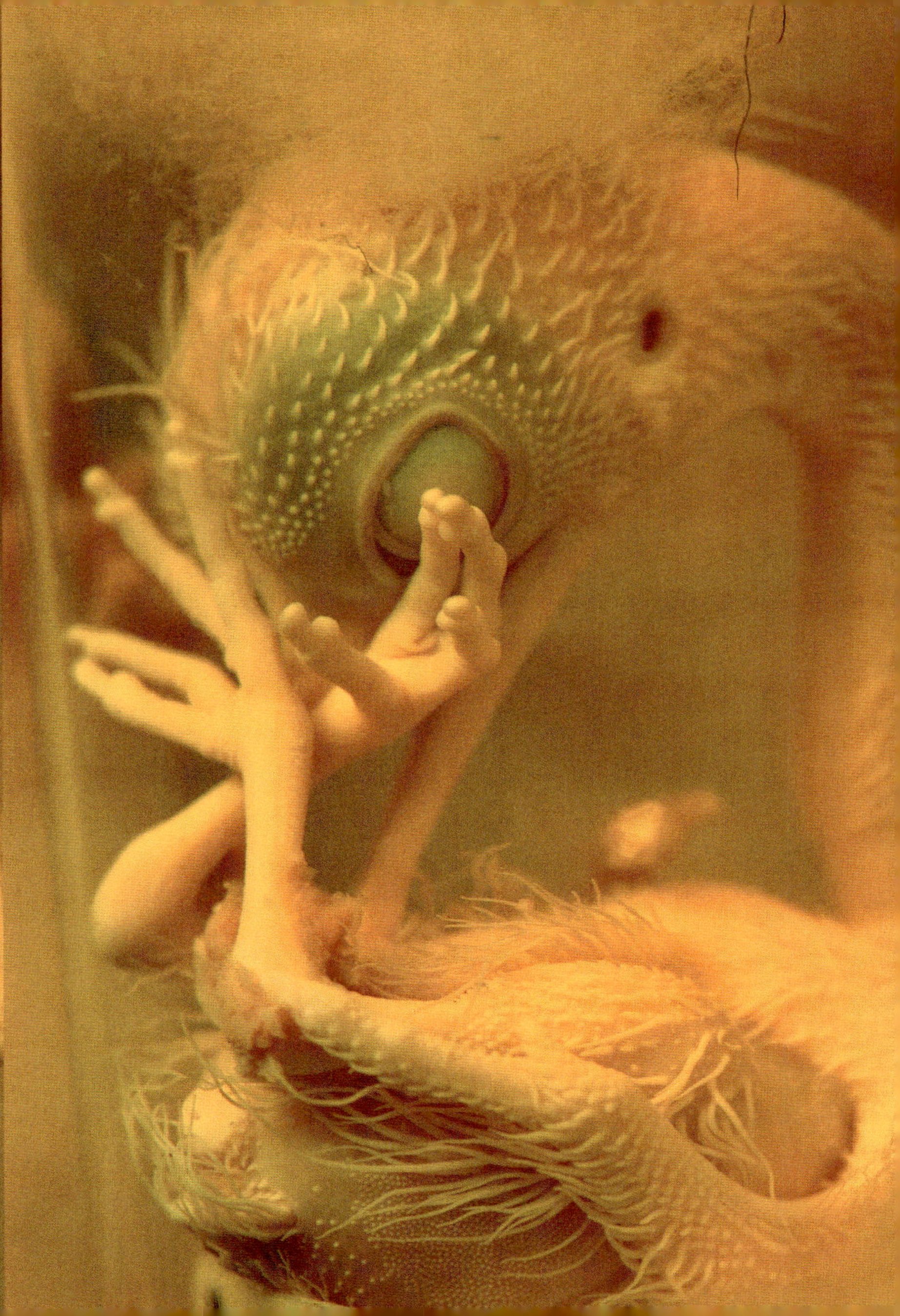

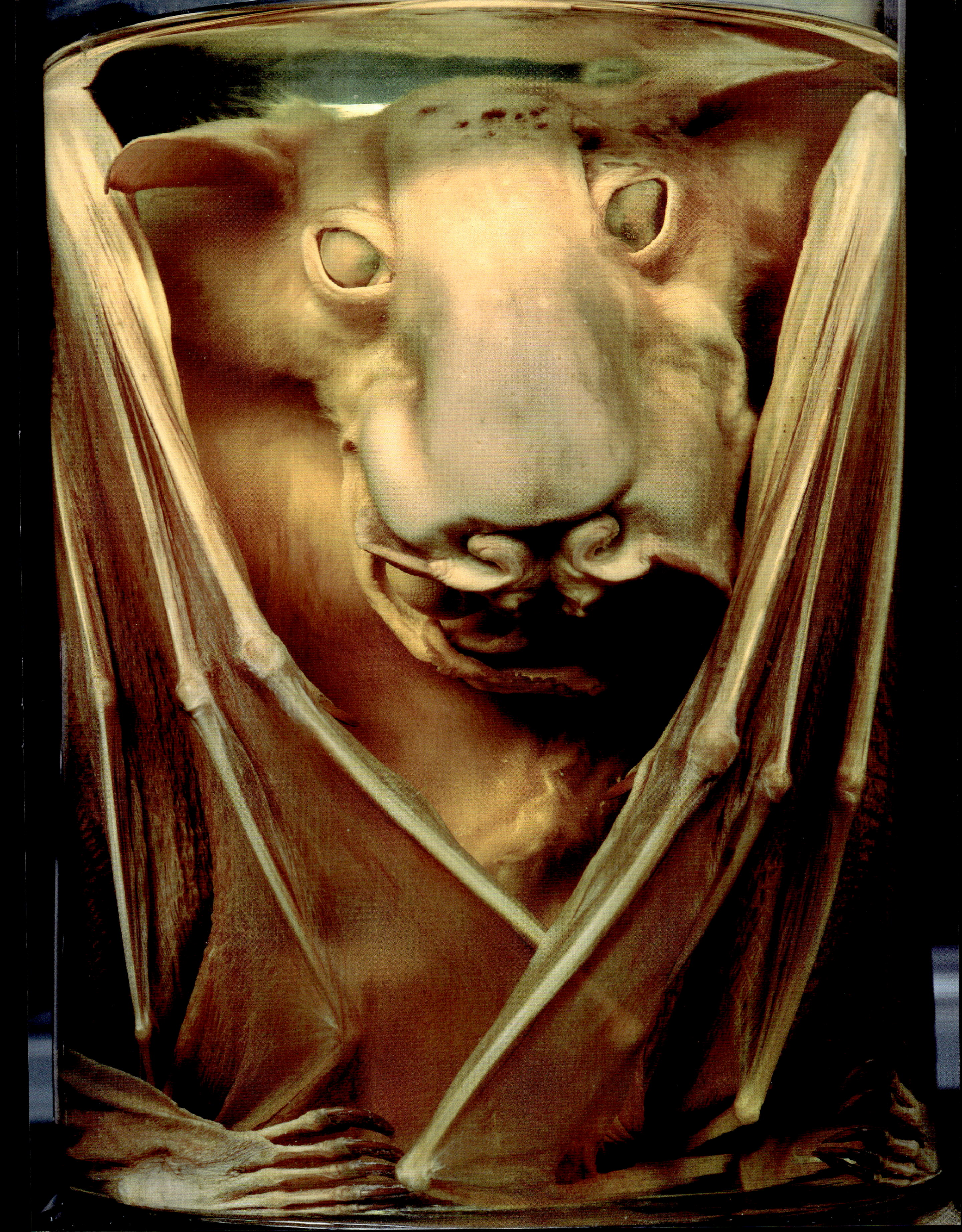

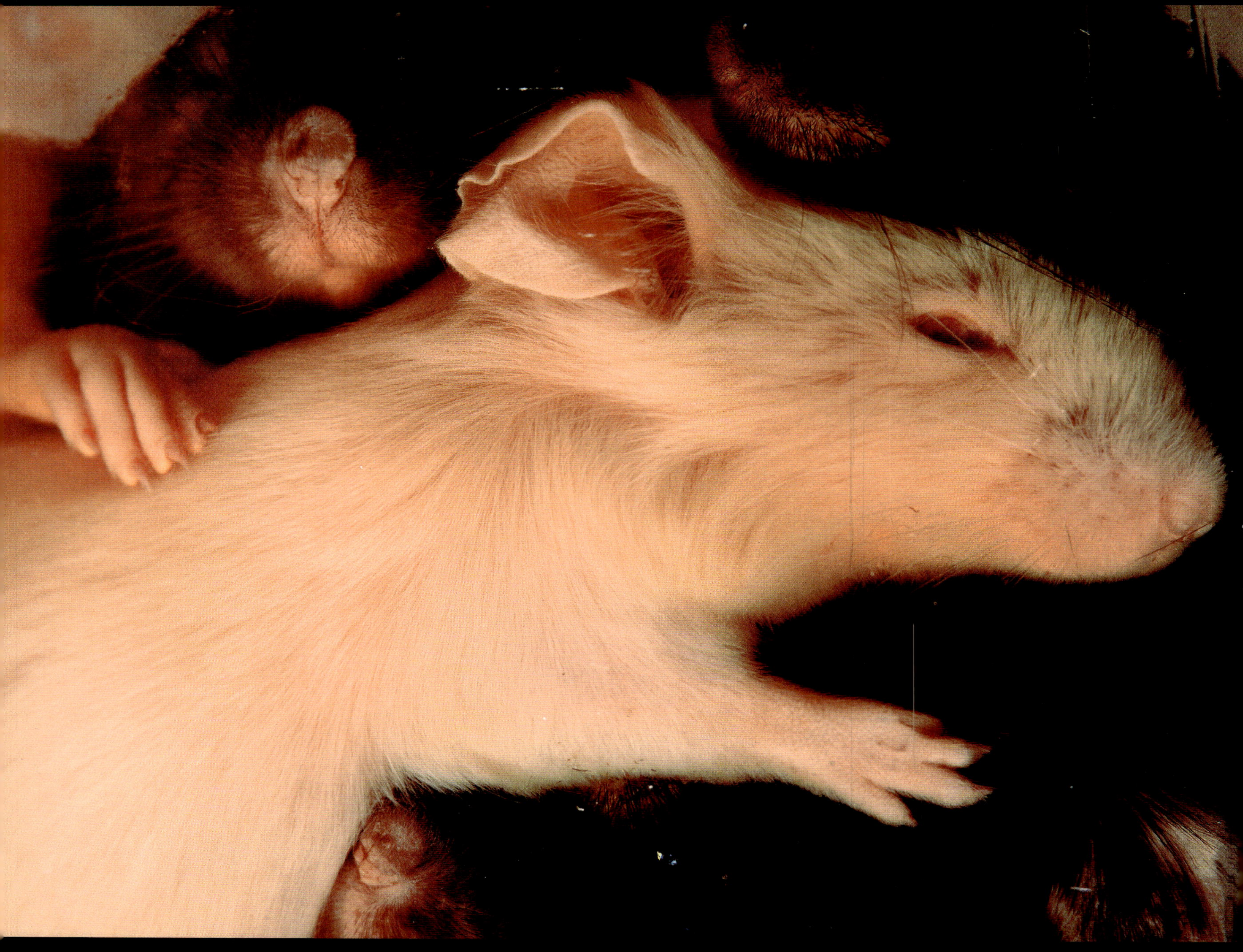

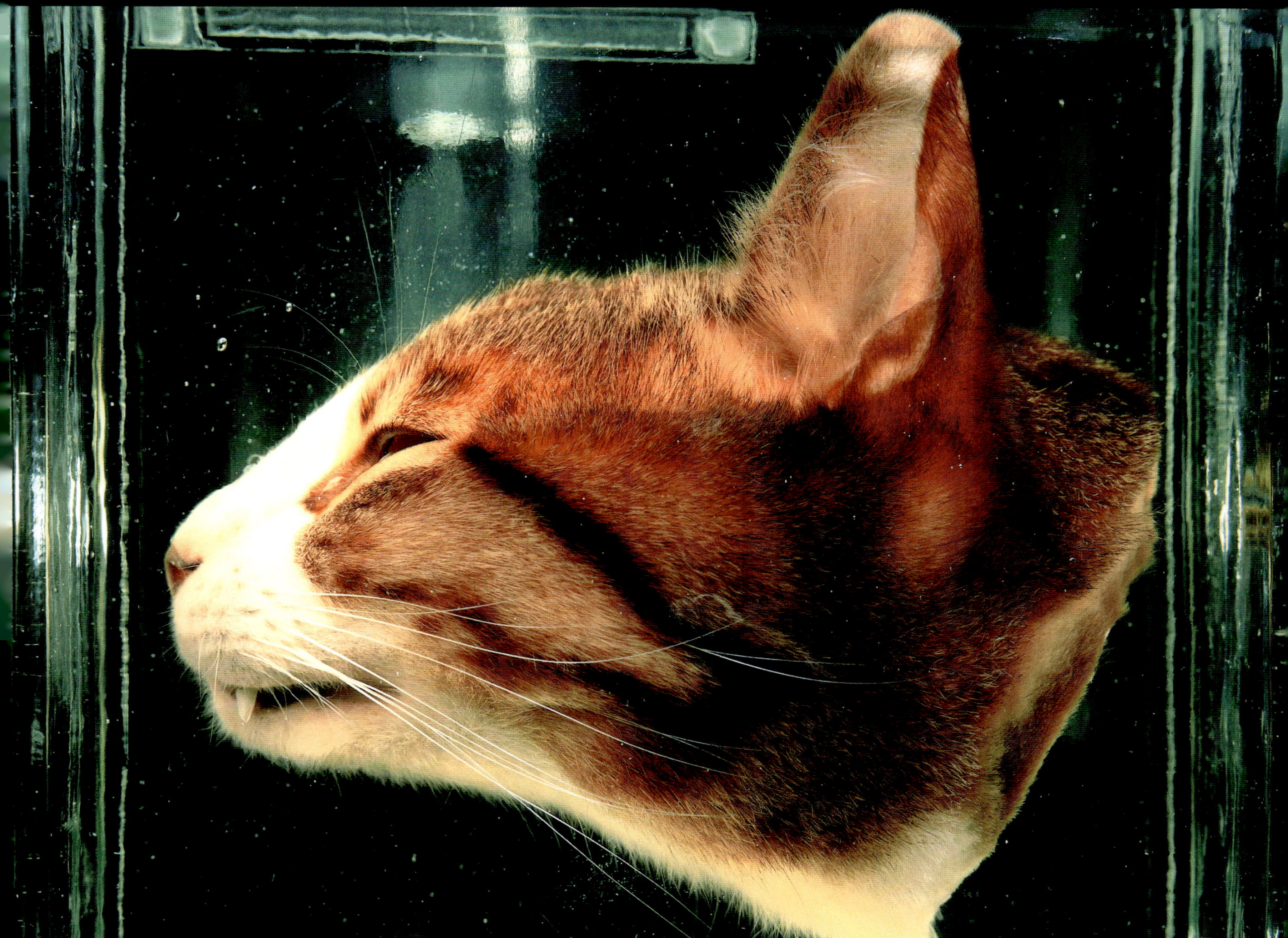

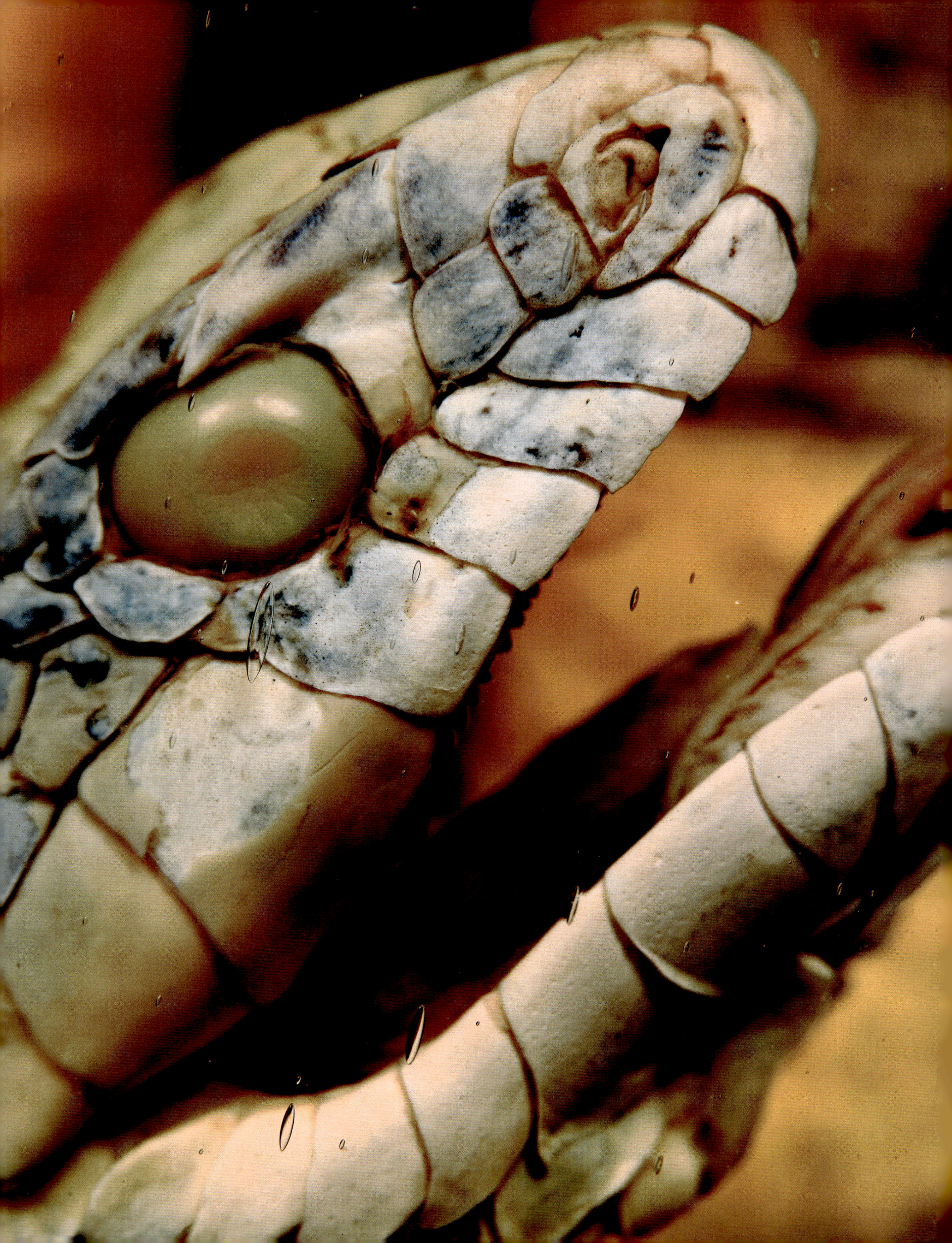

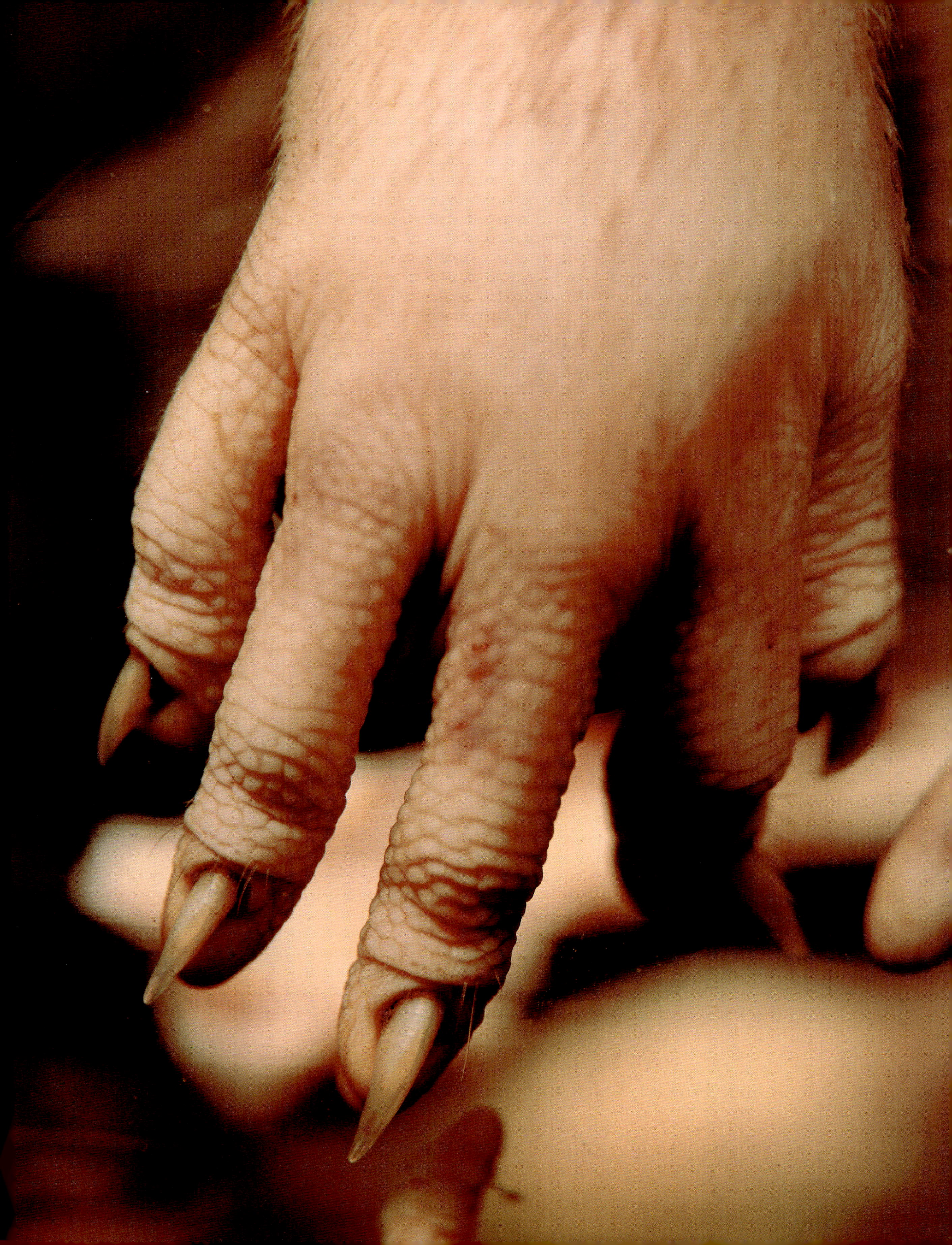

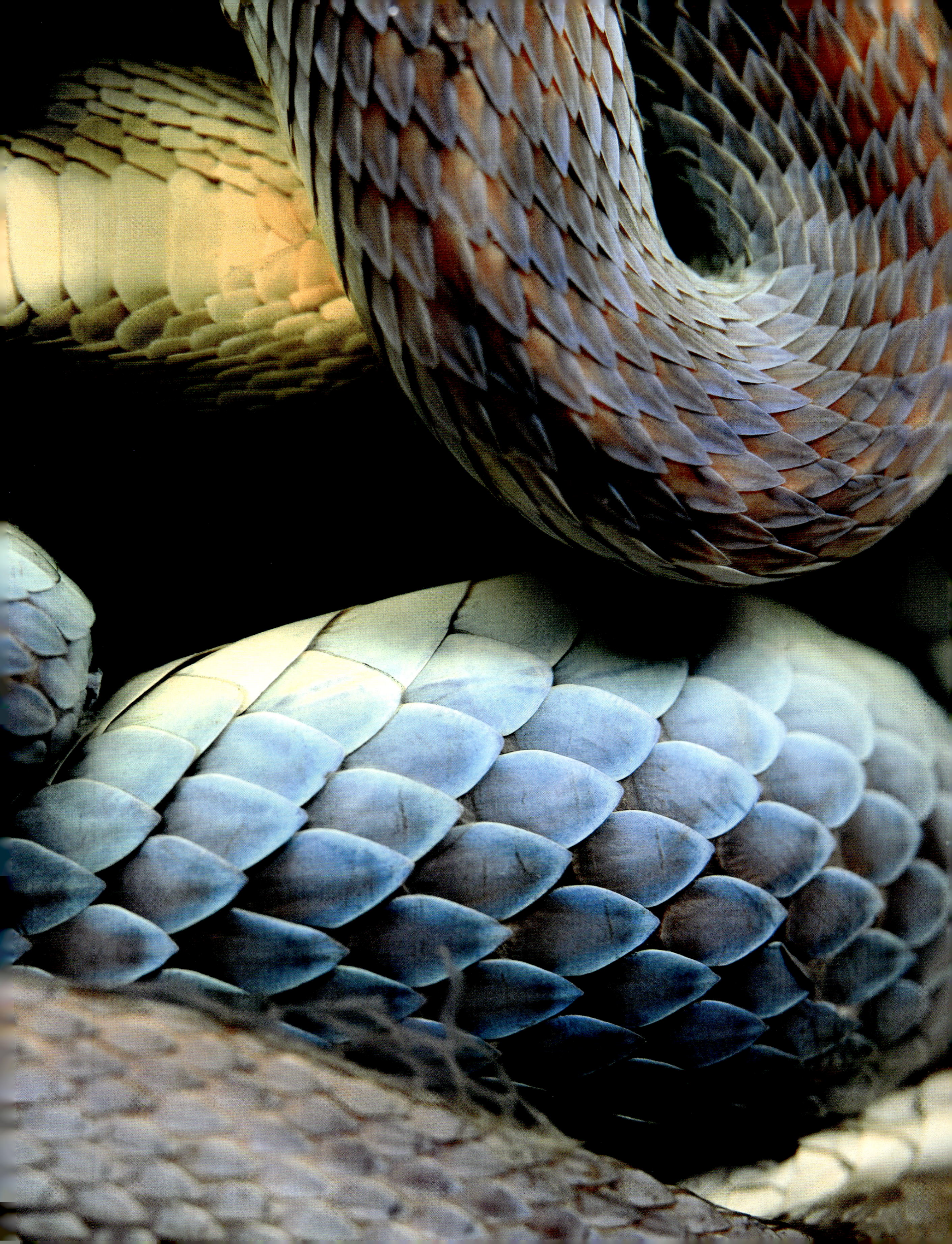

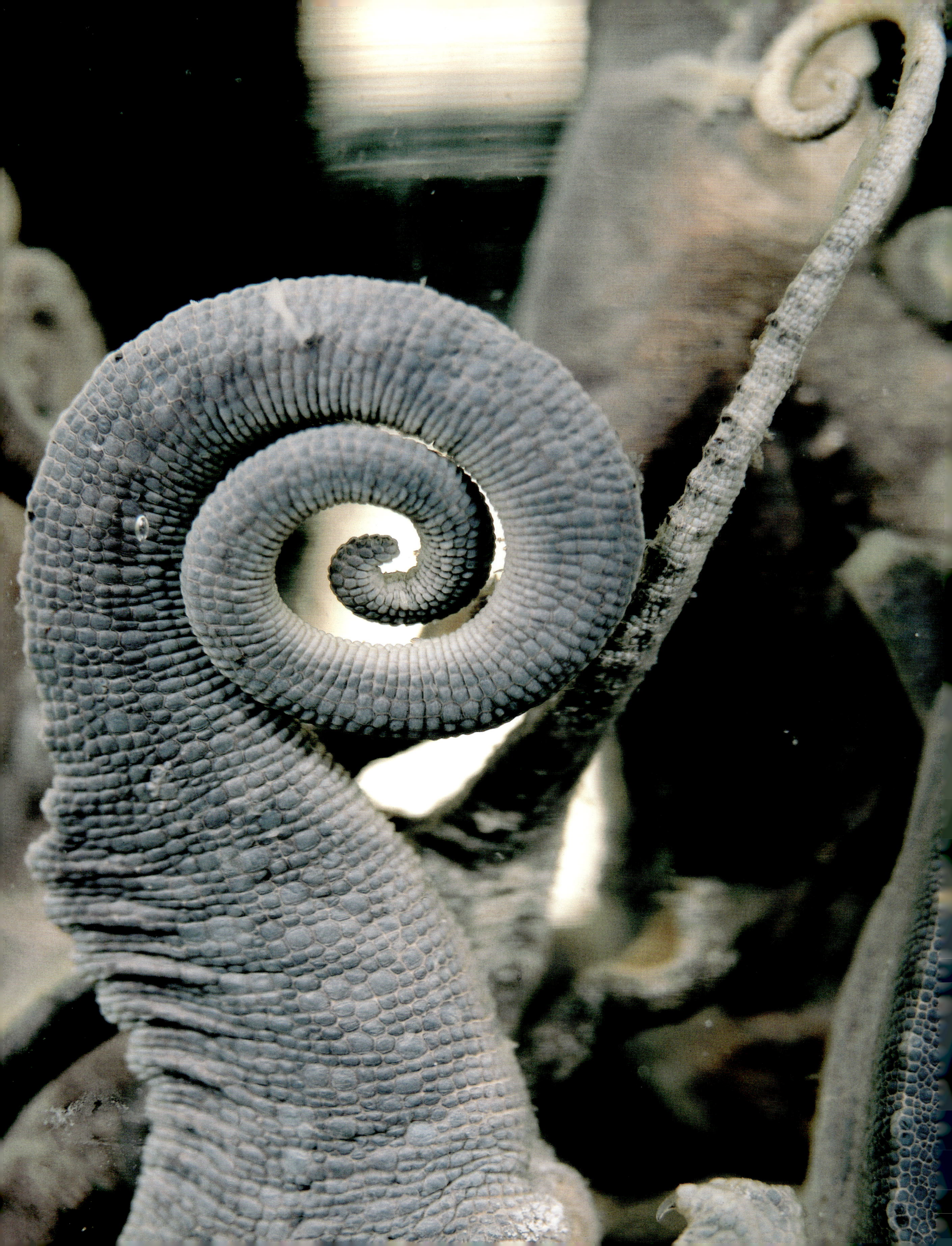

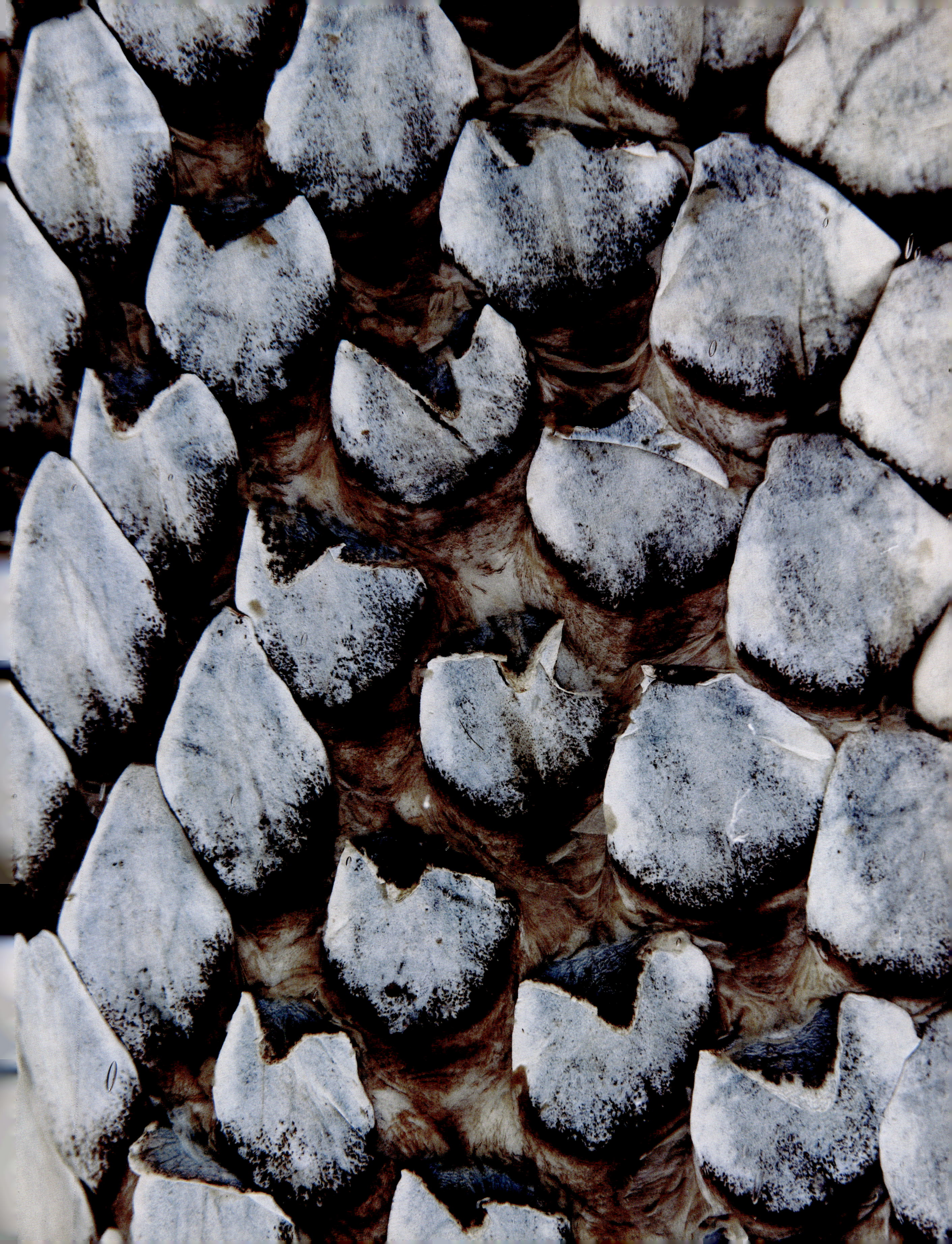

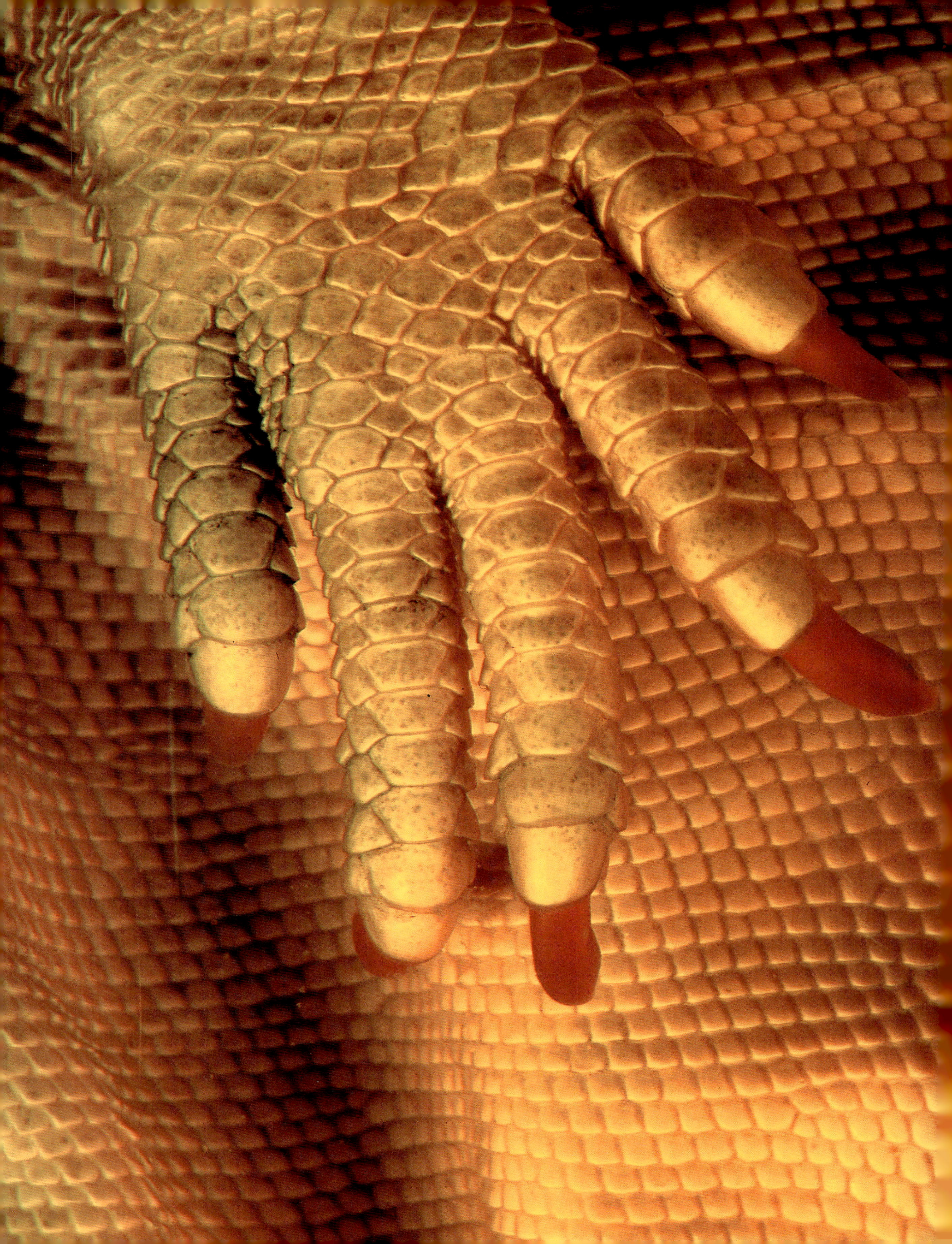

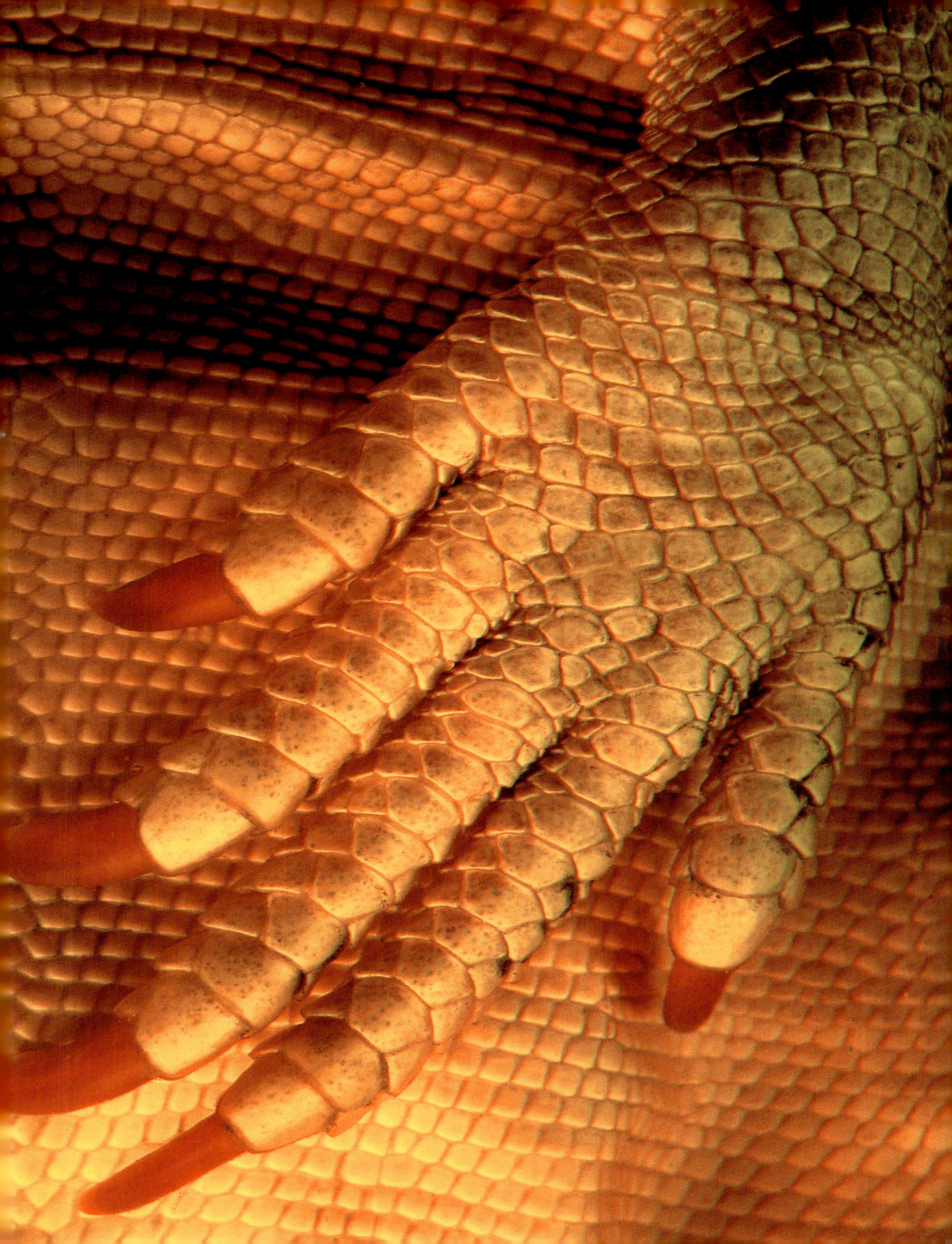

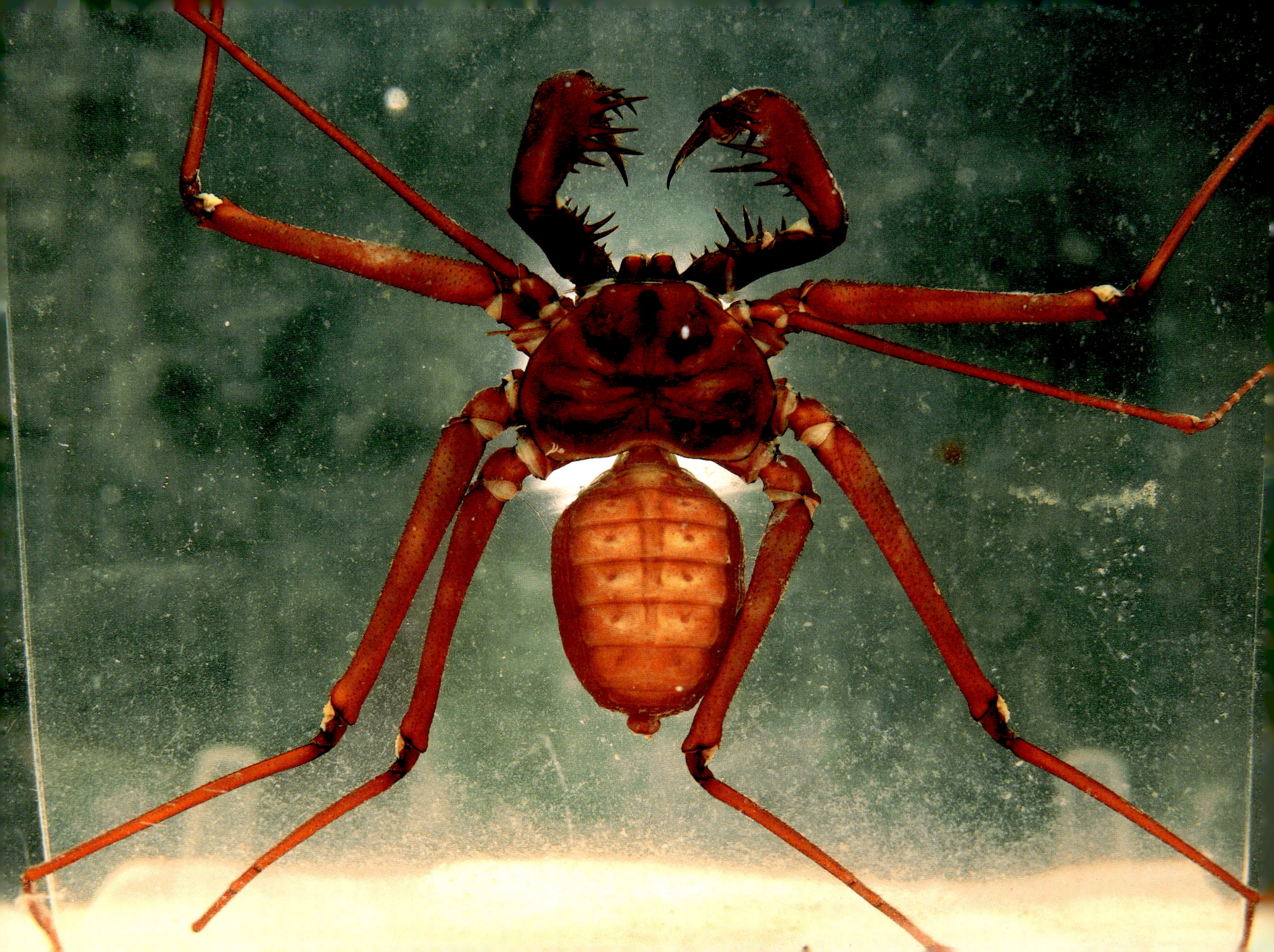

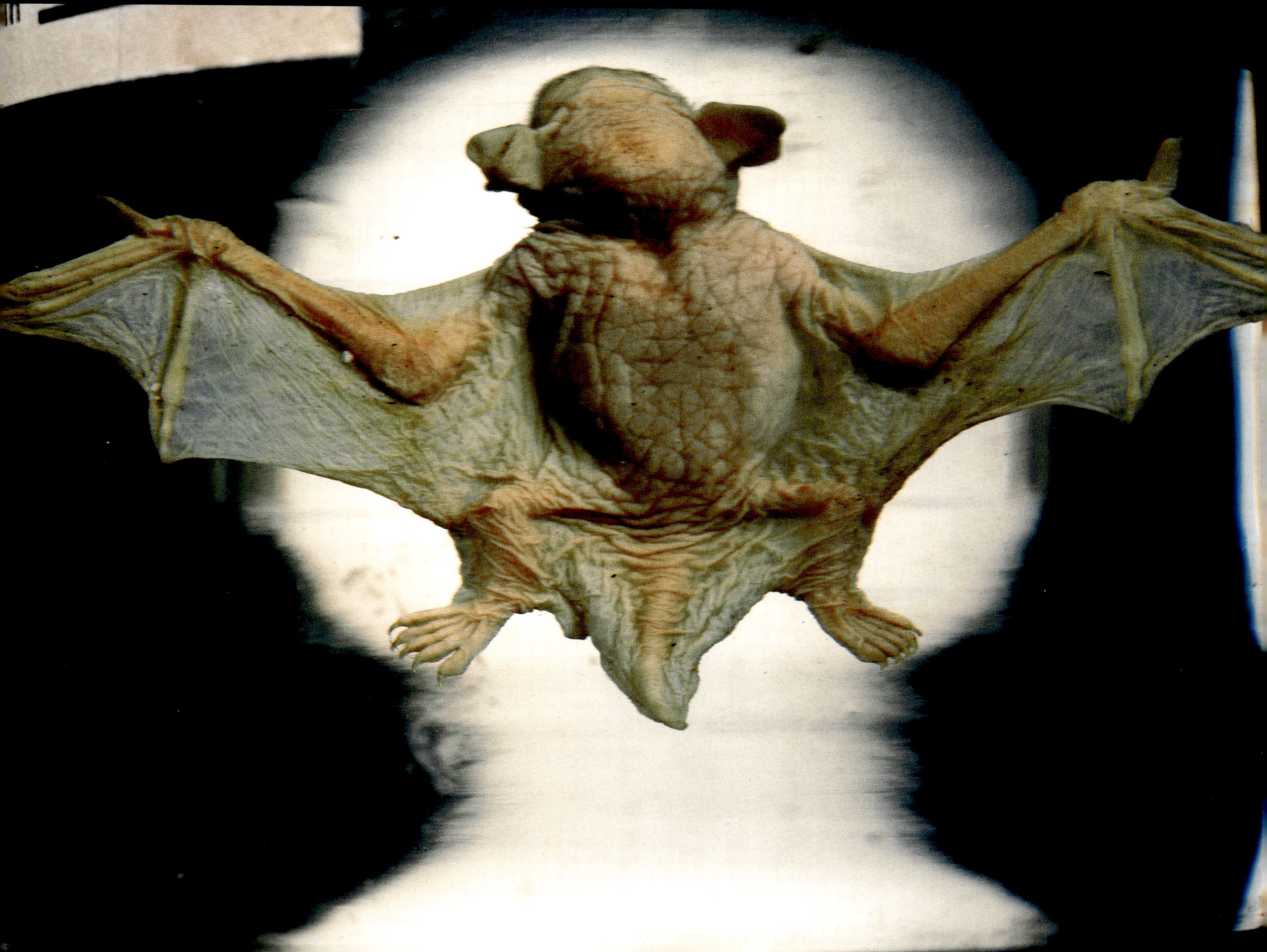

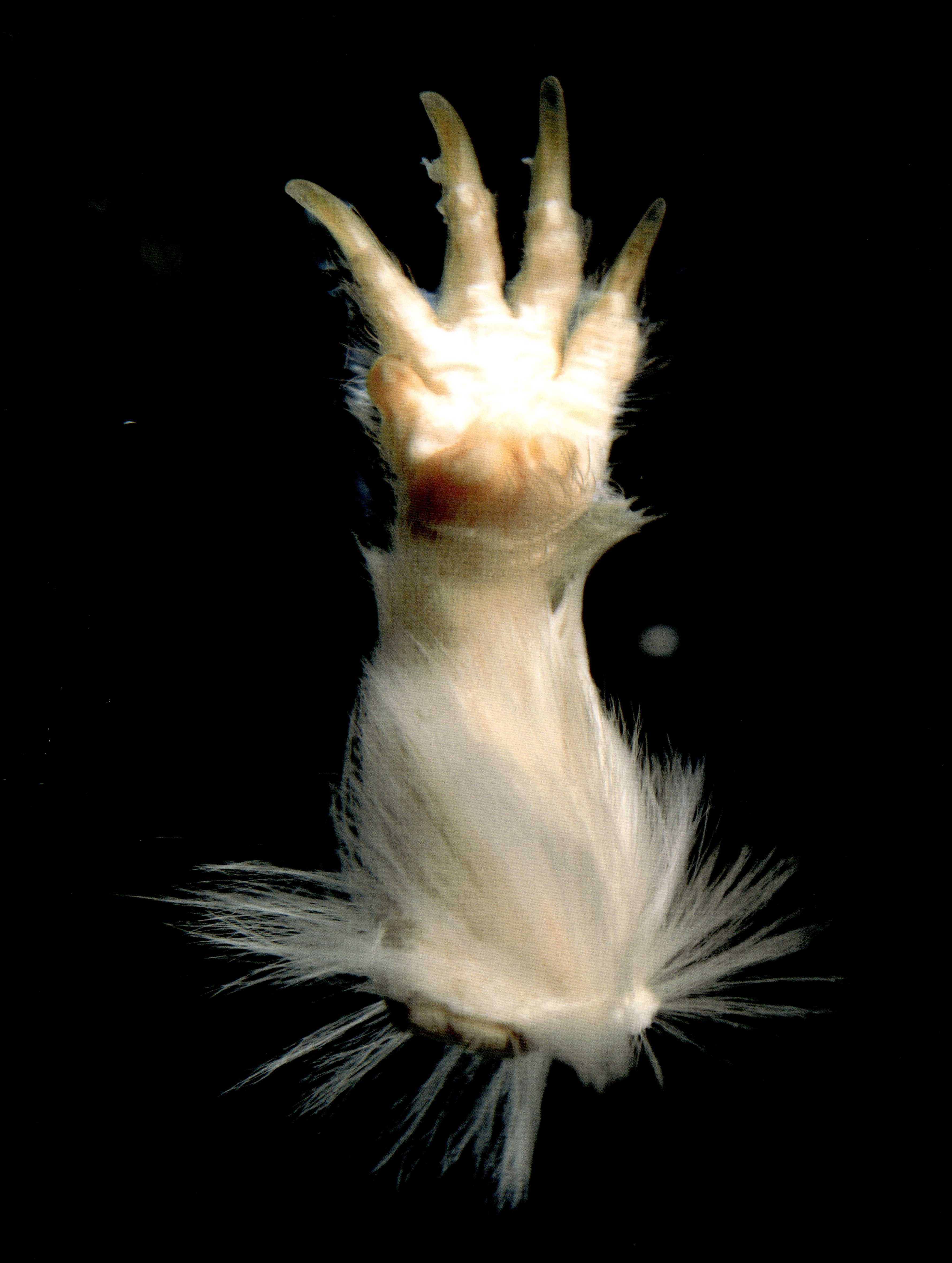

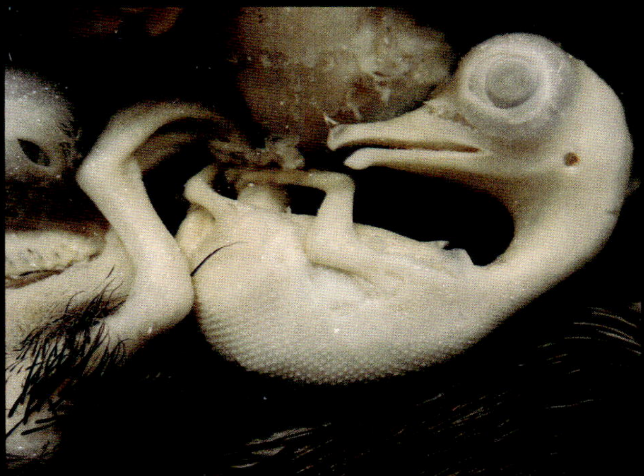

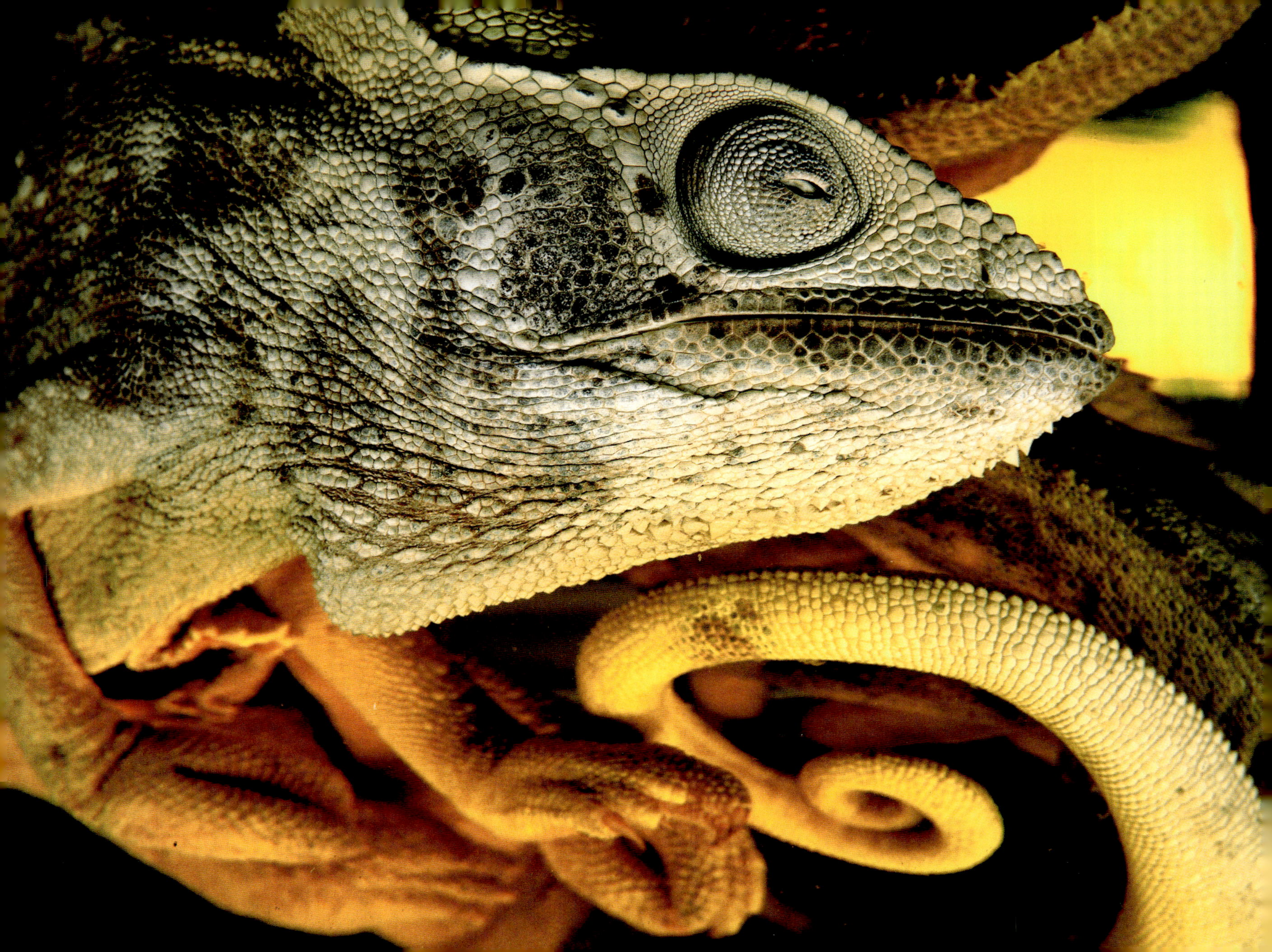

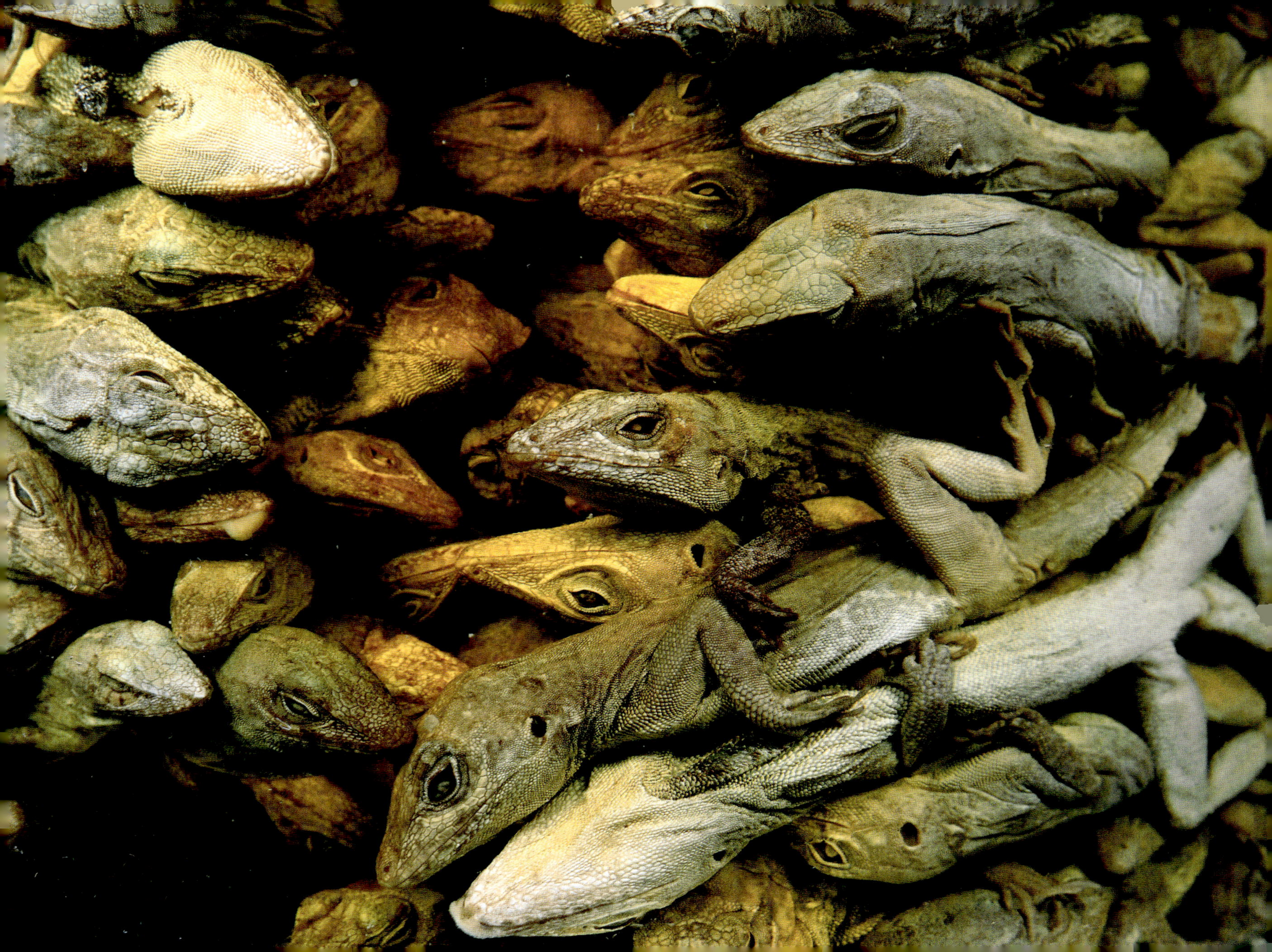

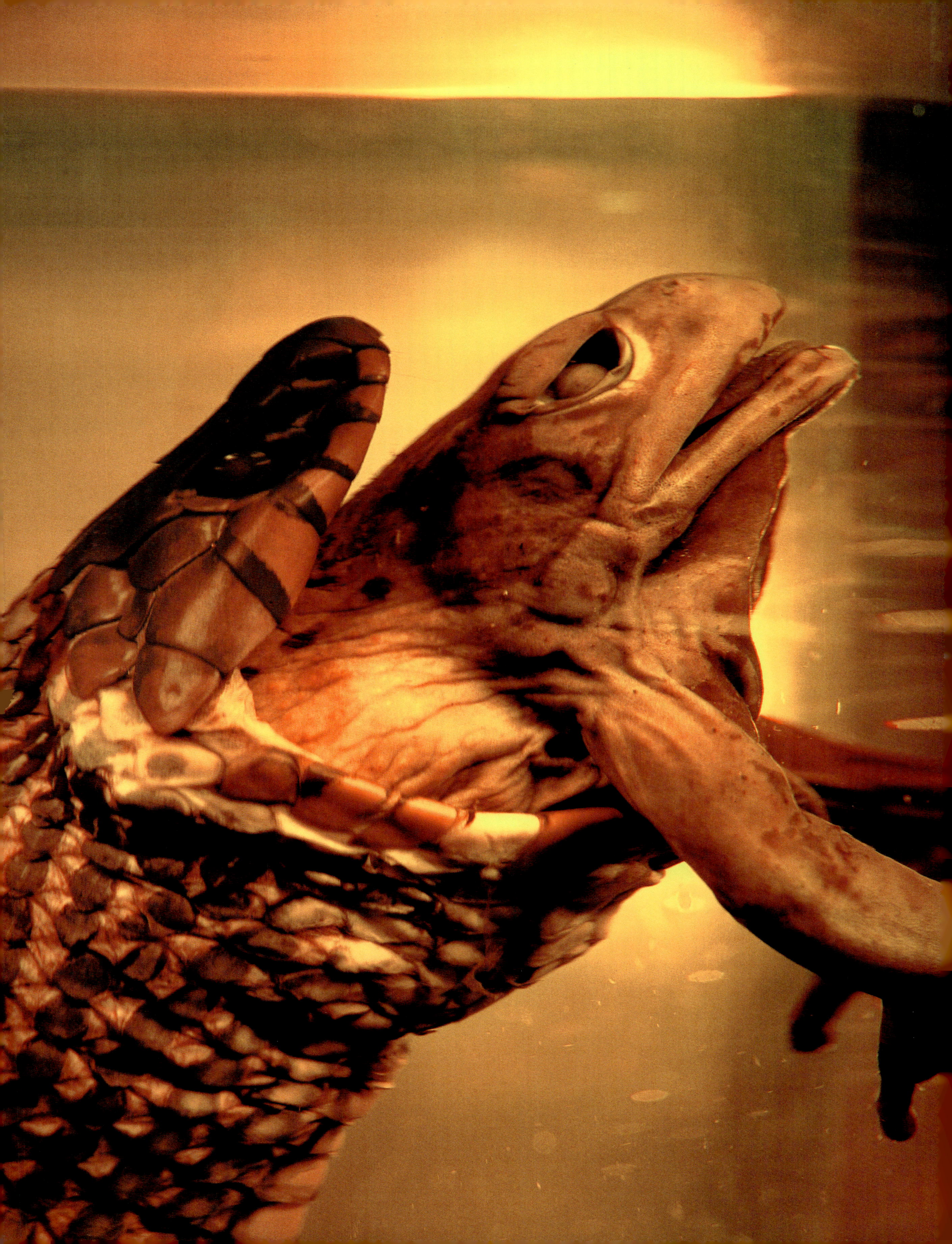

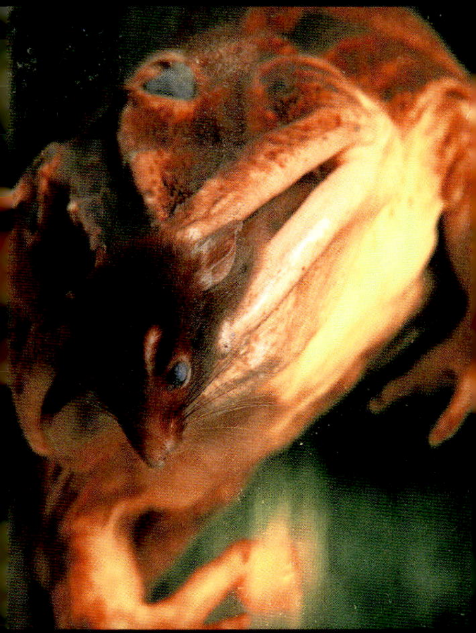

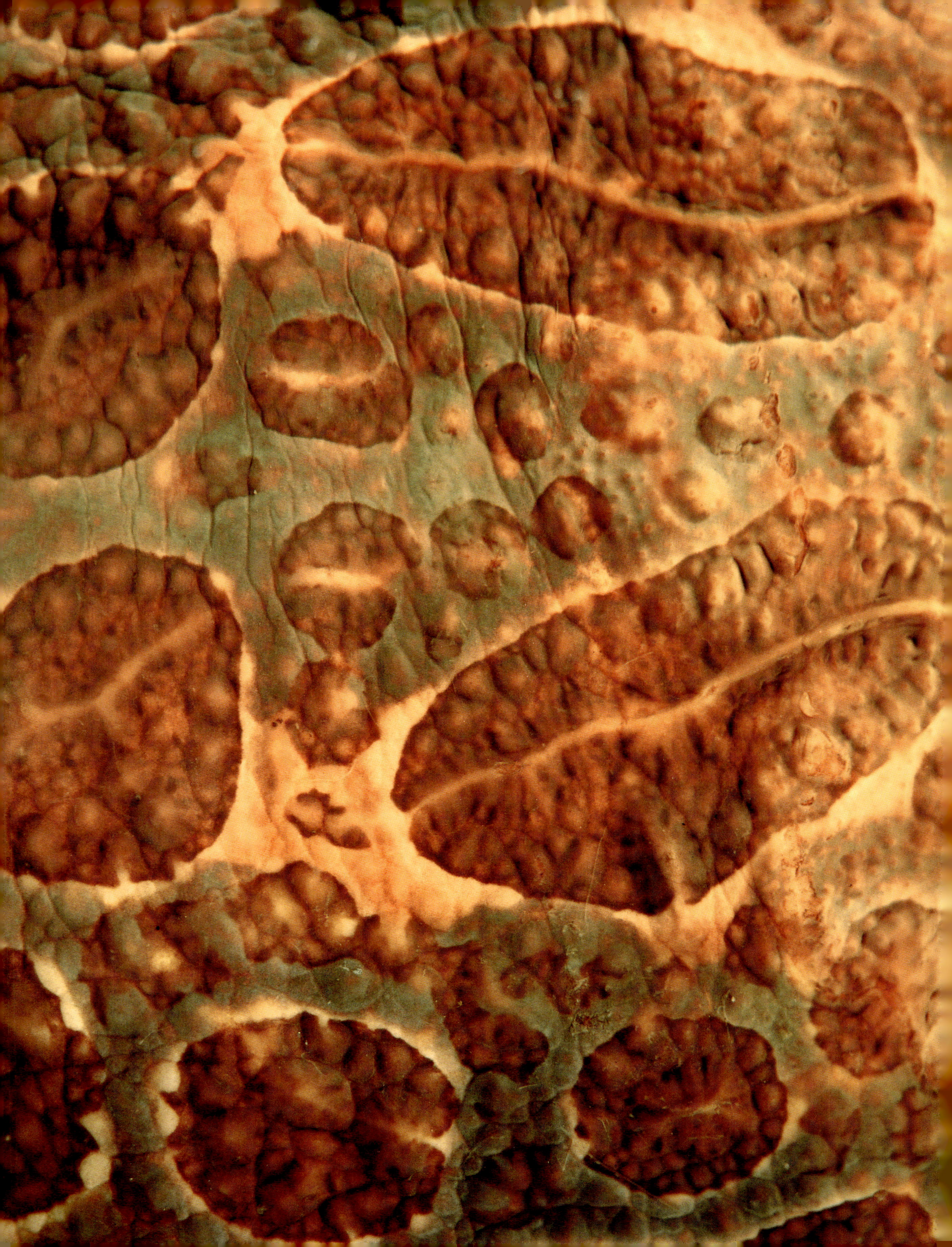

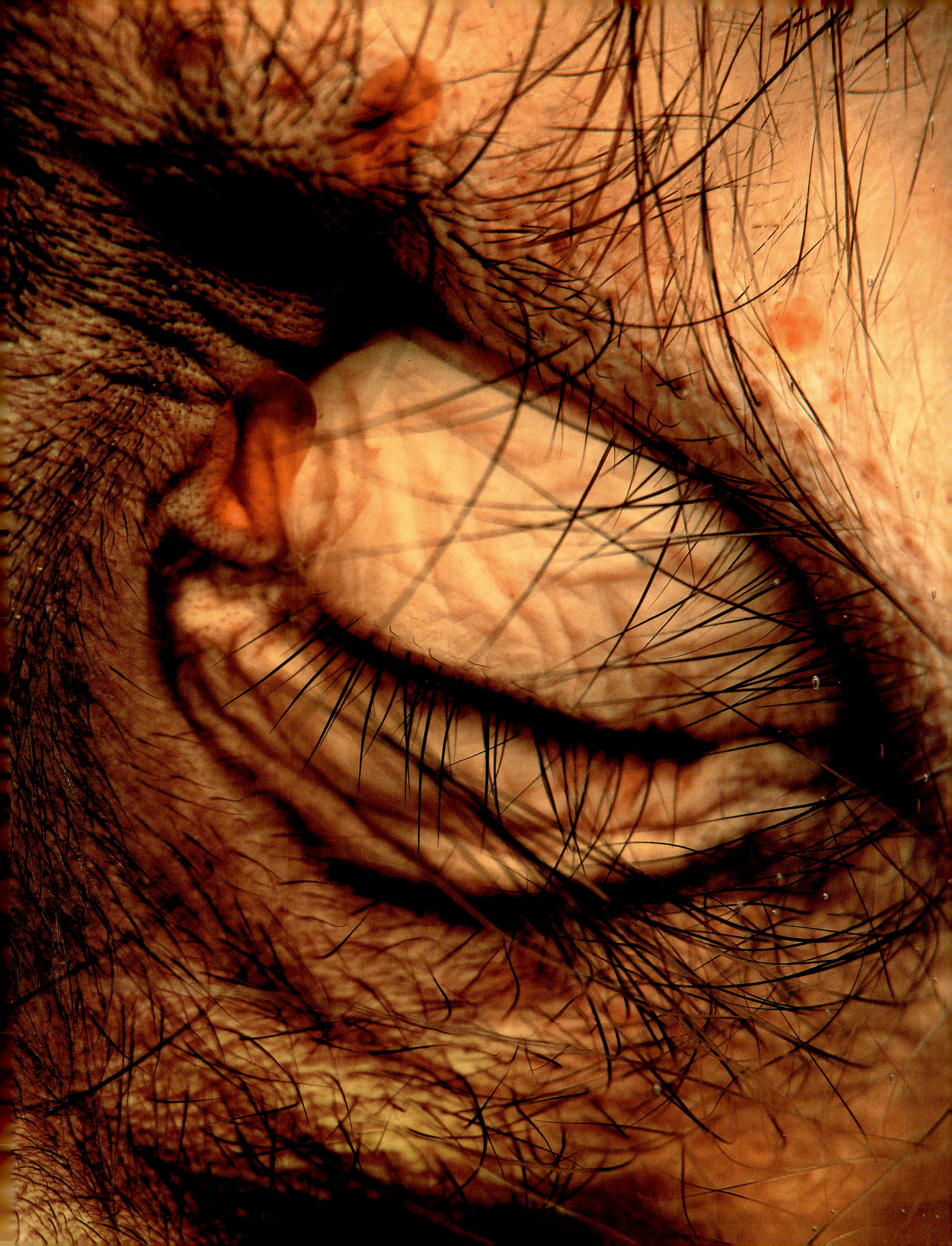

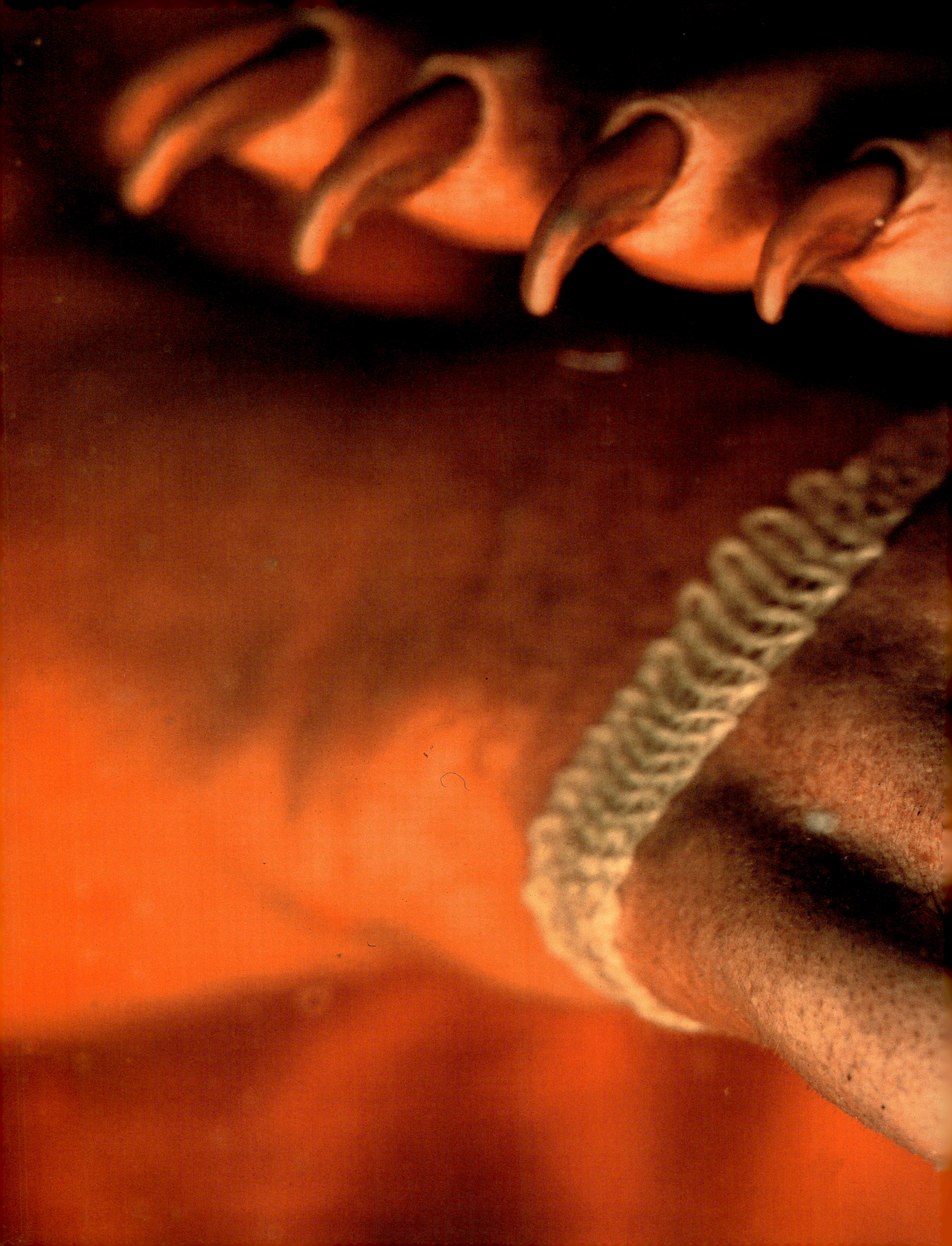

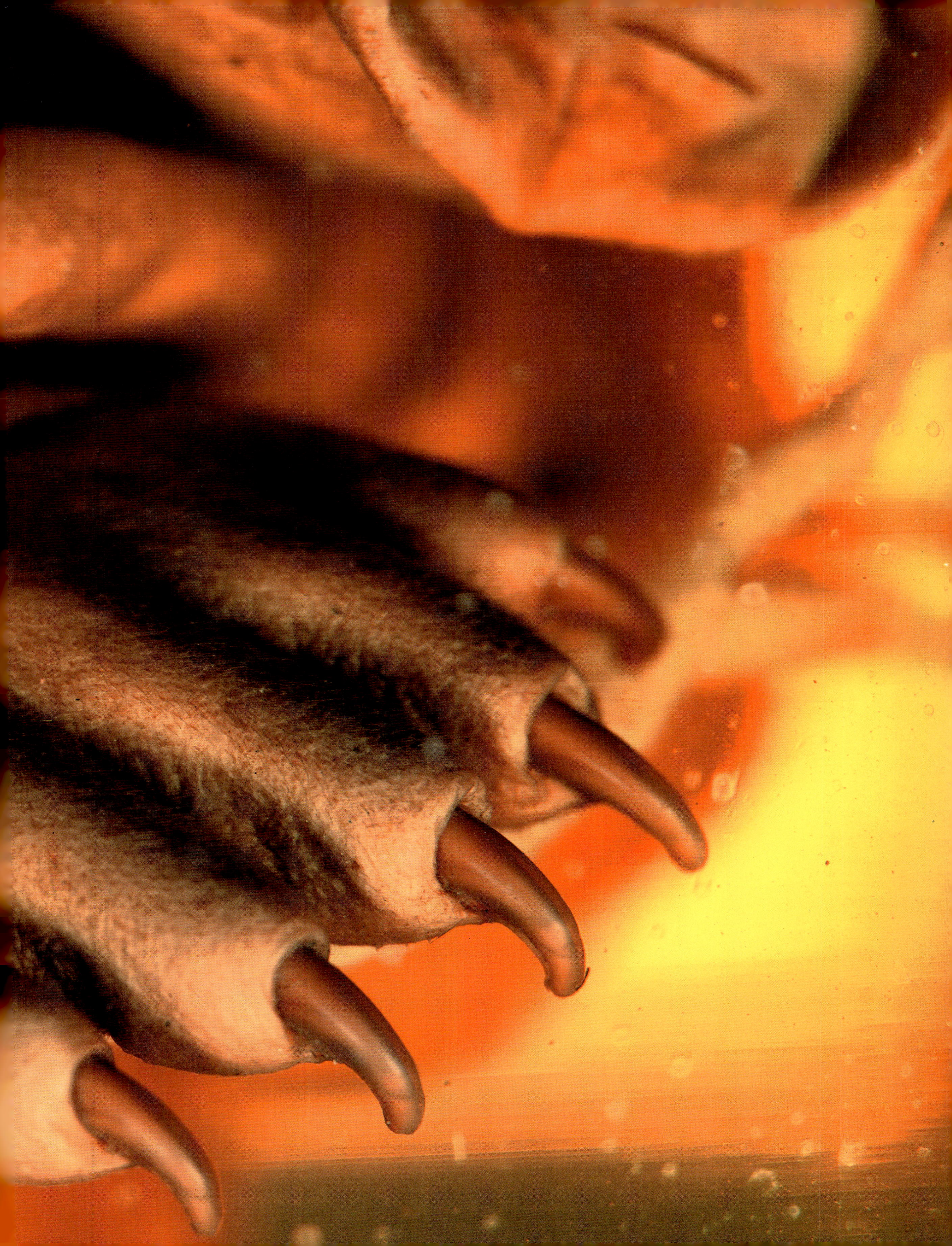

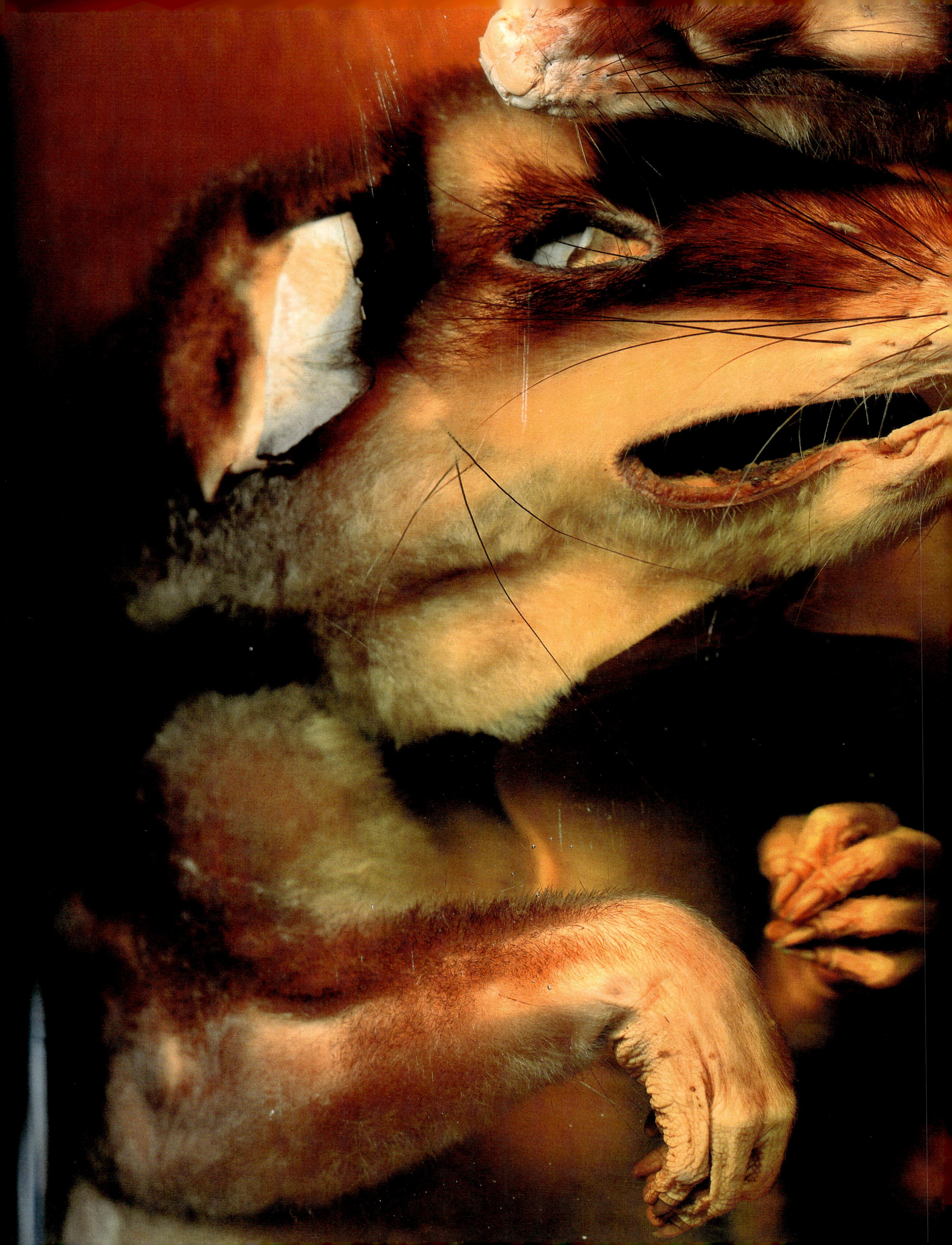

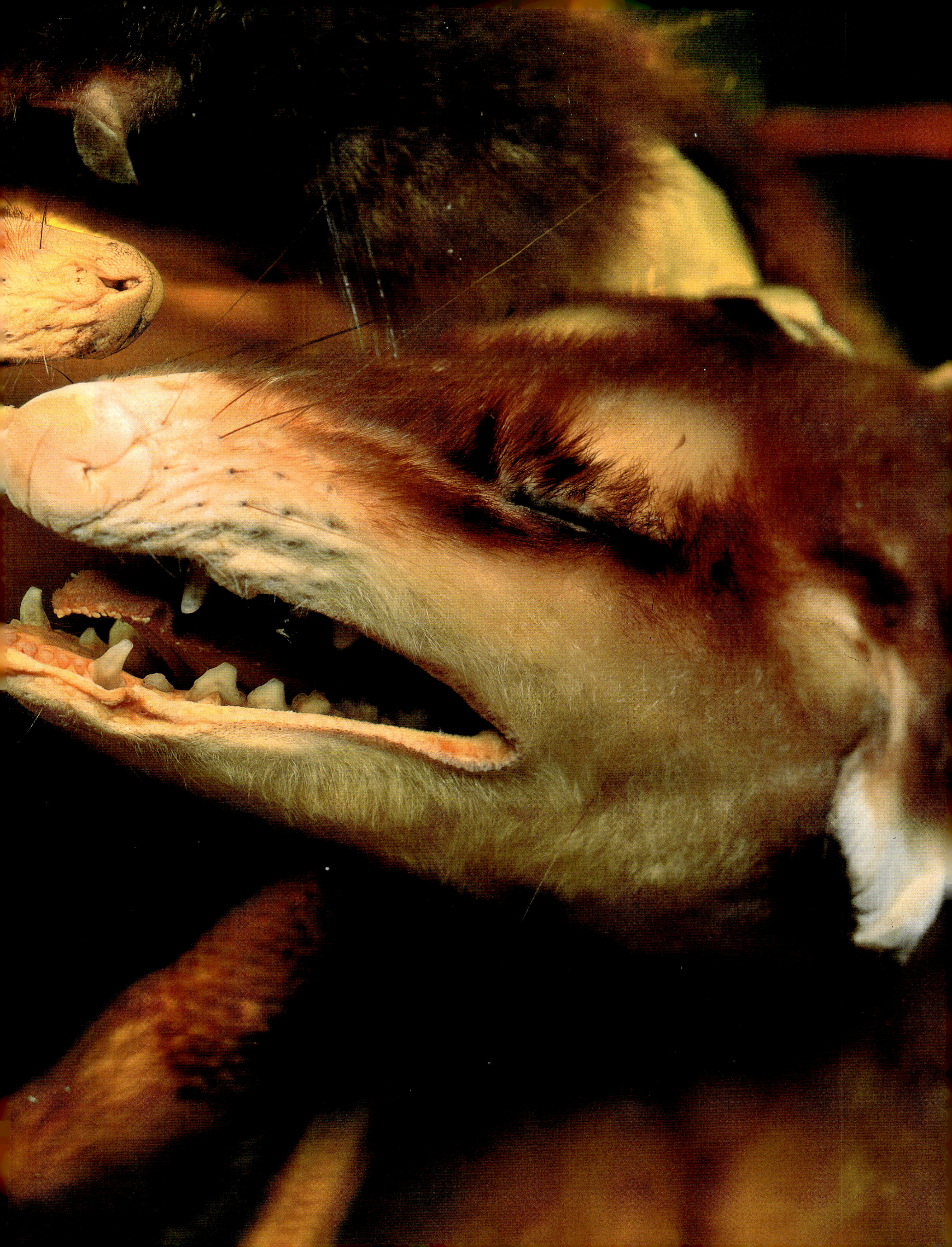

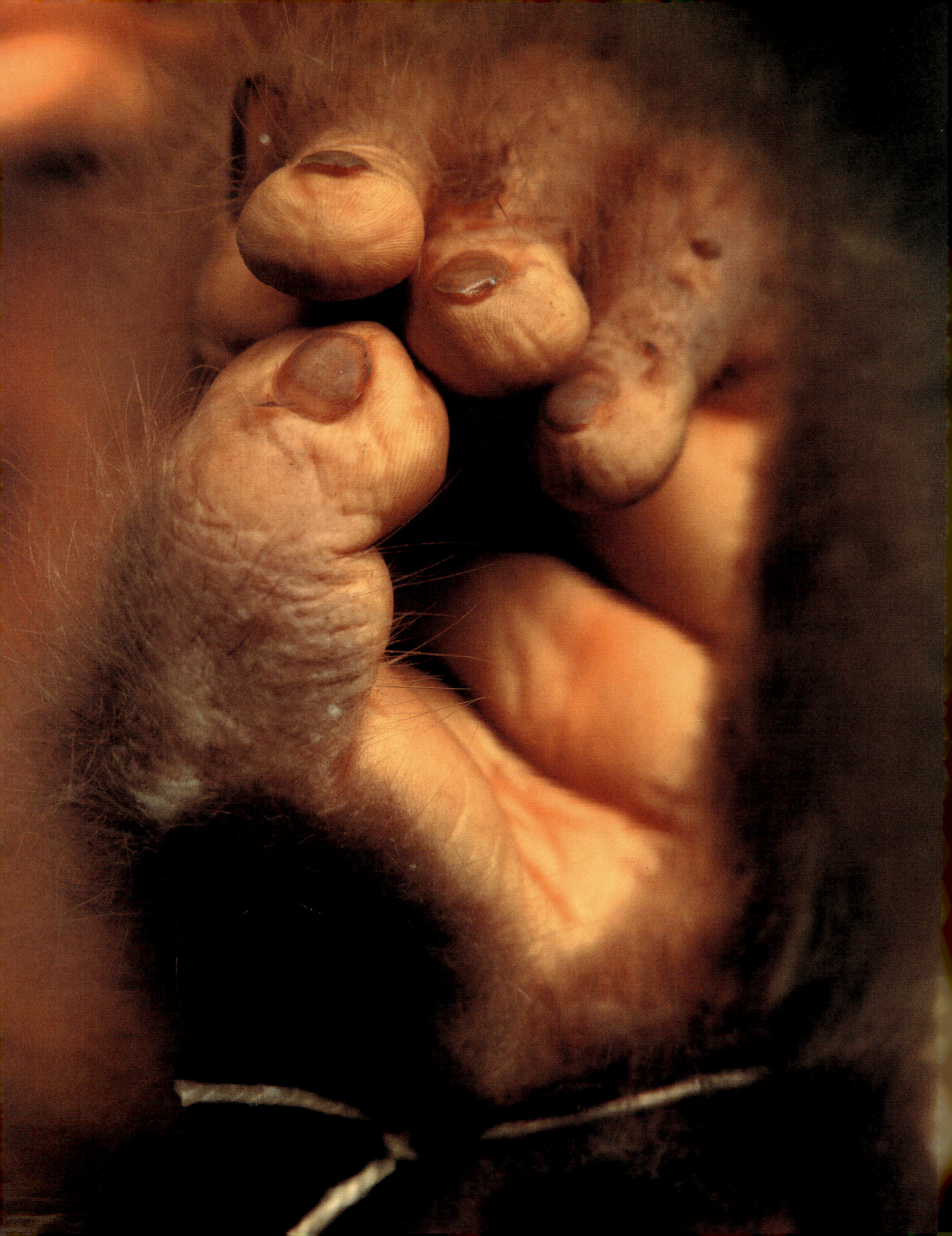

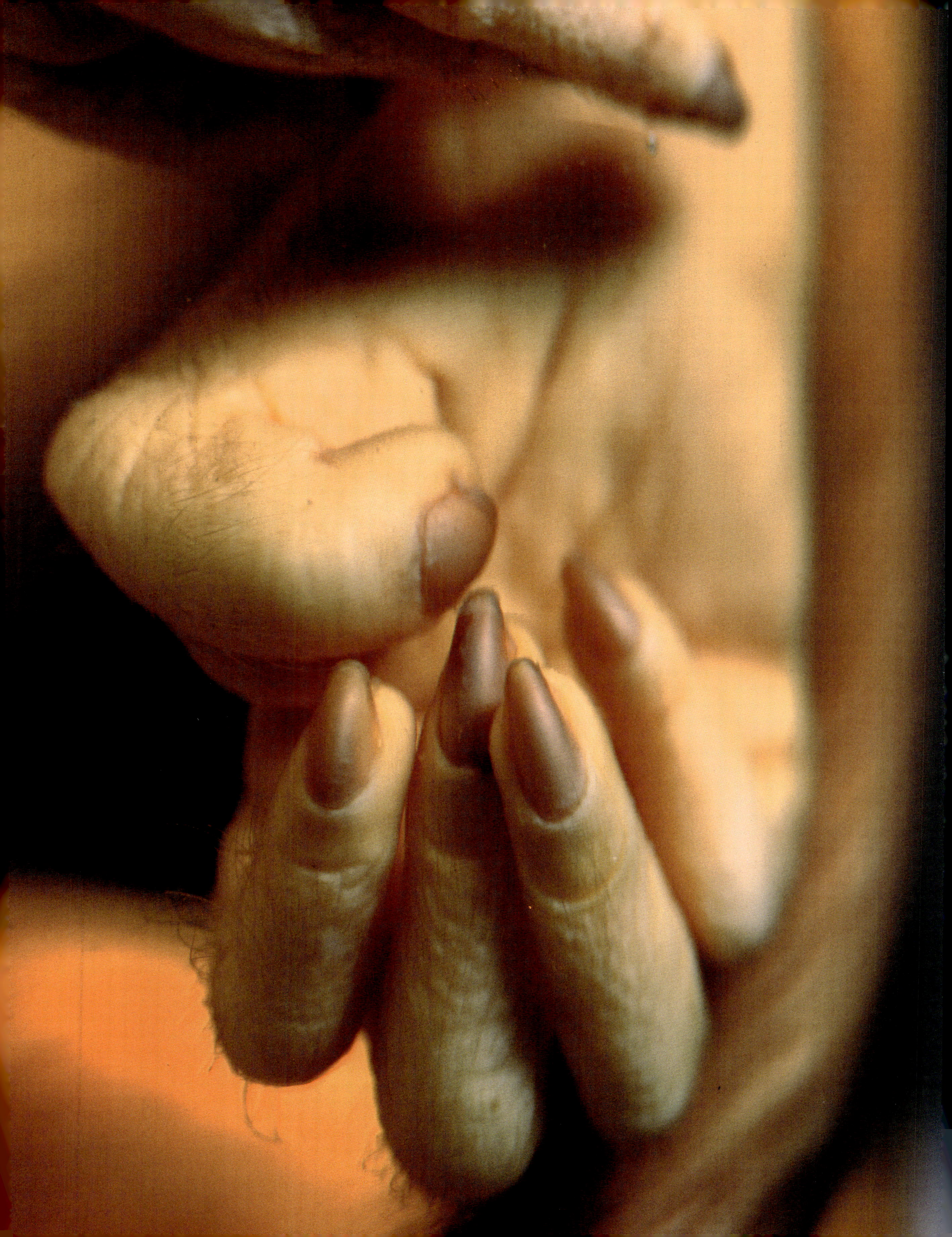

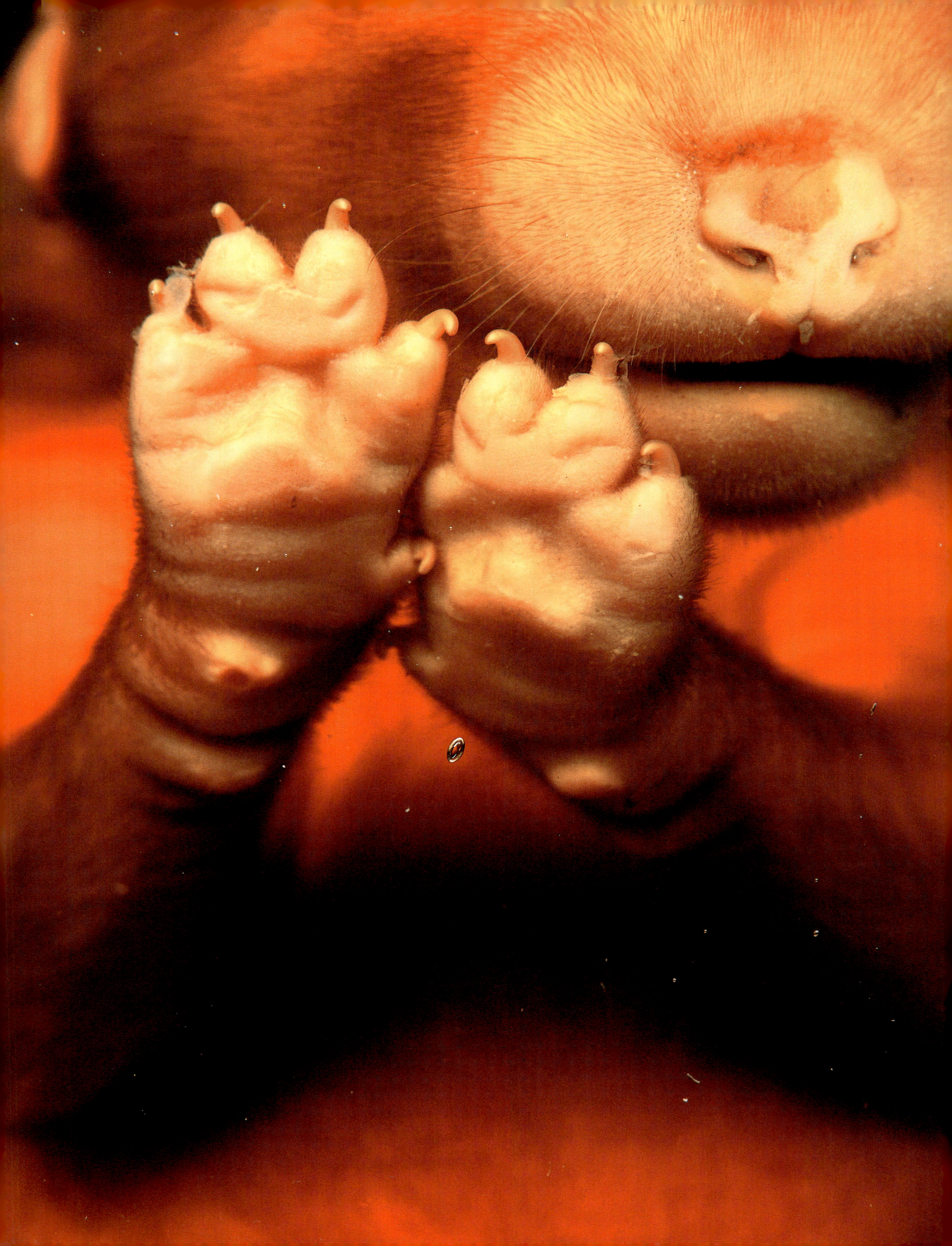

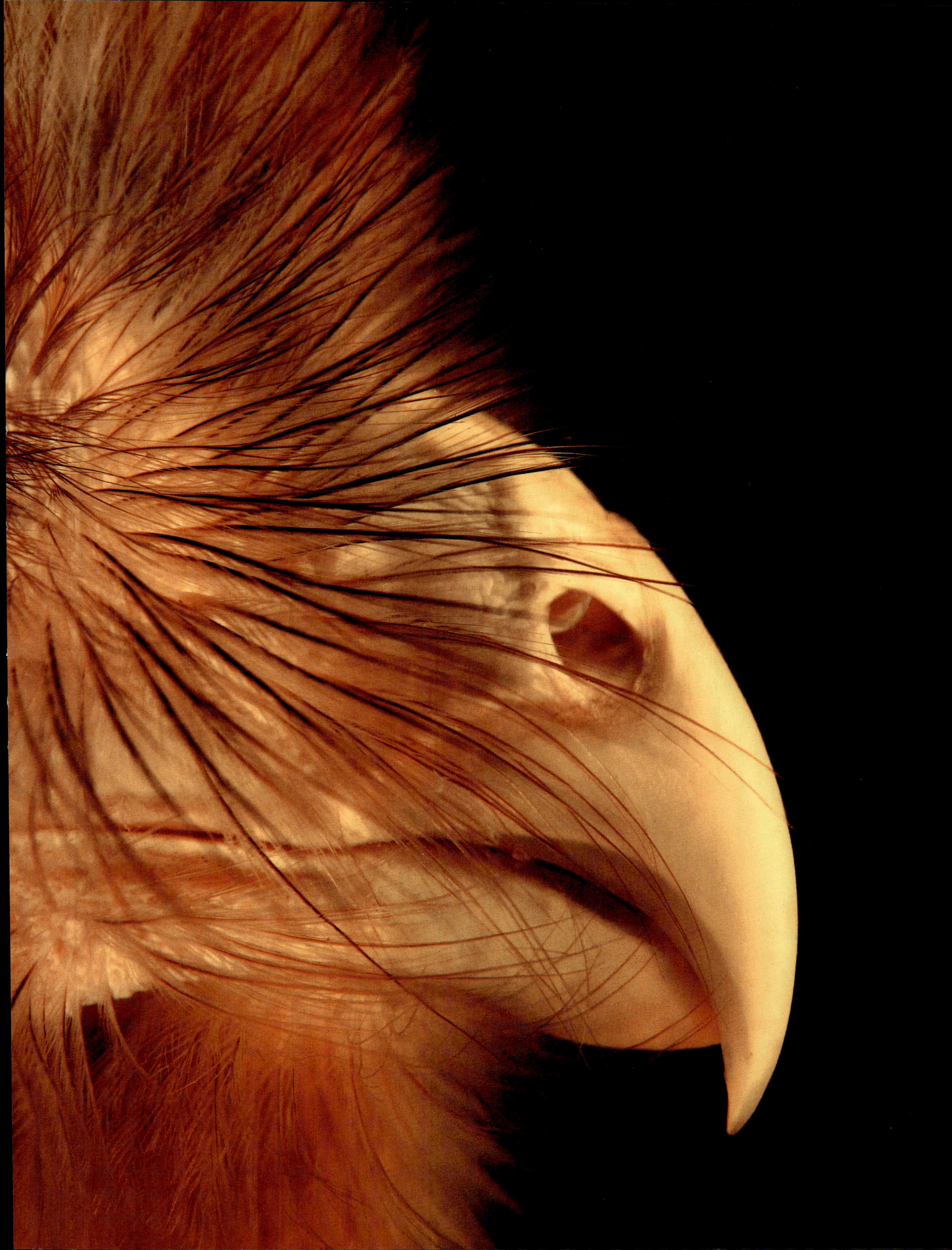

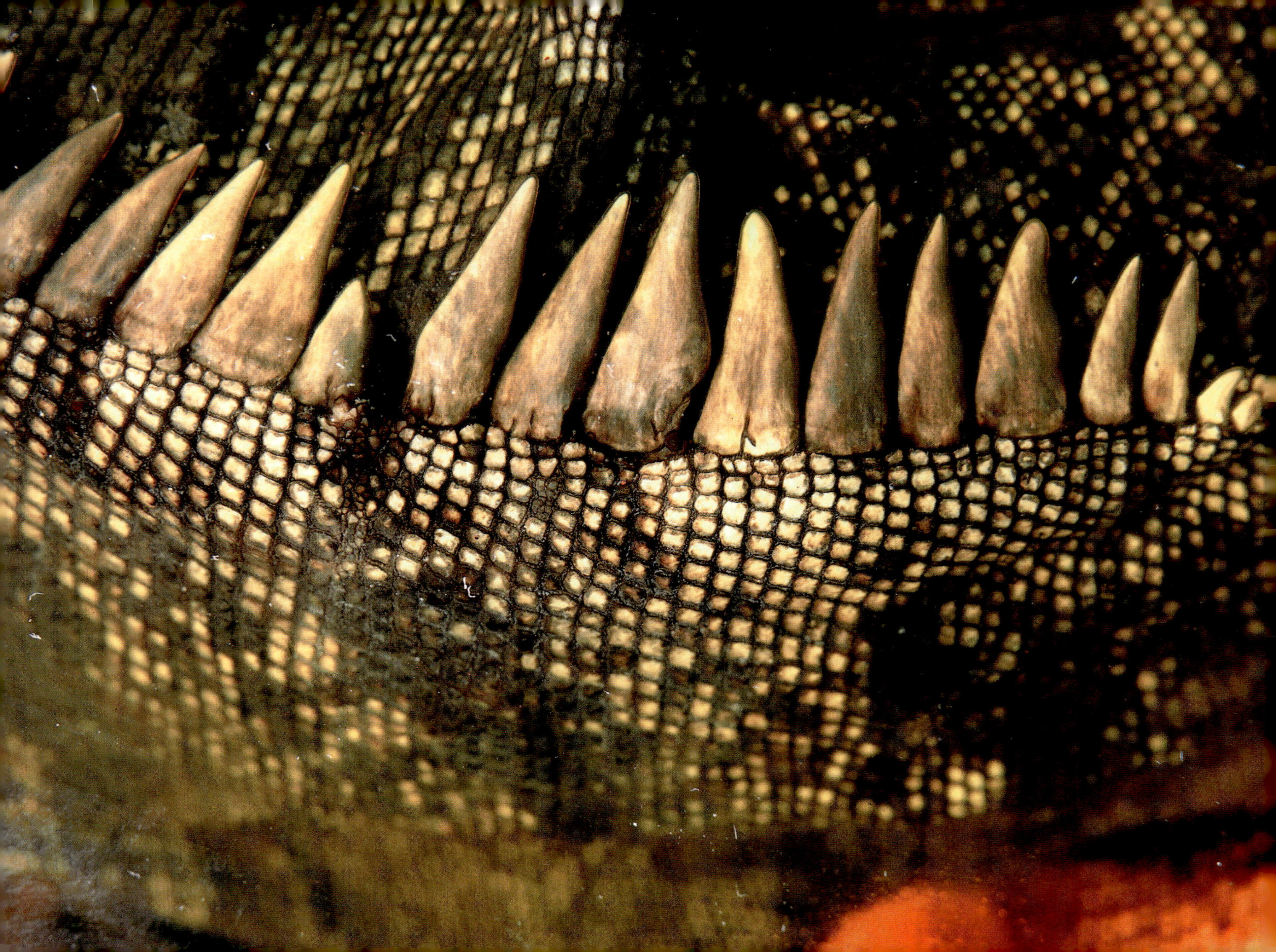

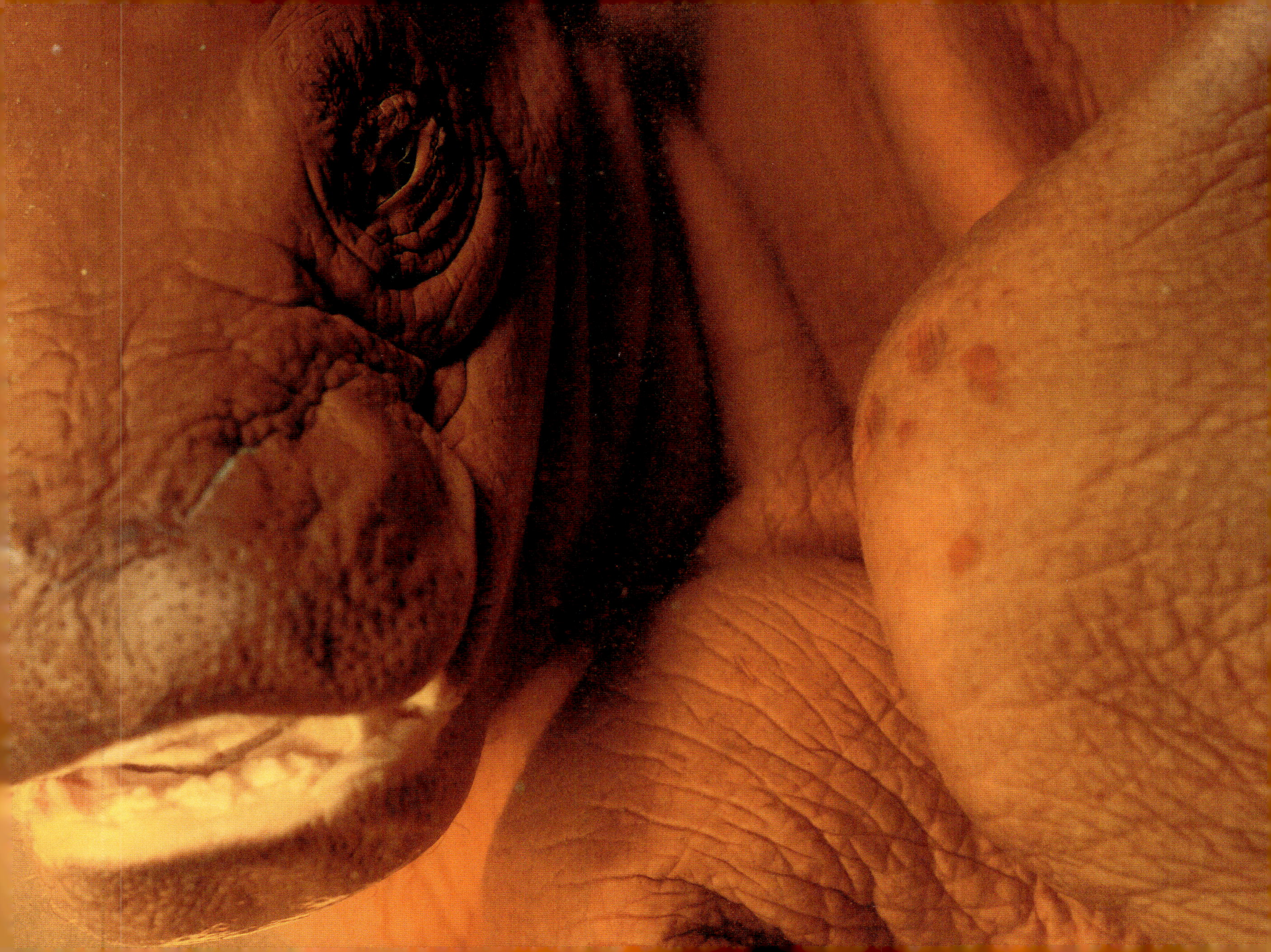

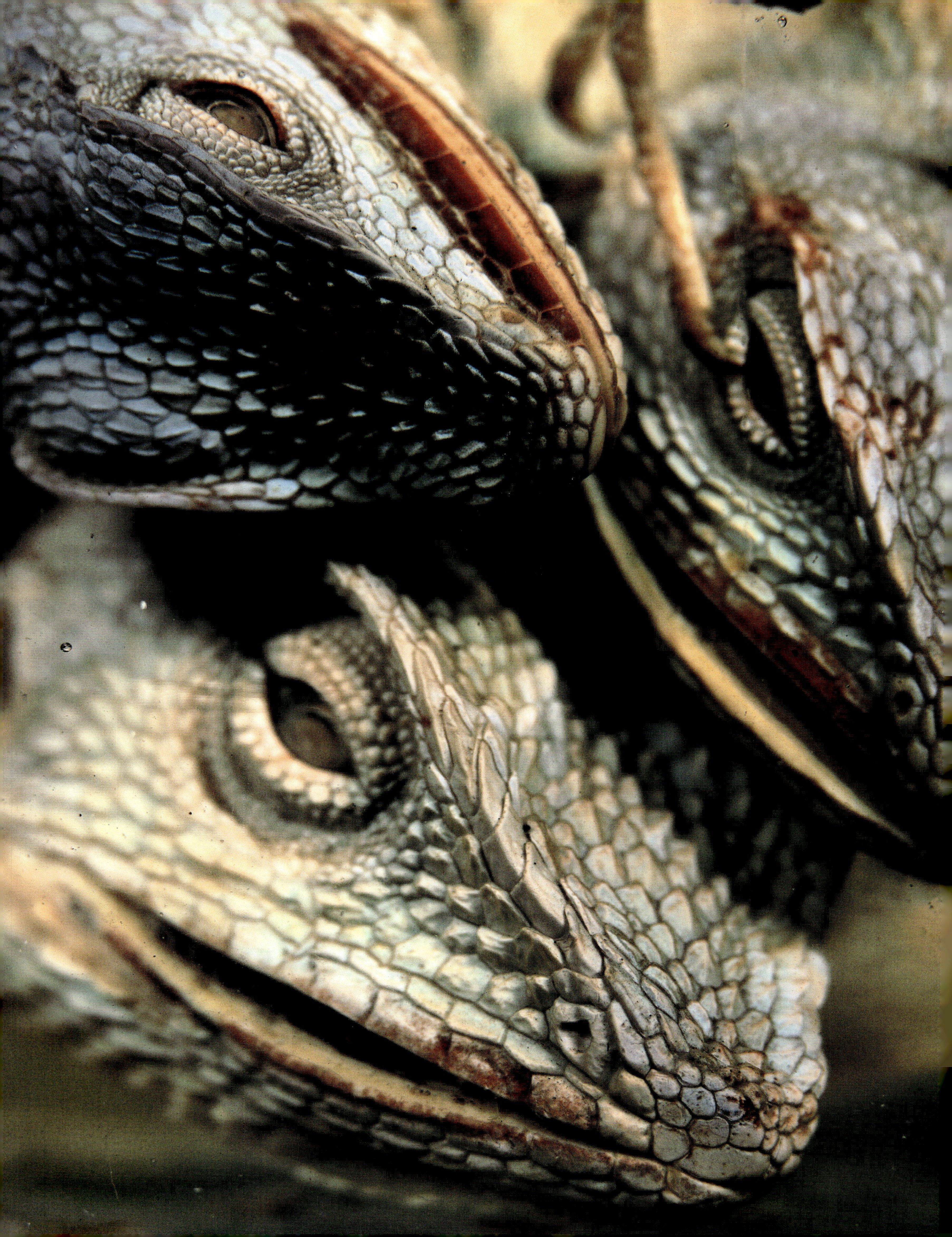

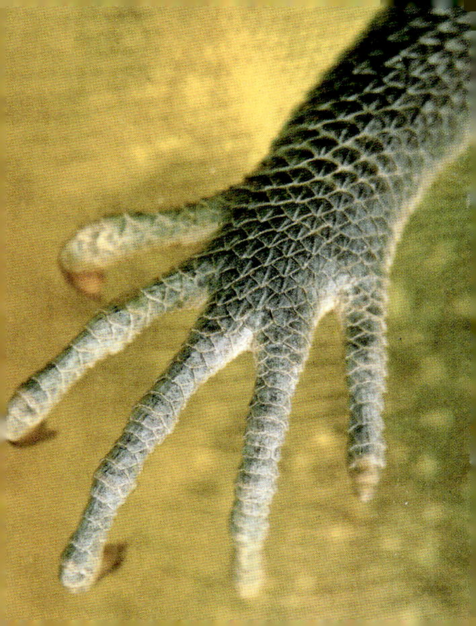

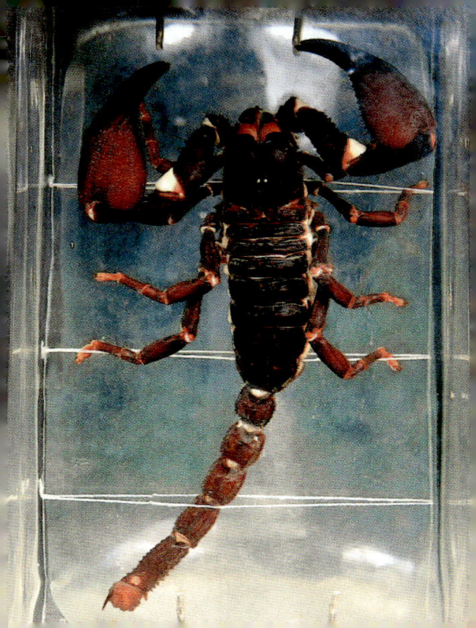

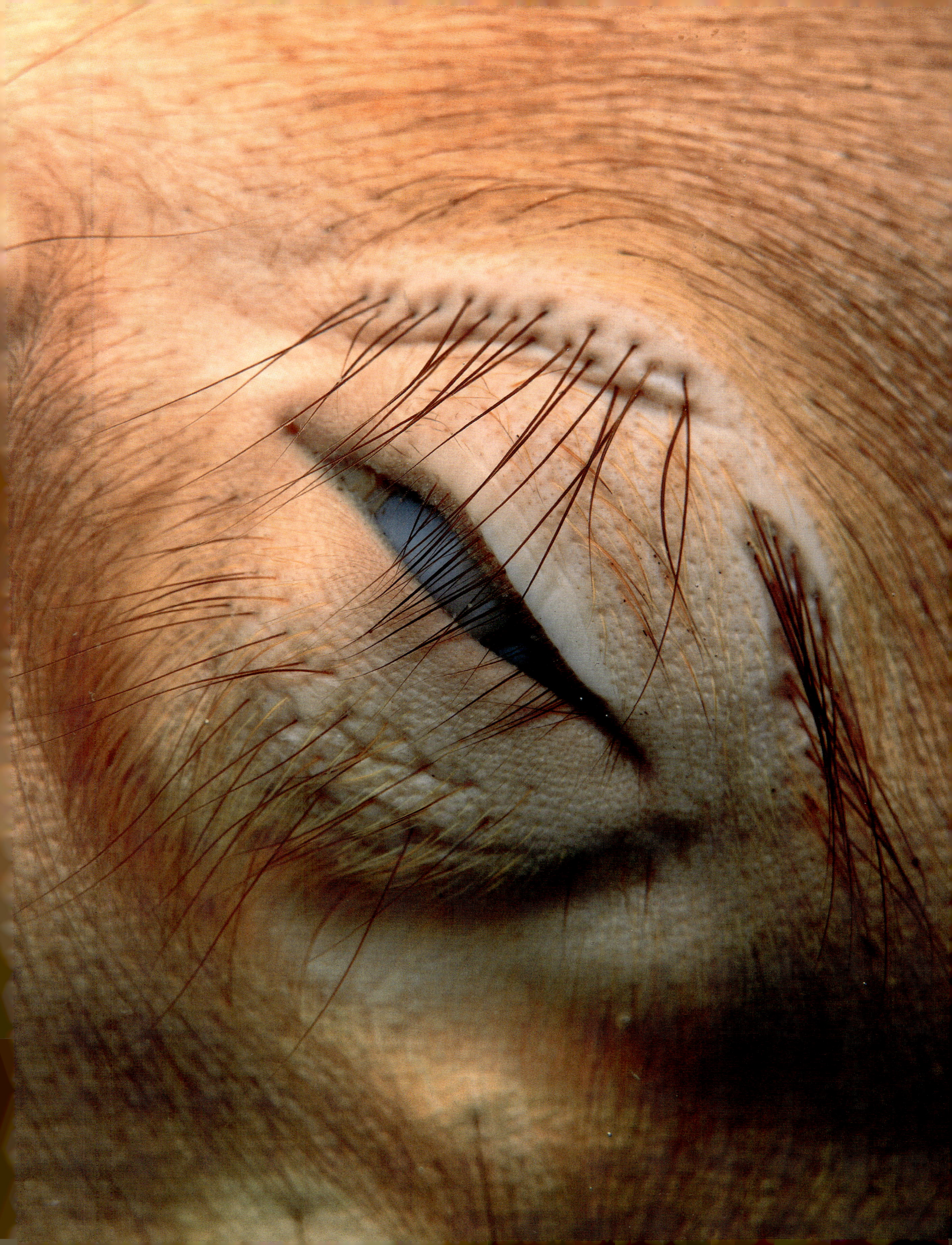

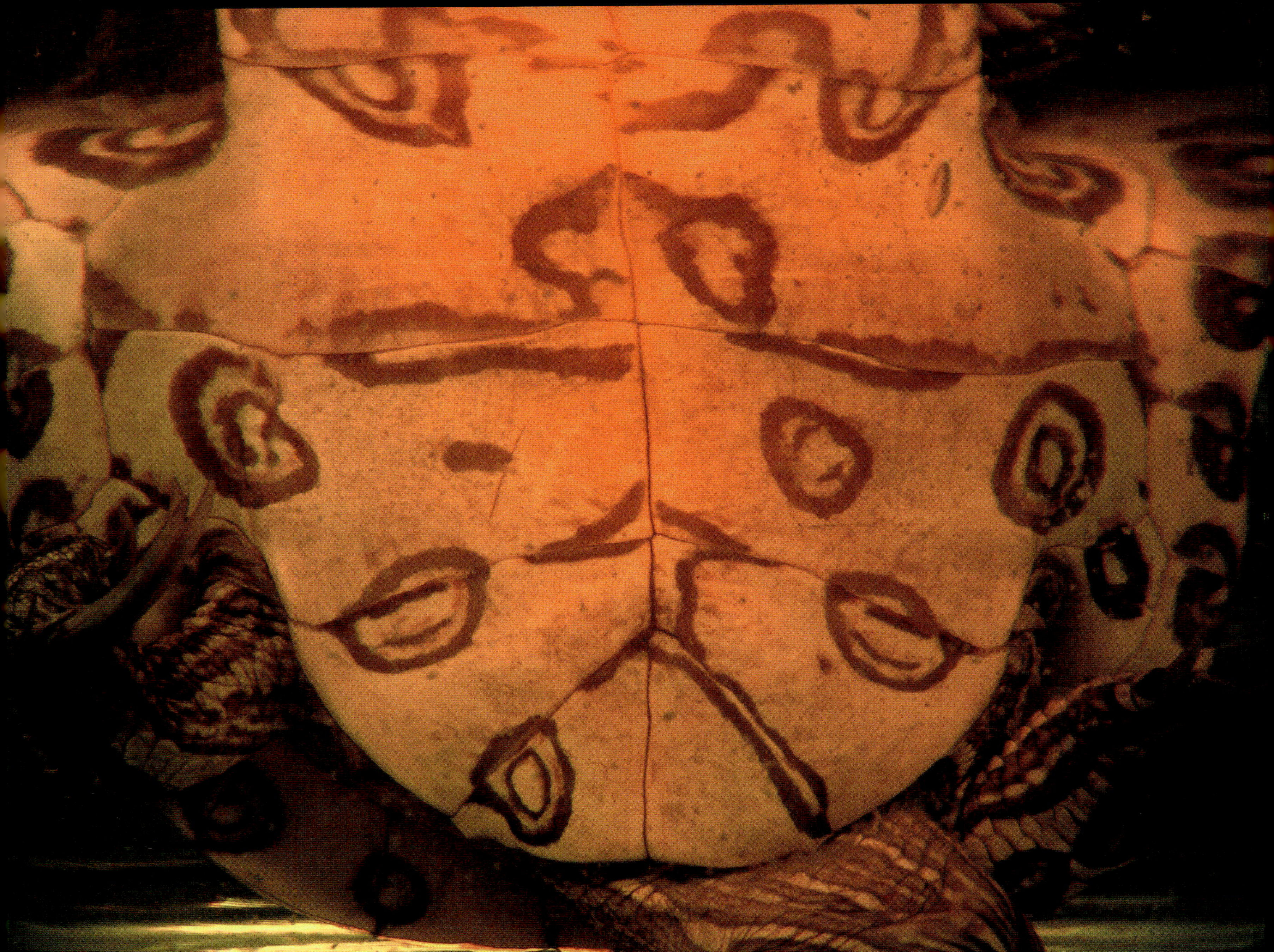

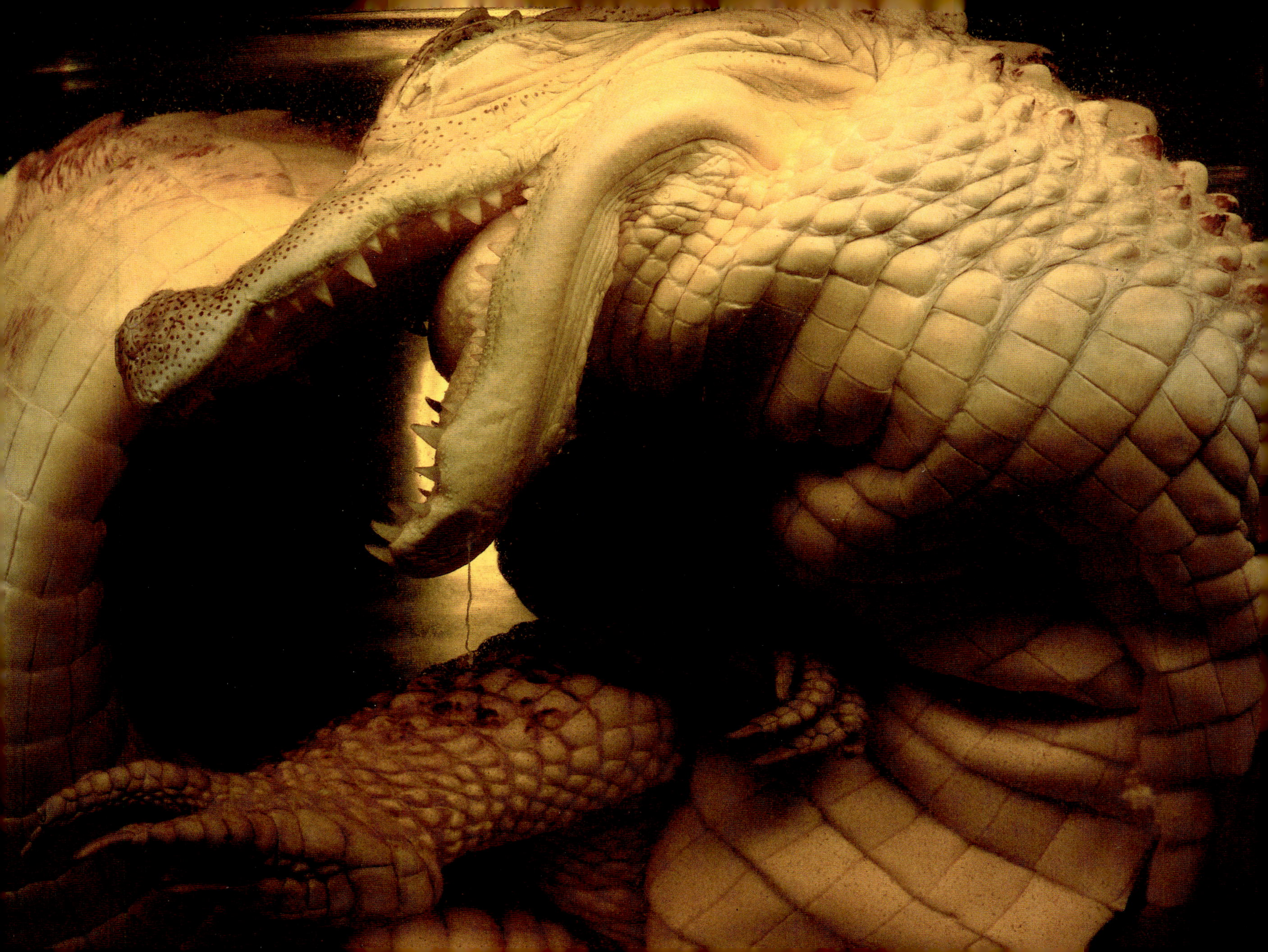

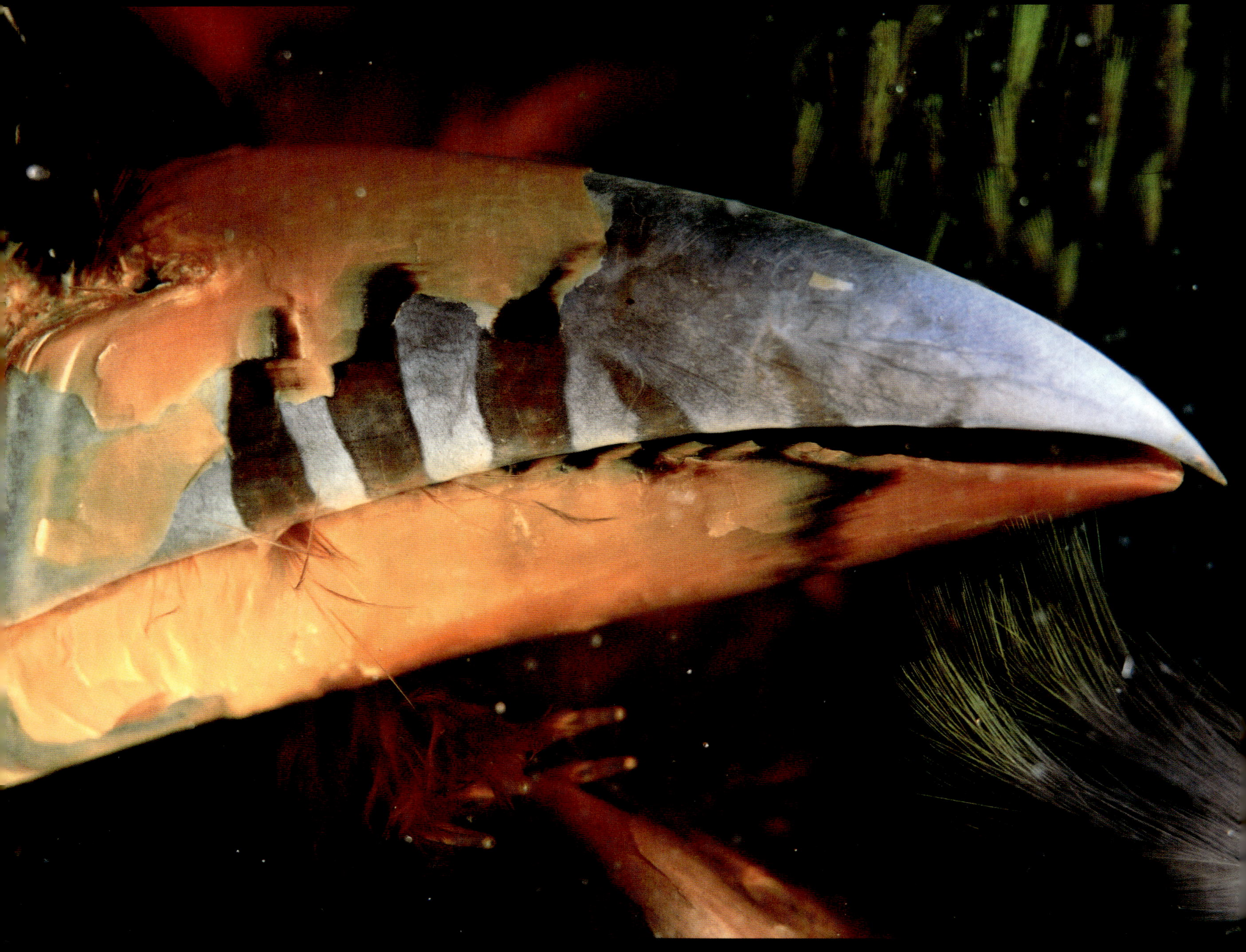

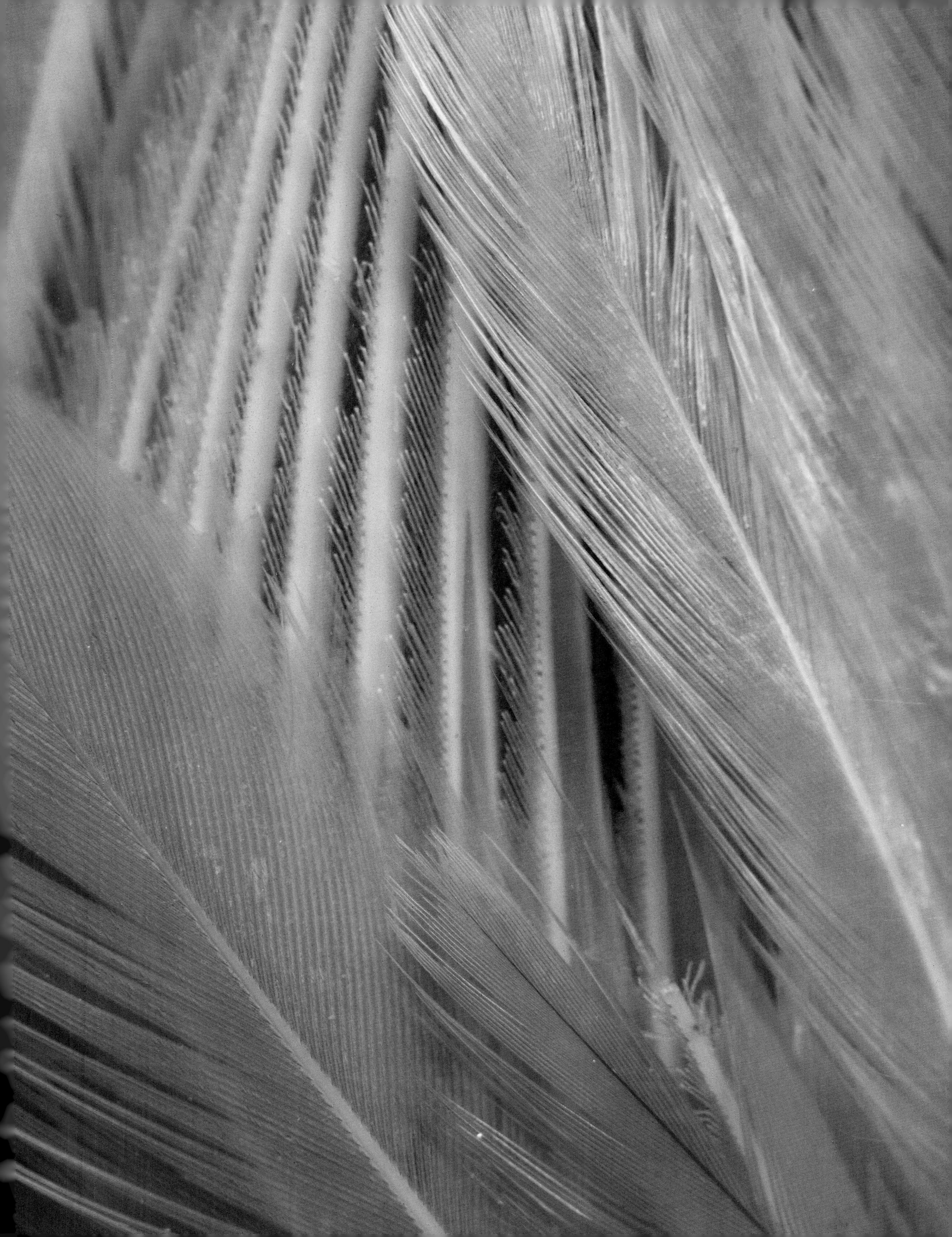

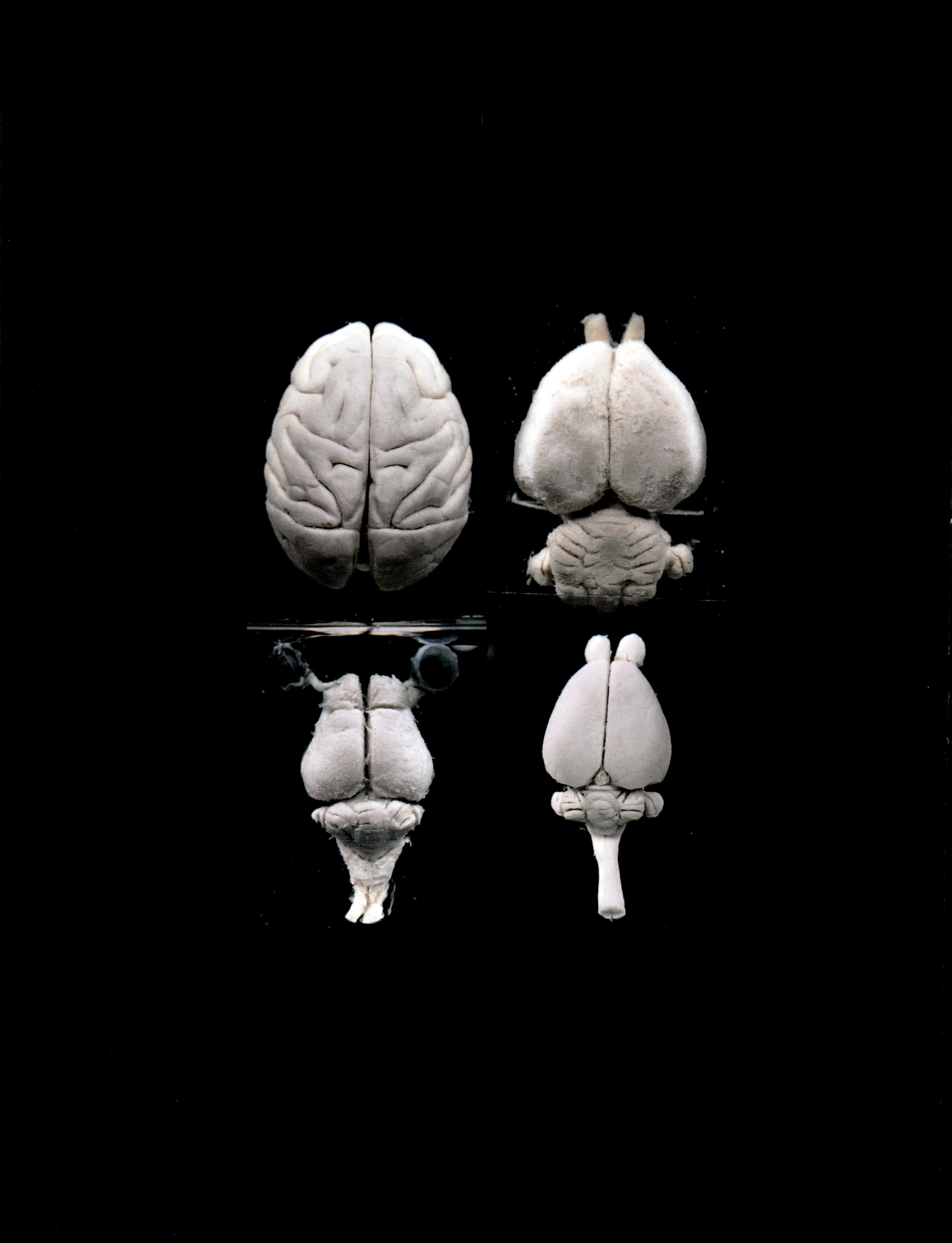

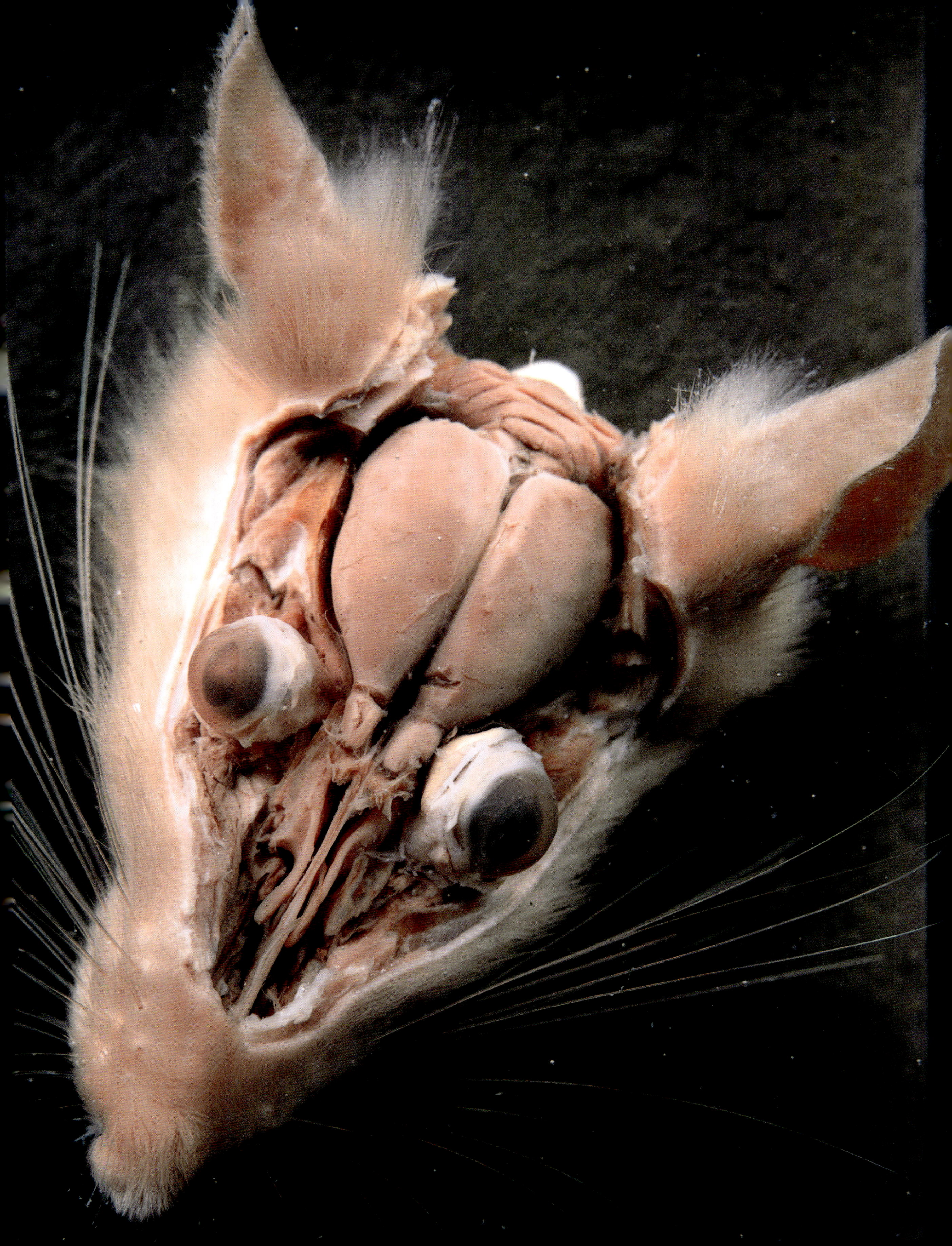

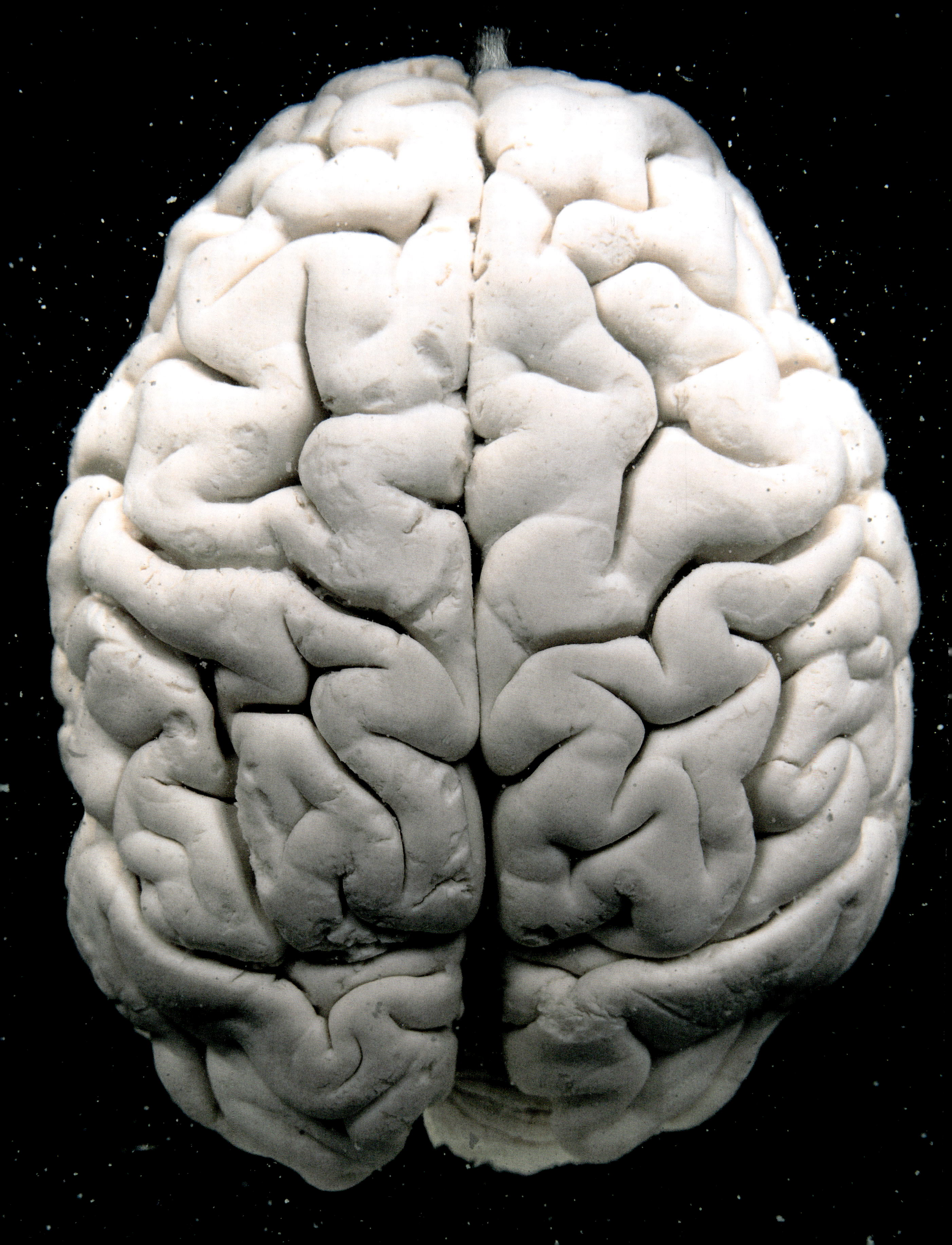

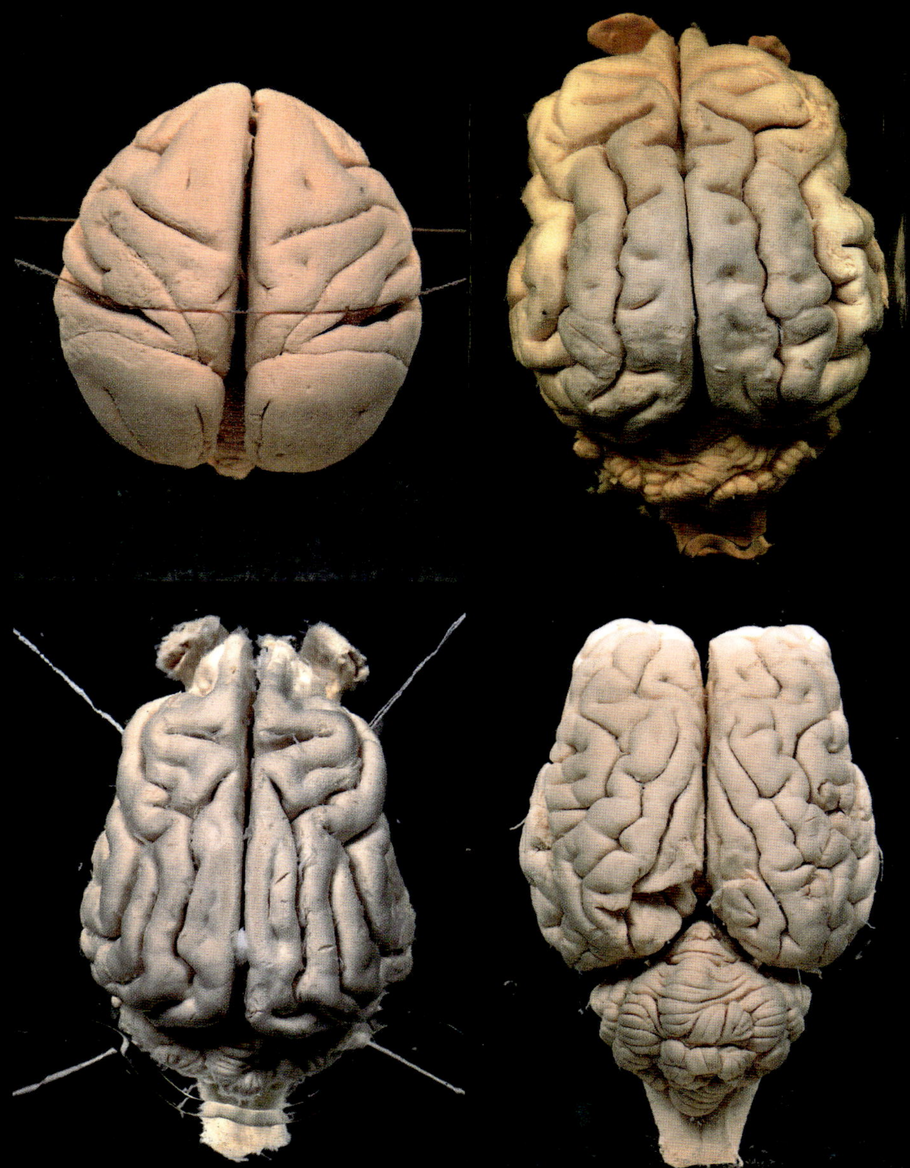

CONSERVING HUMANS

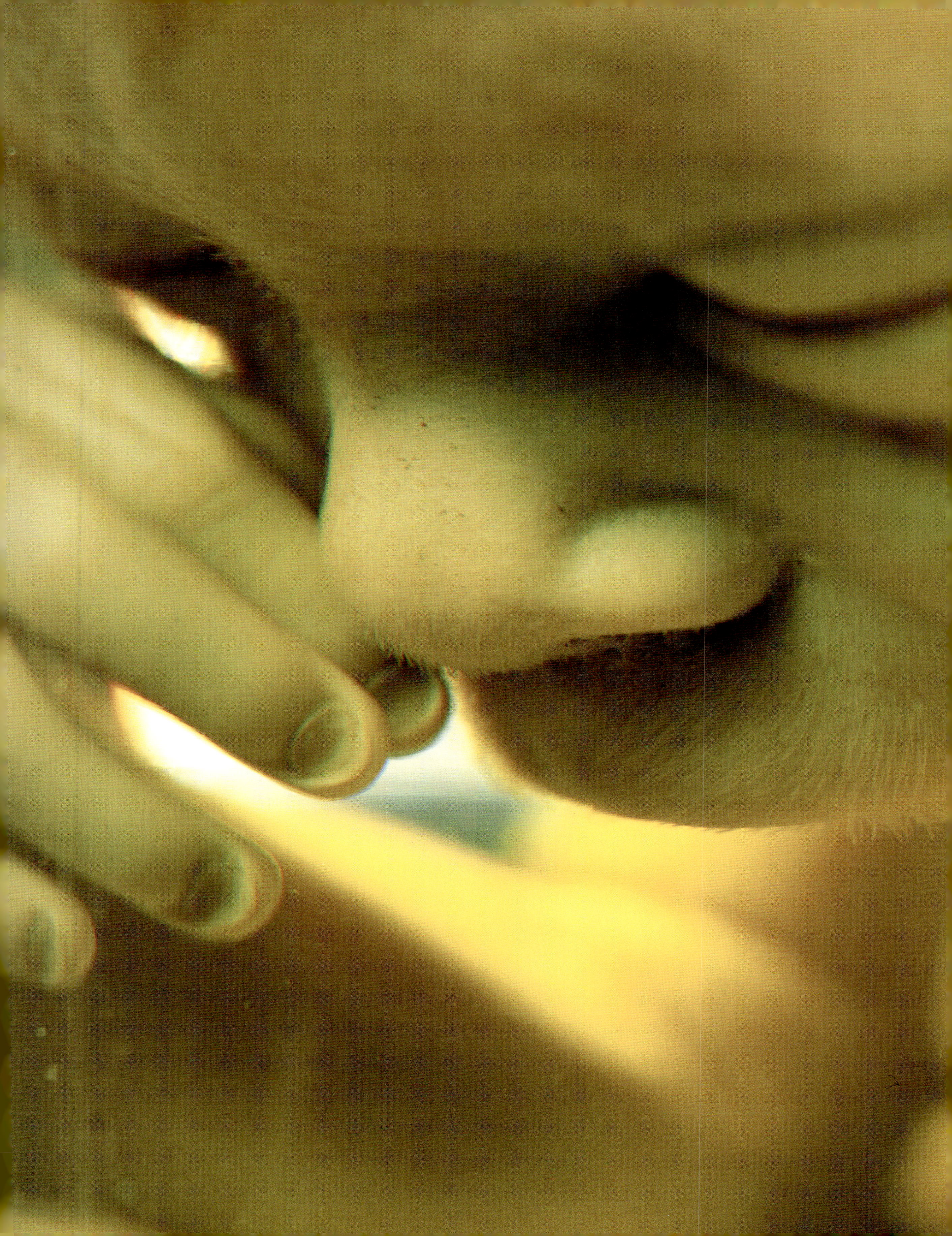

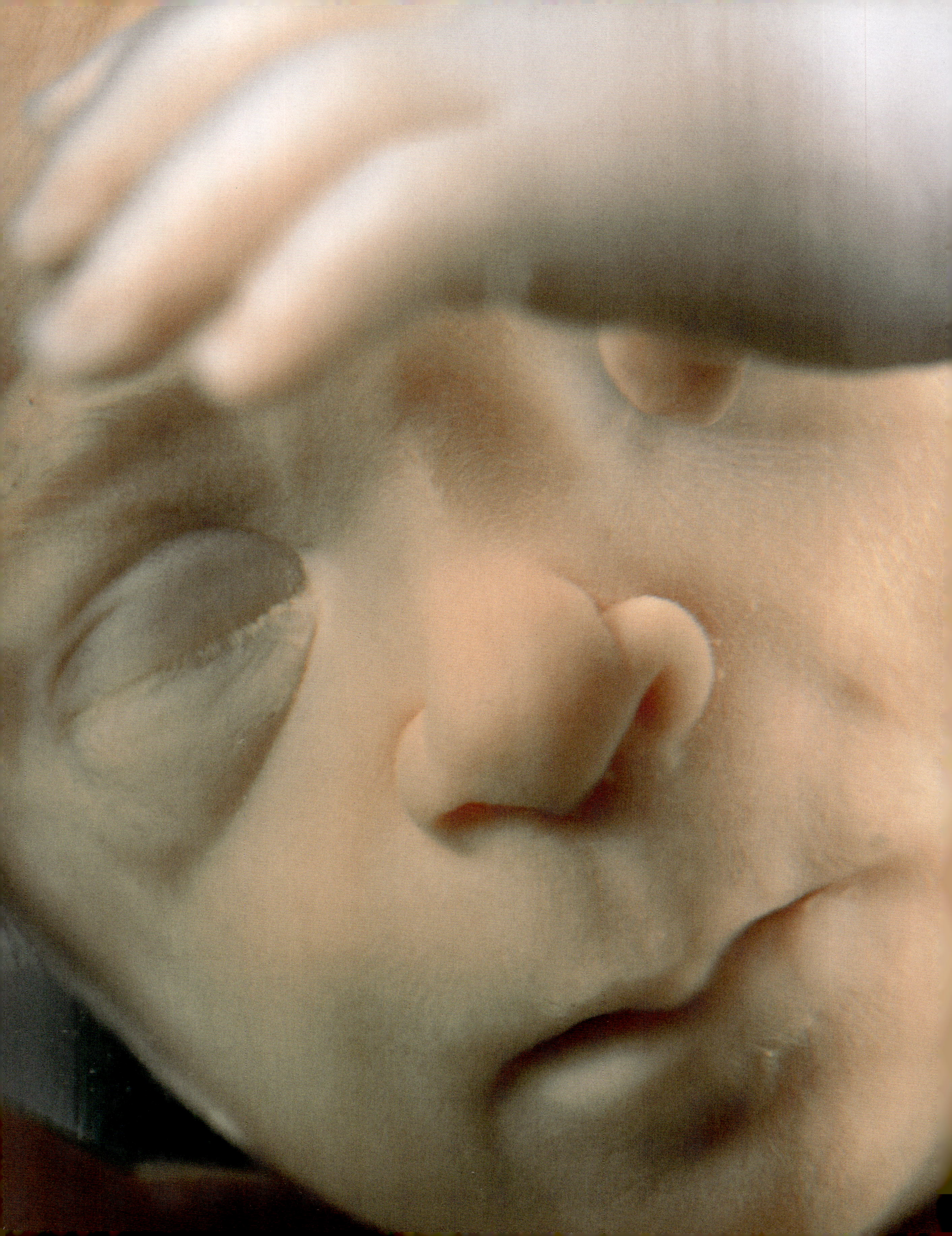

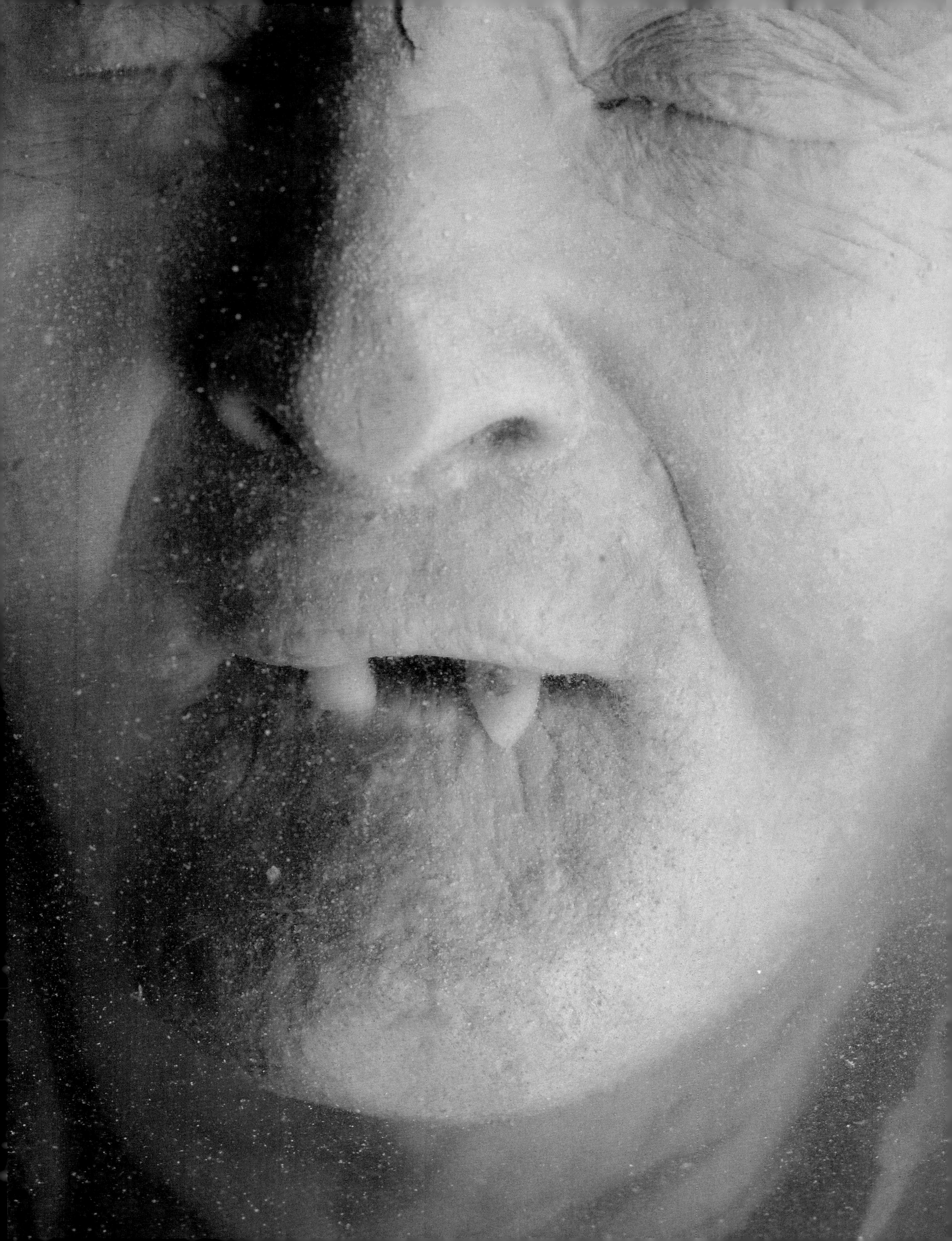

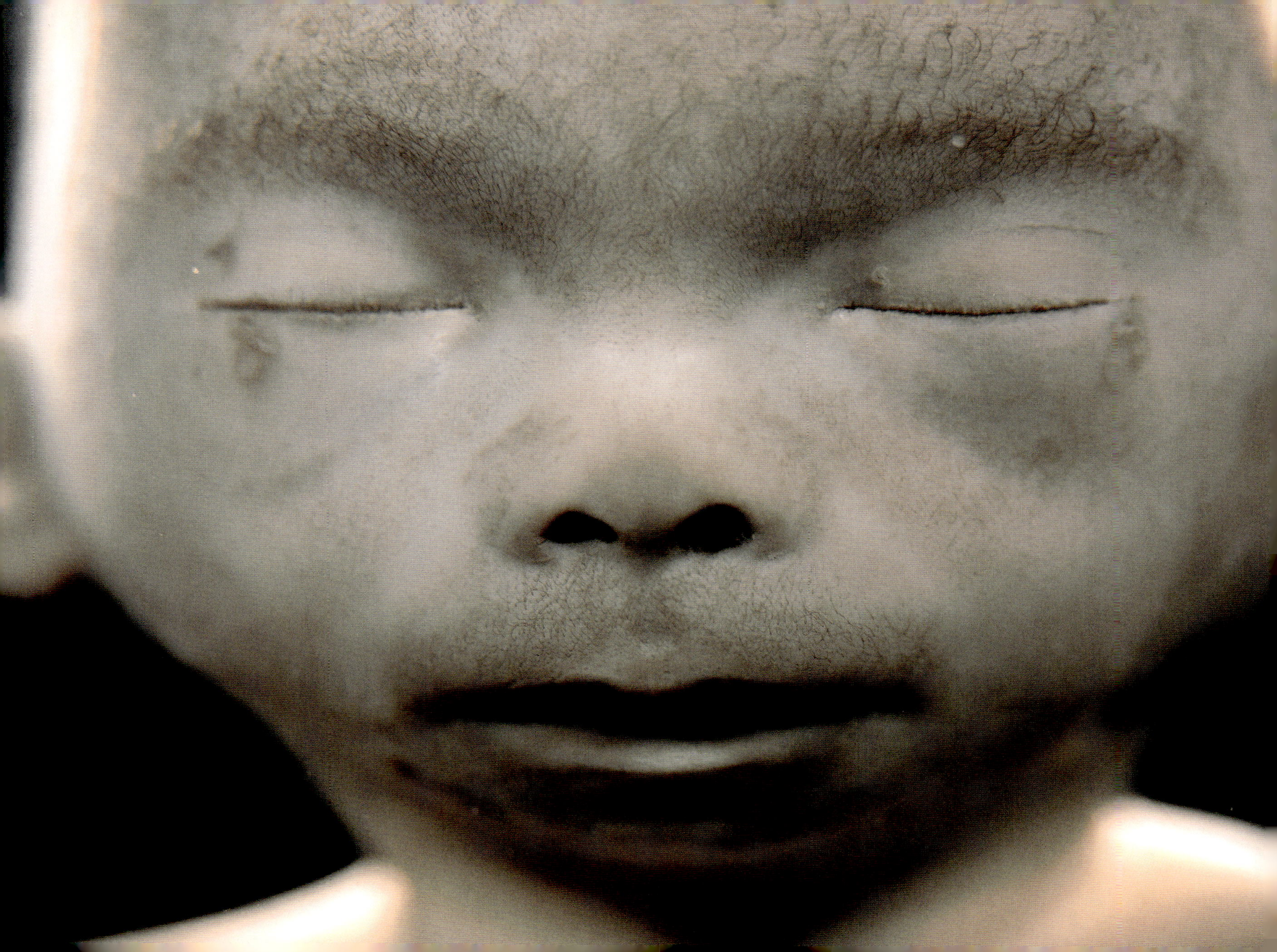

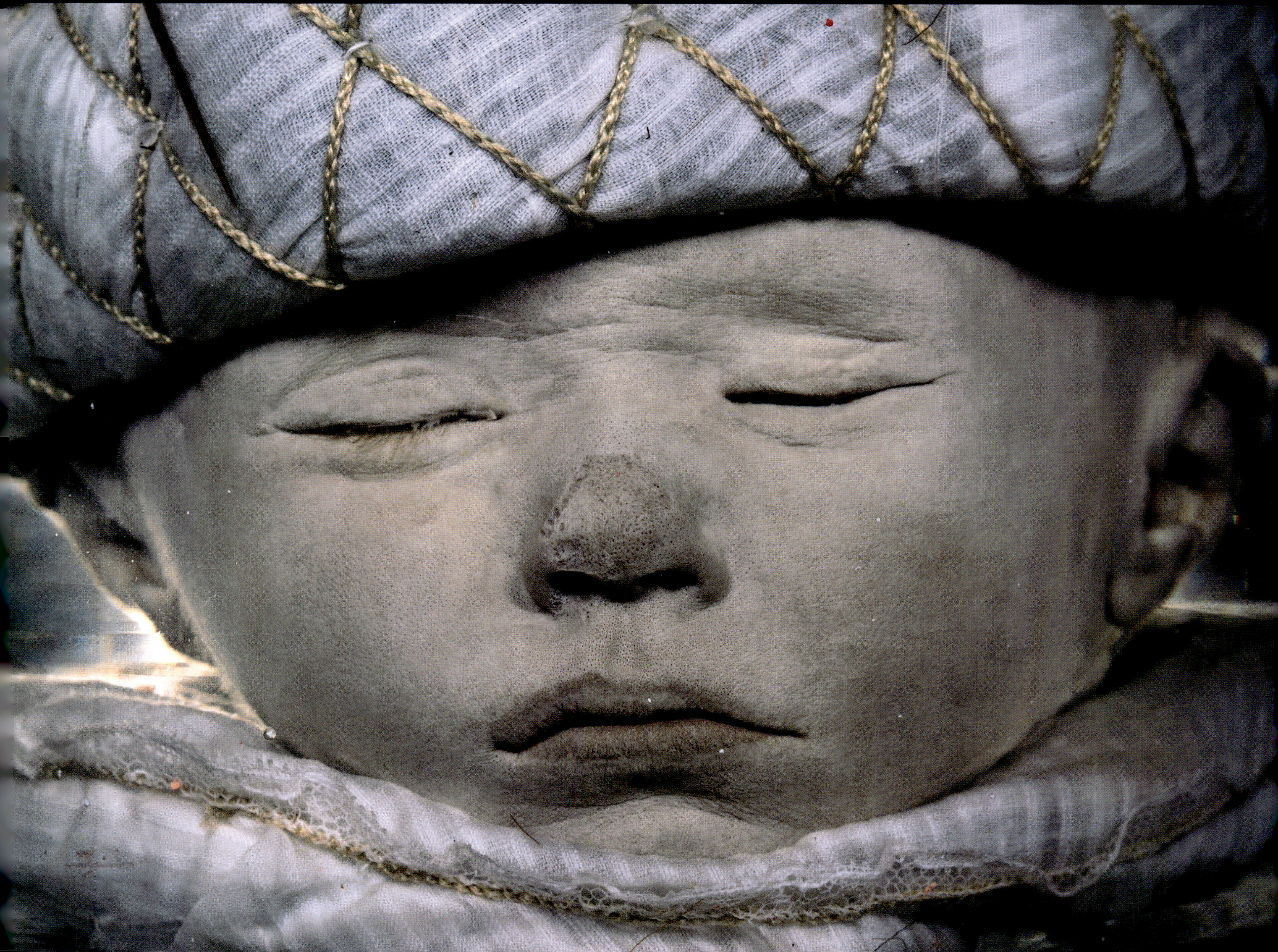

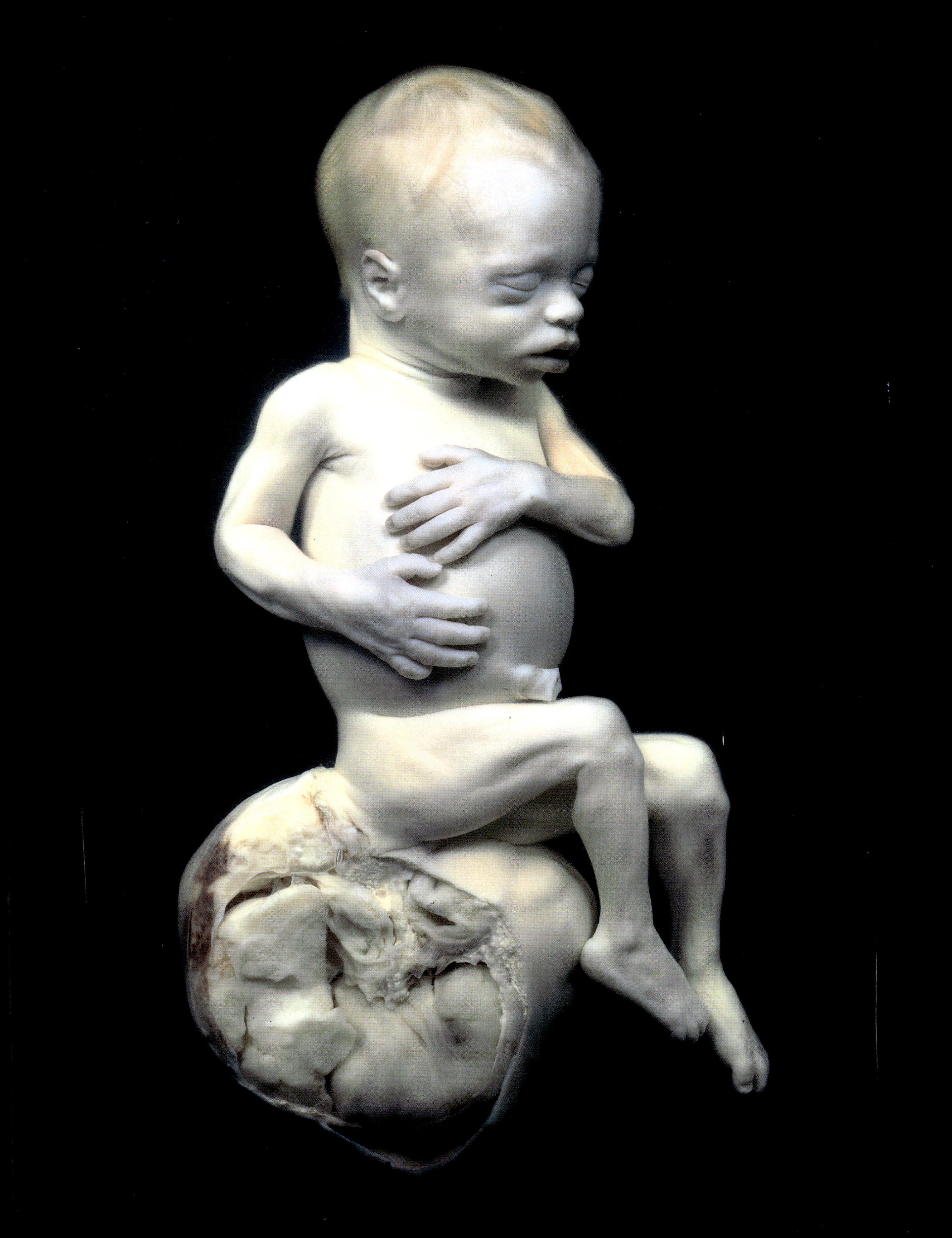

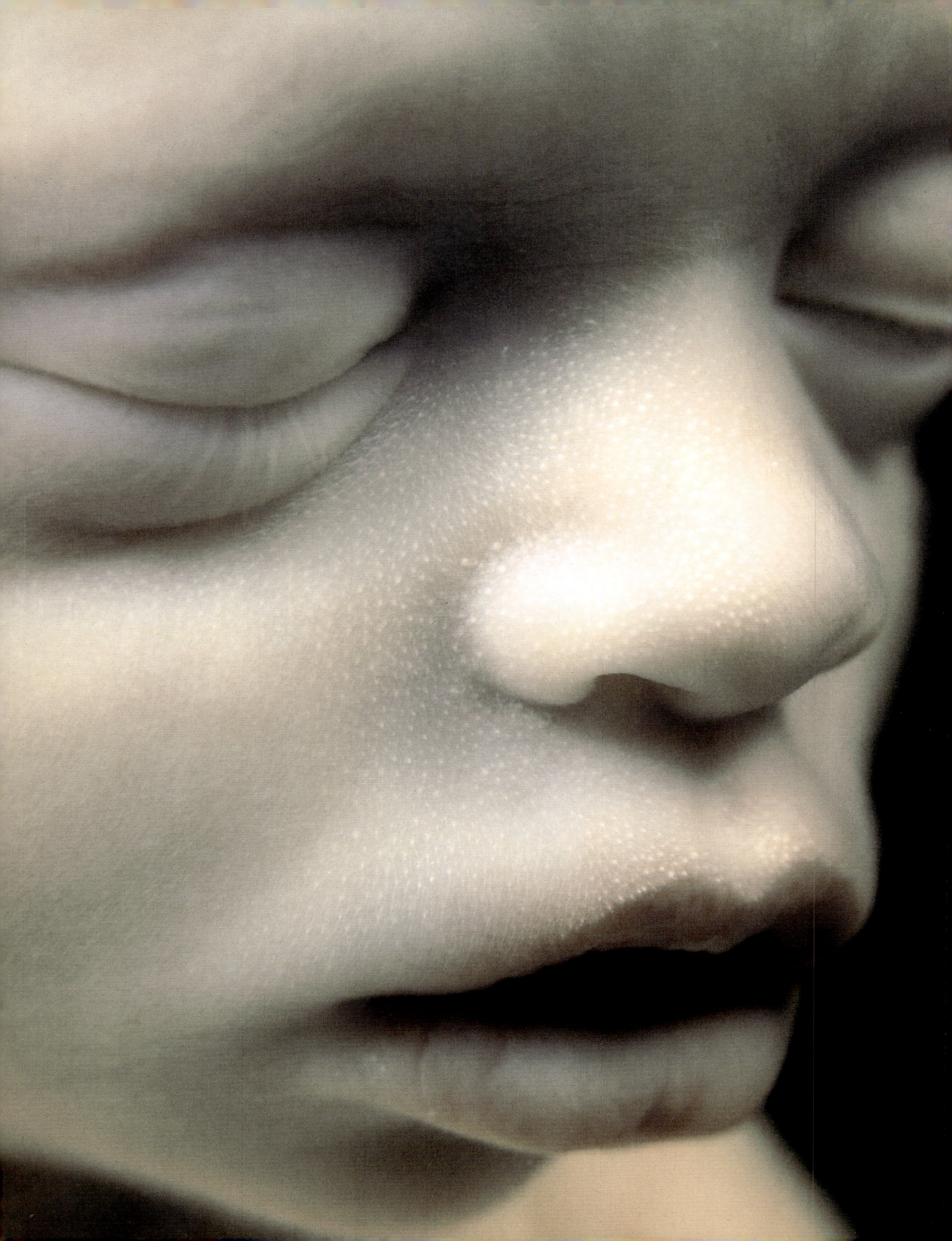

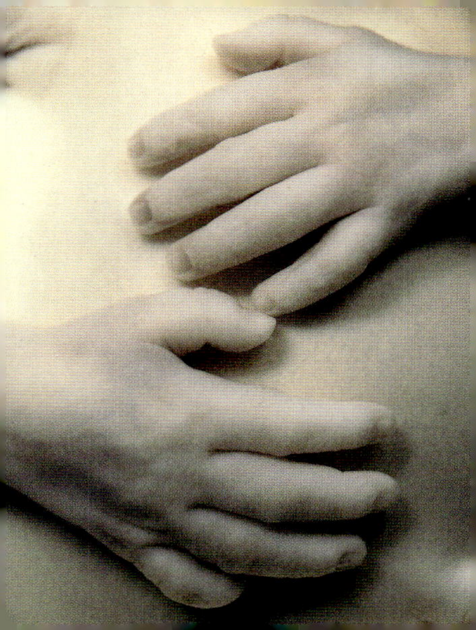

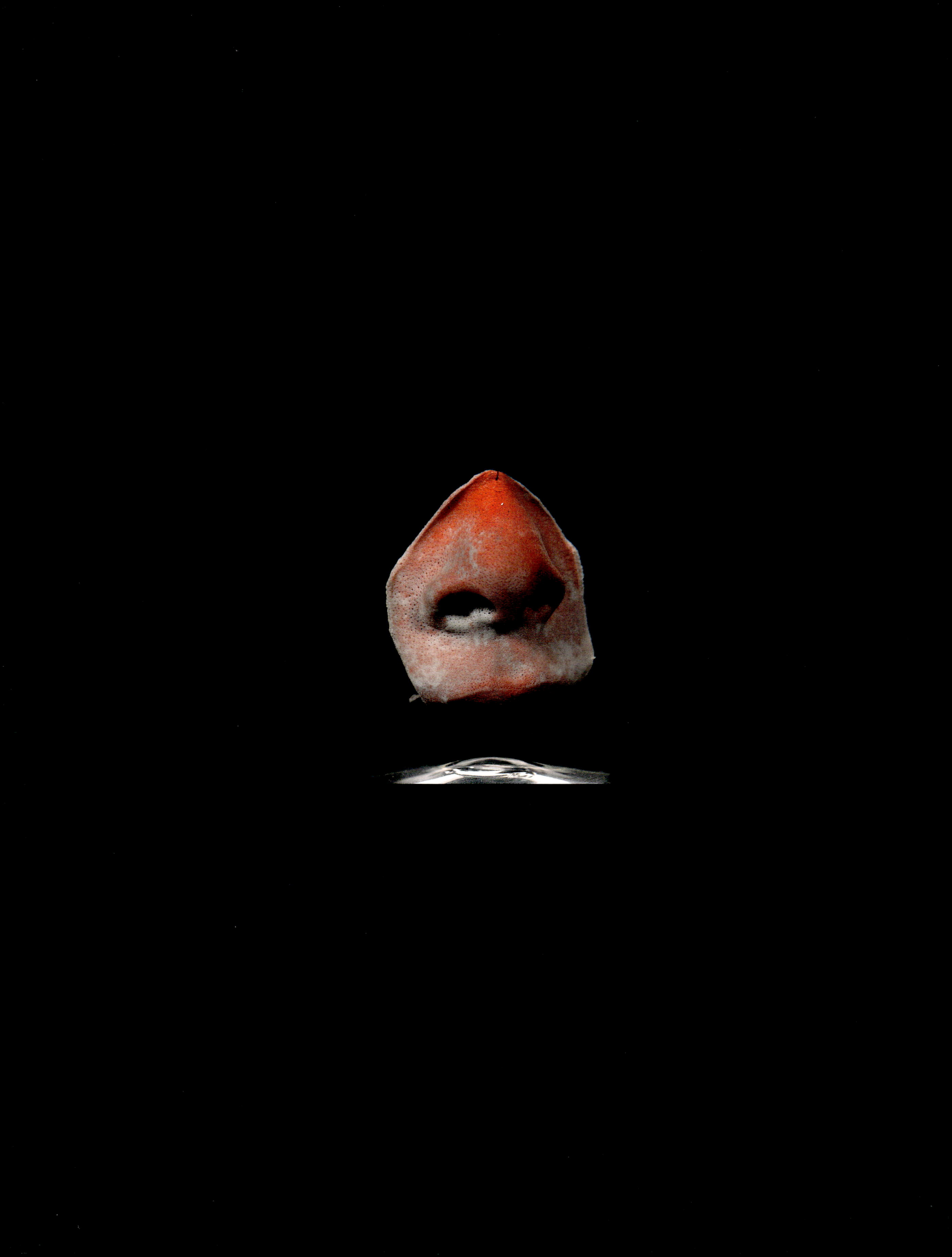

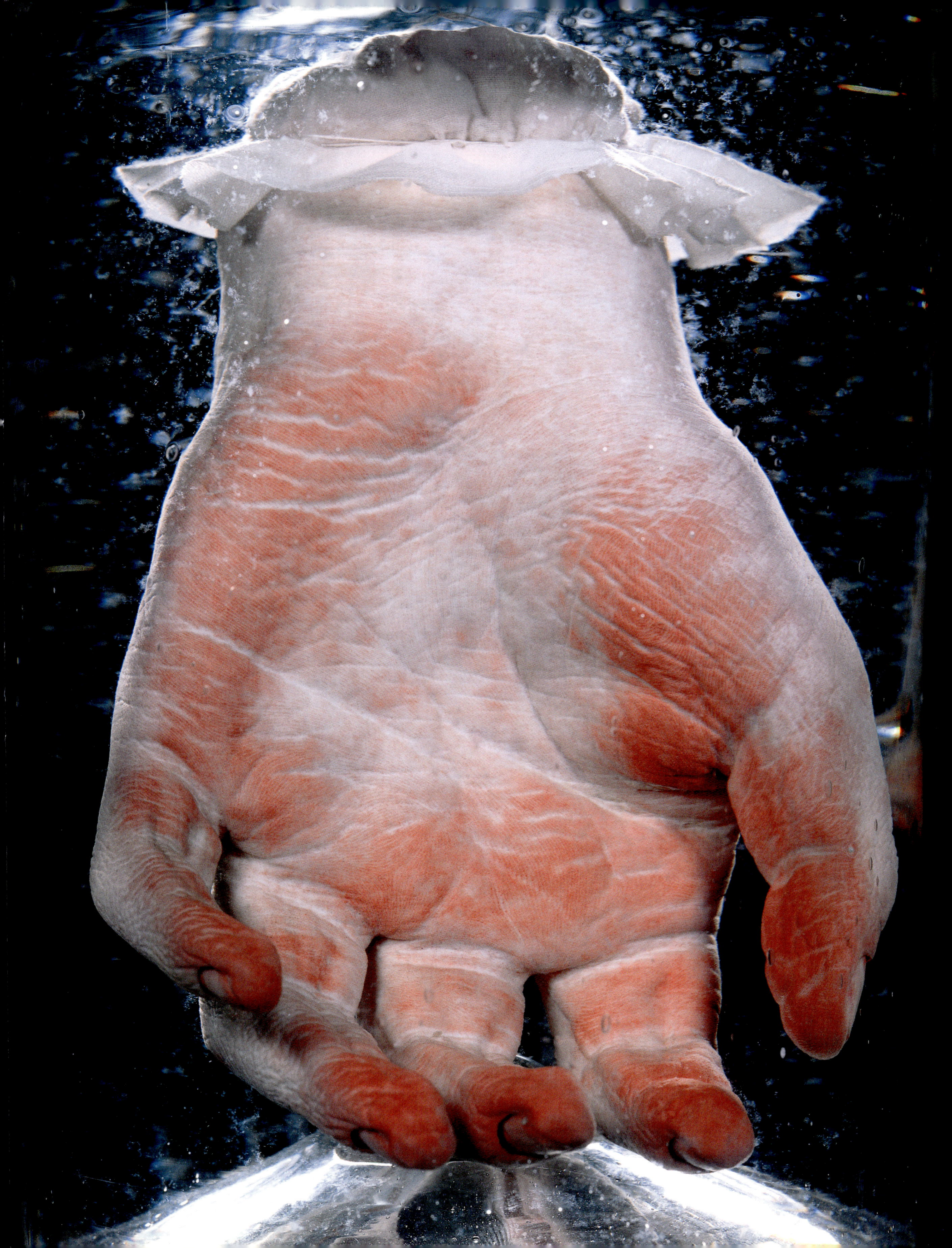

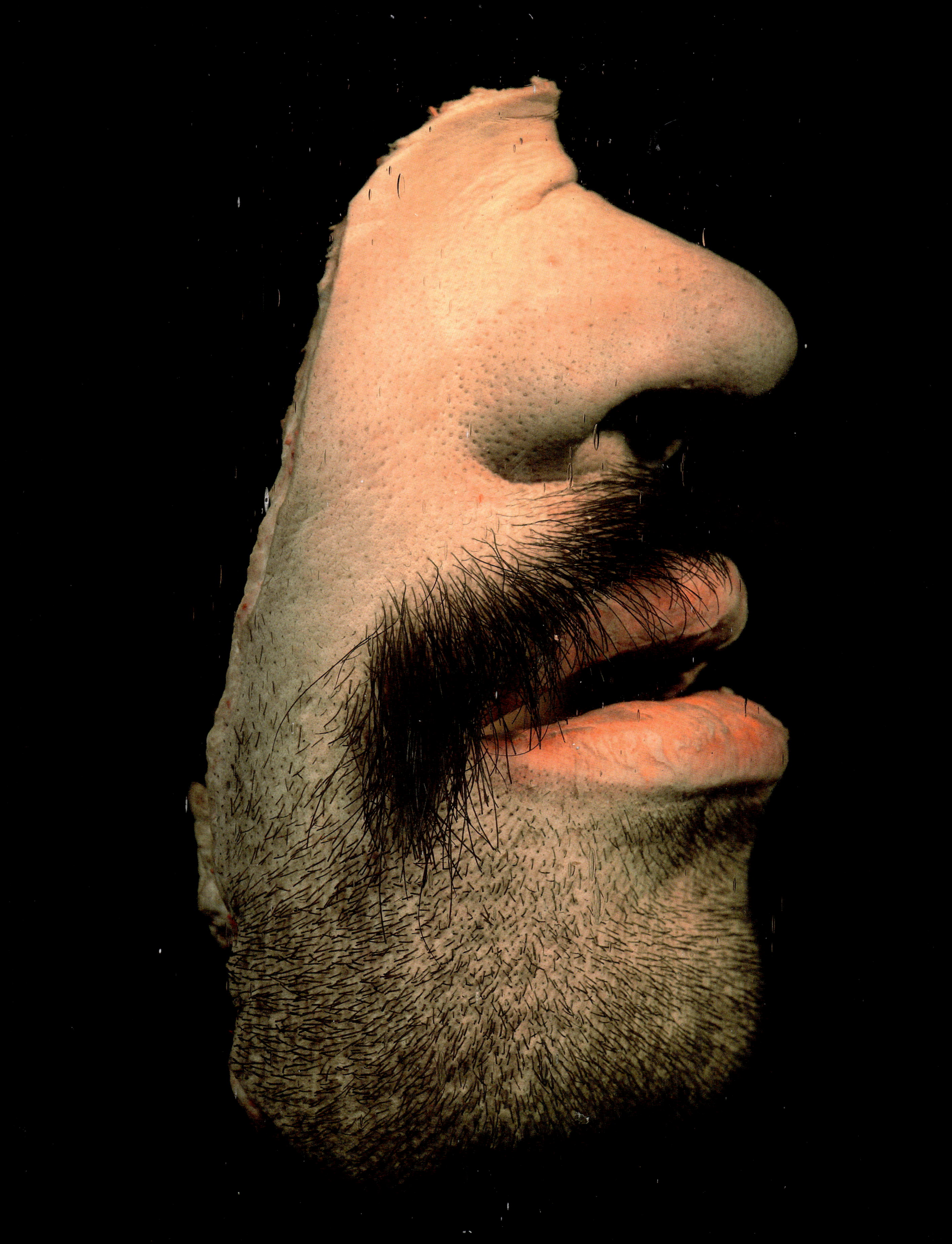

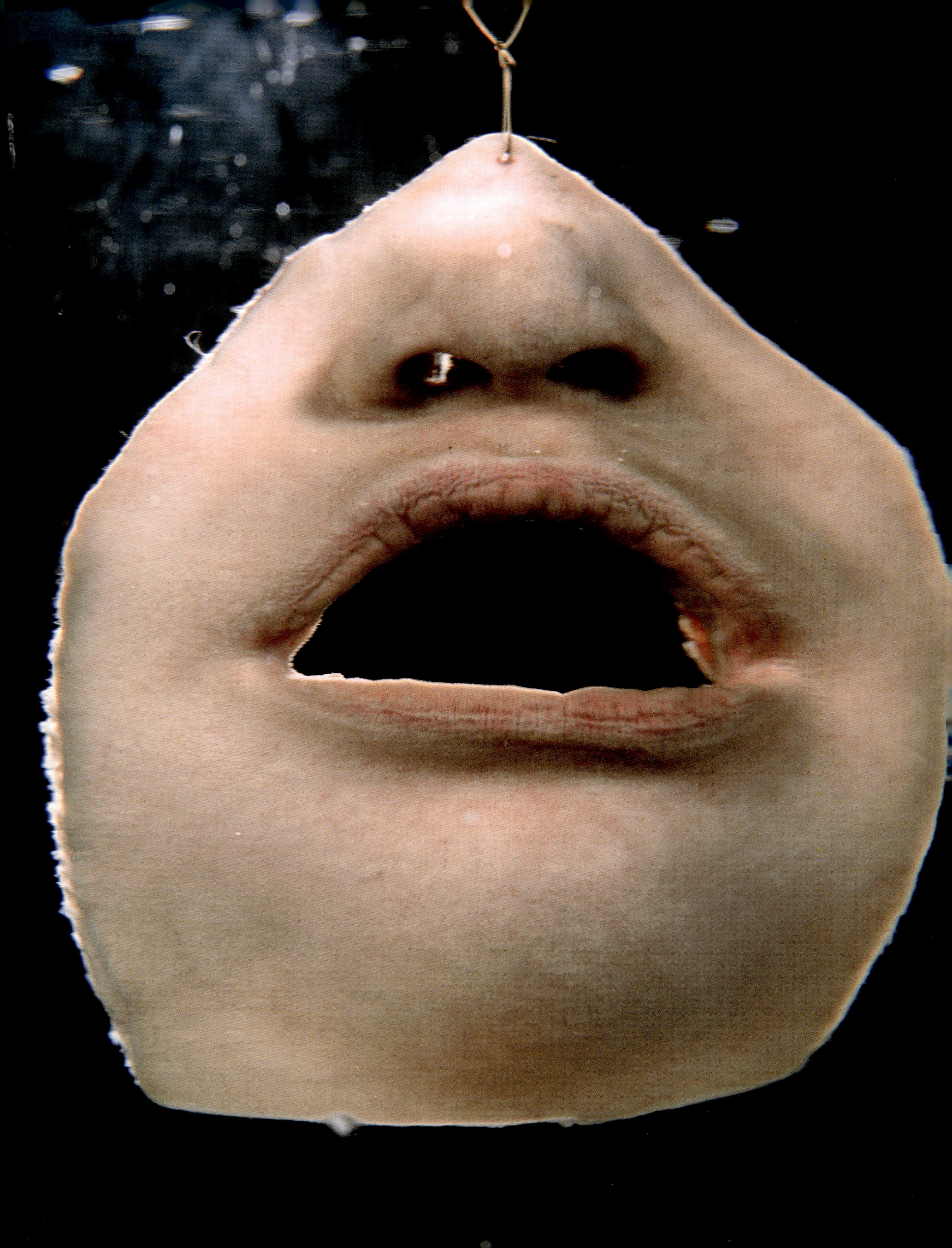

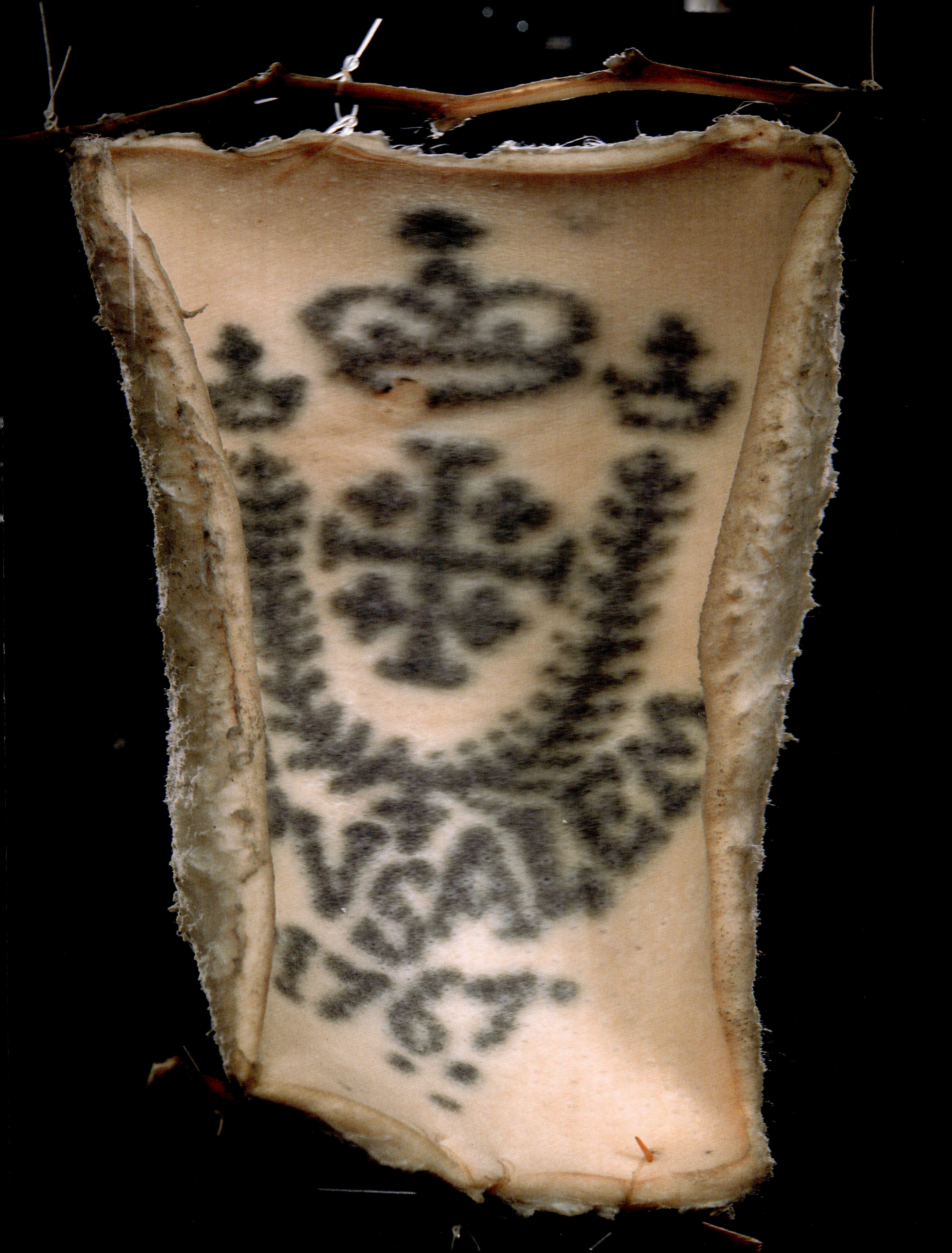

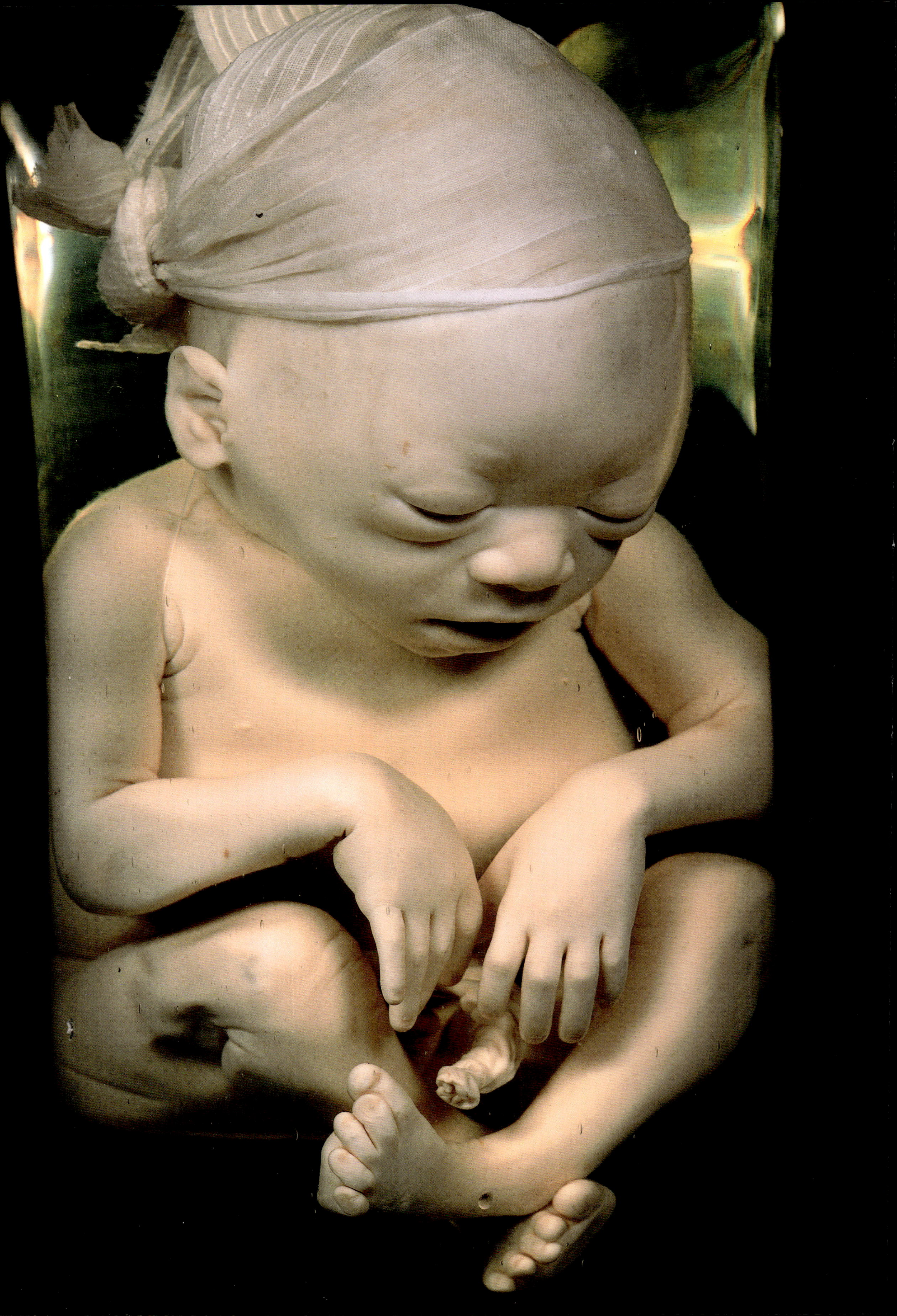

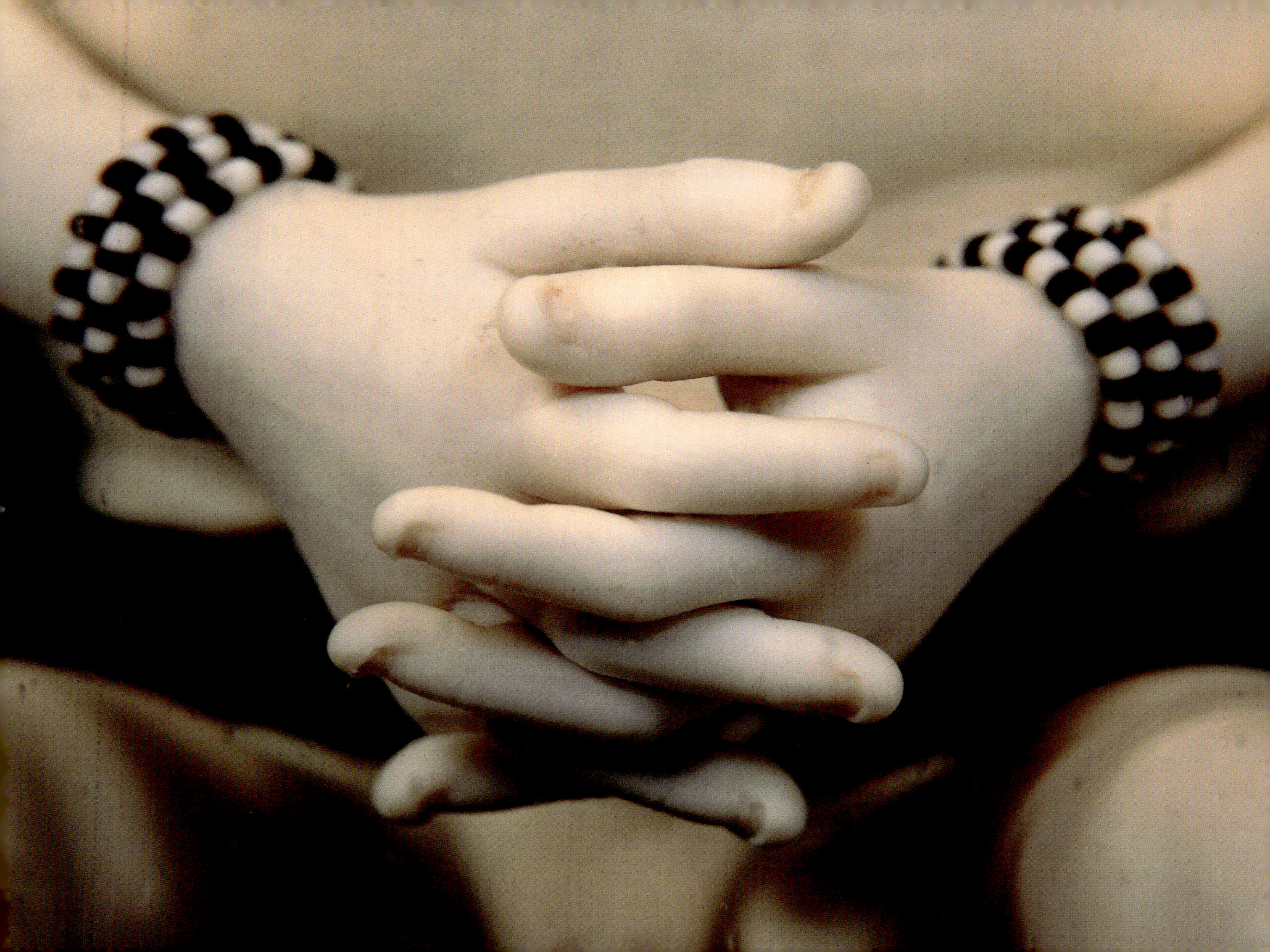

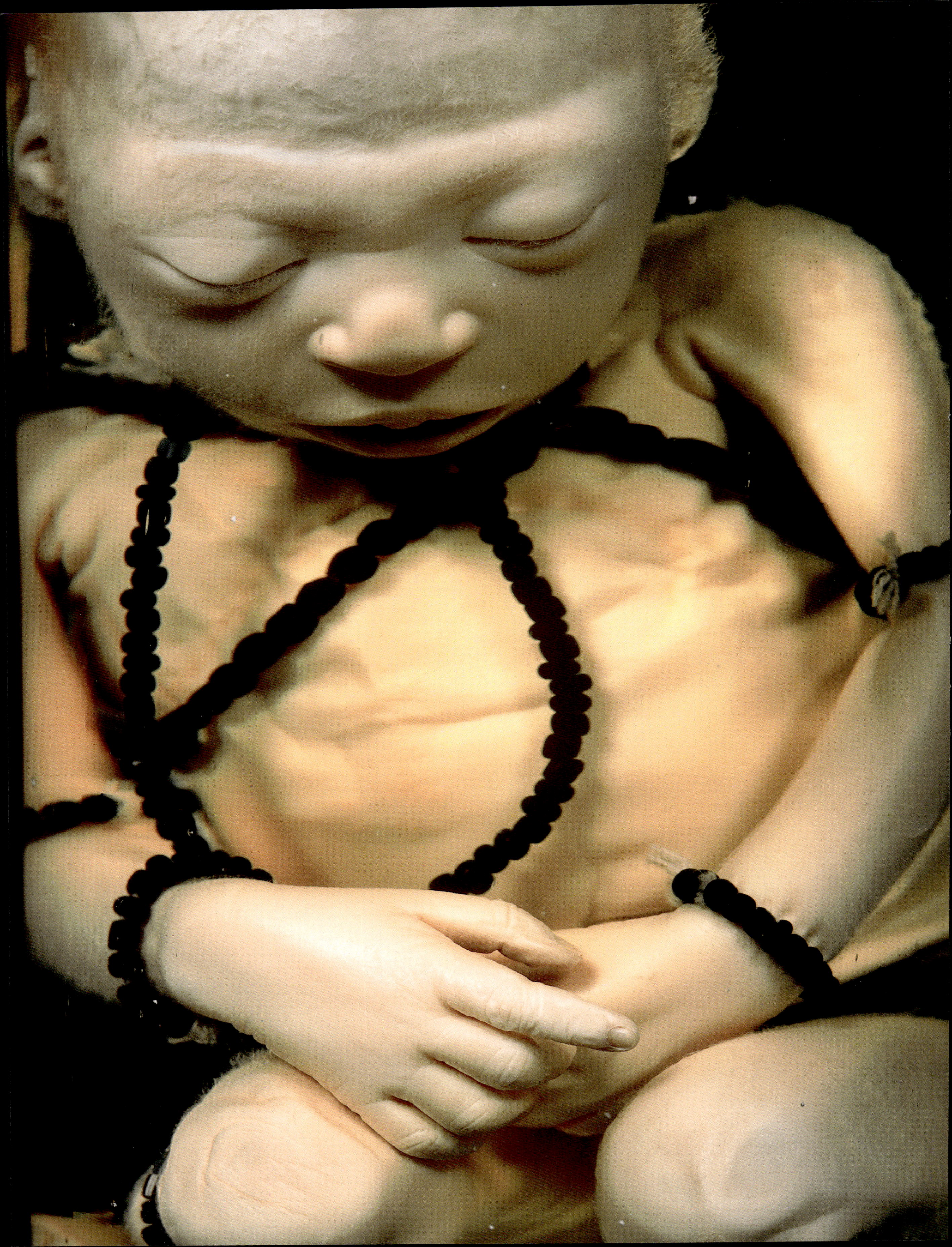

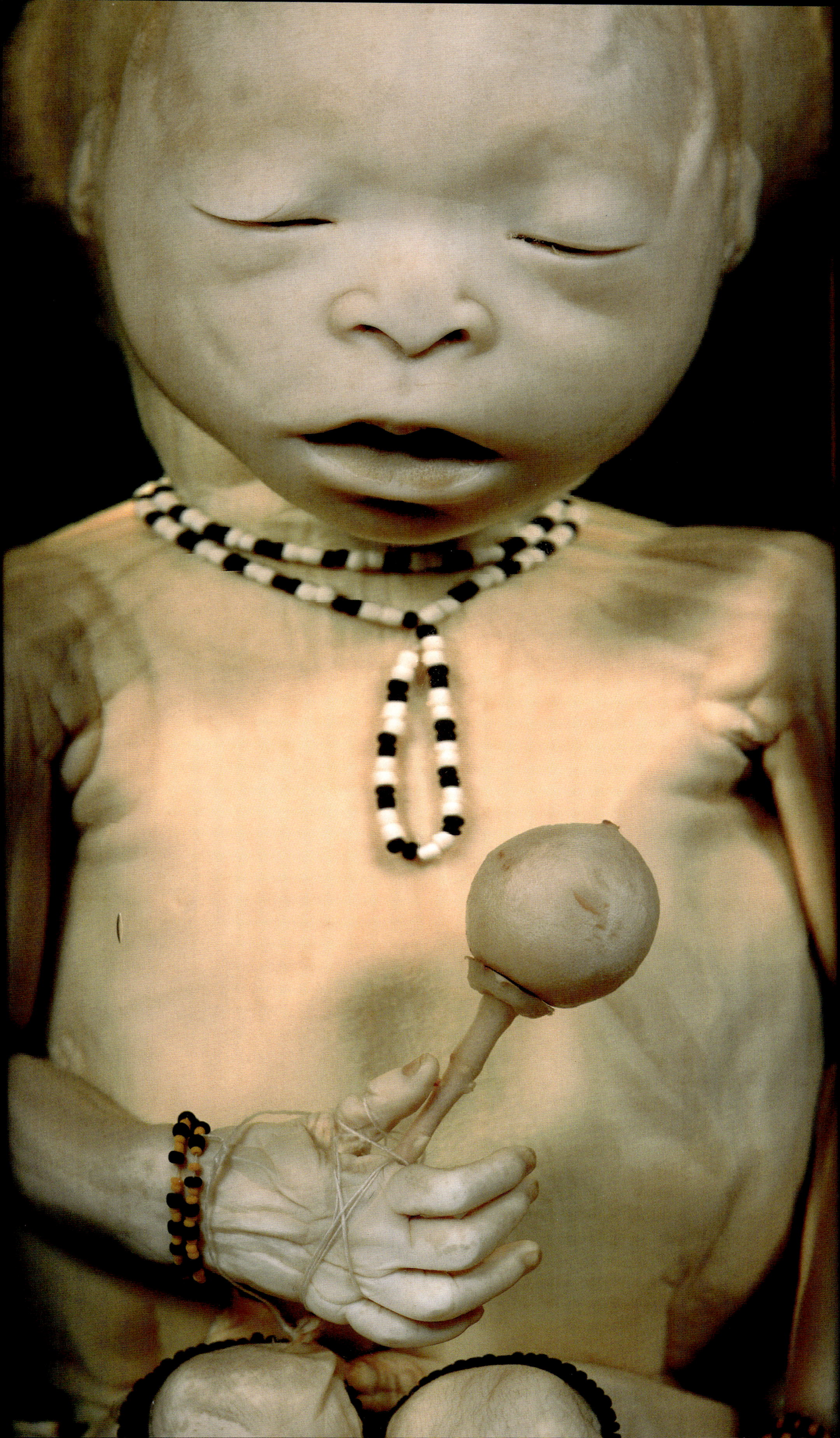

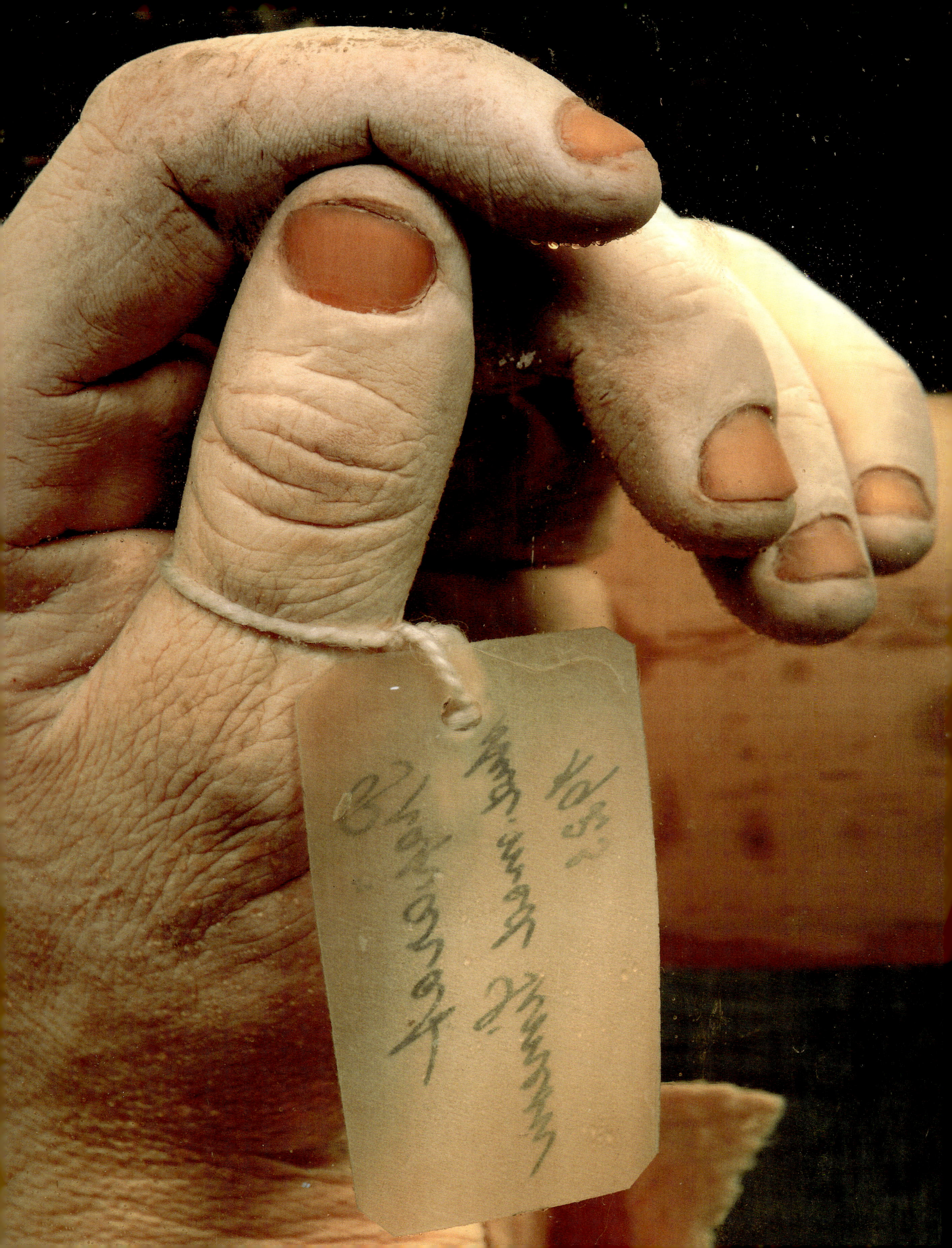

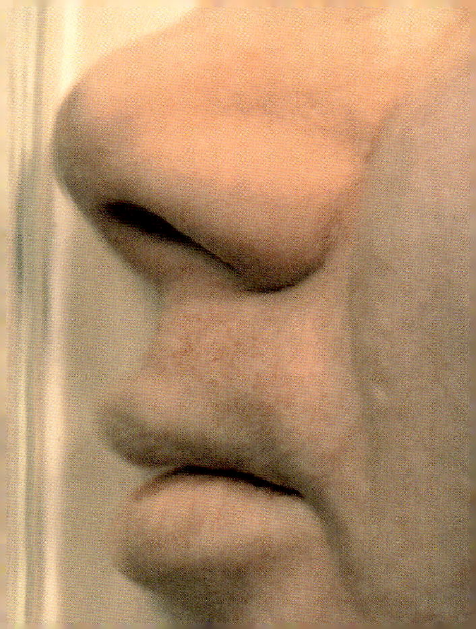

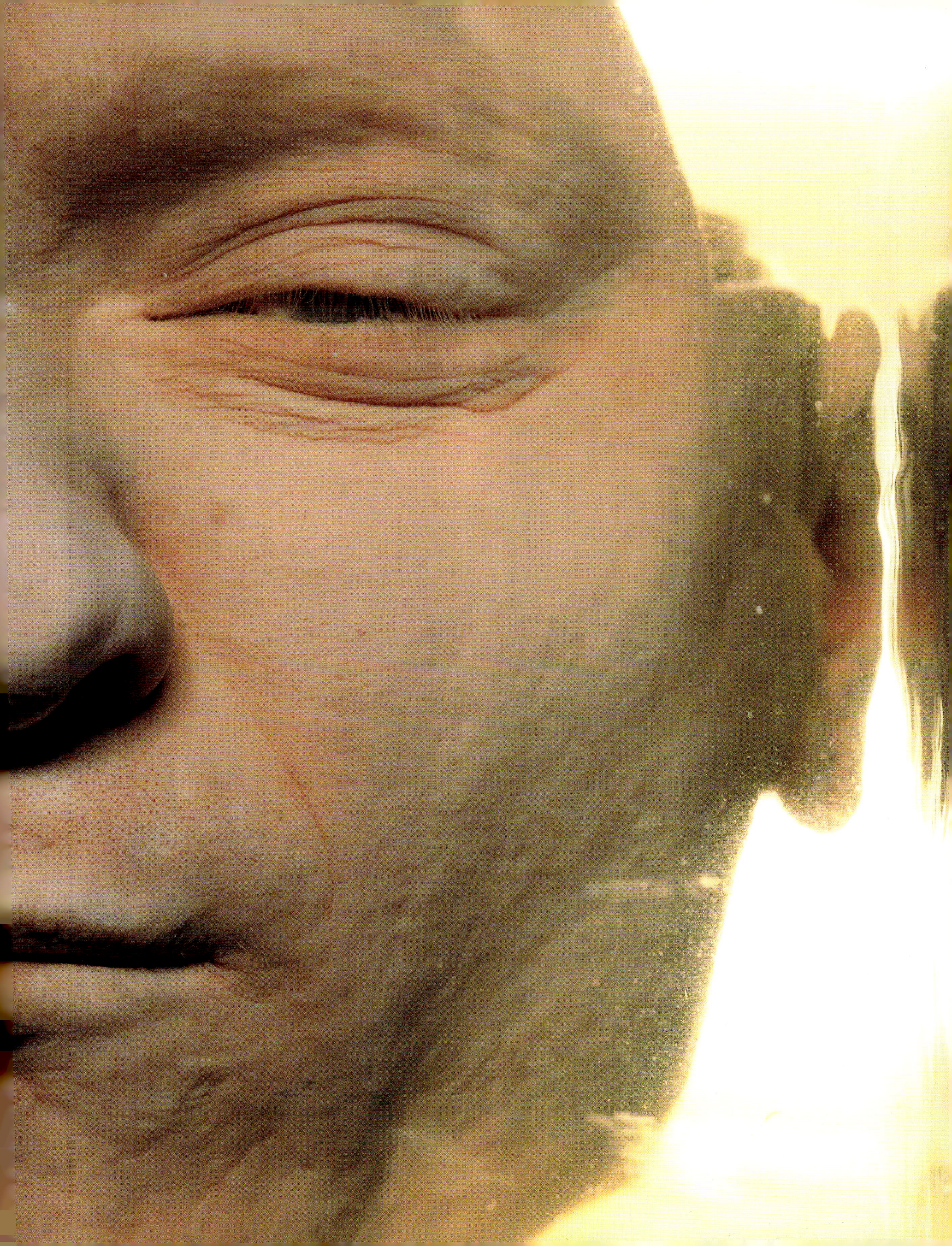

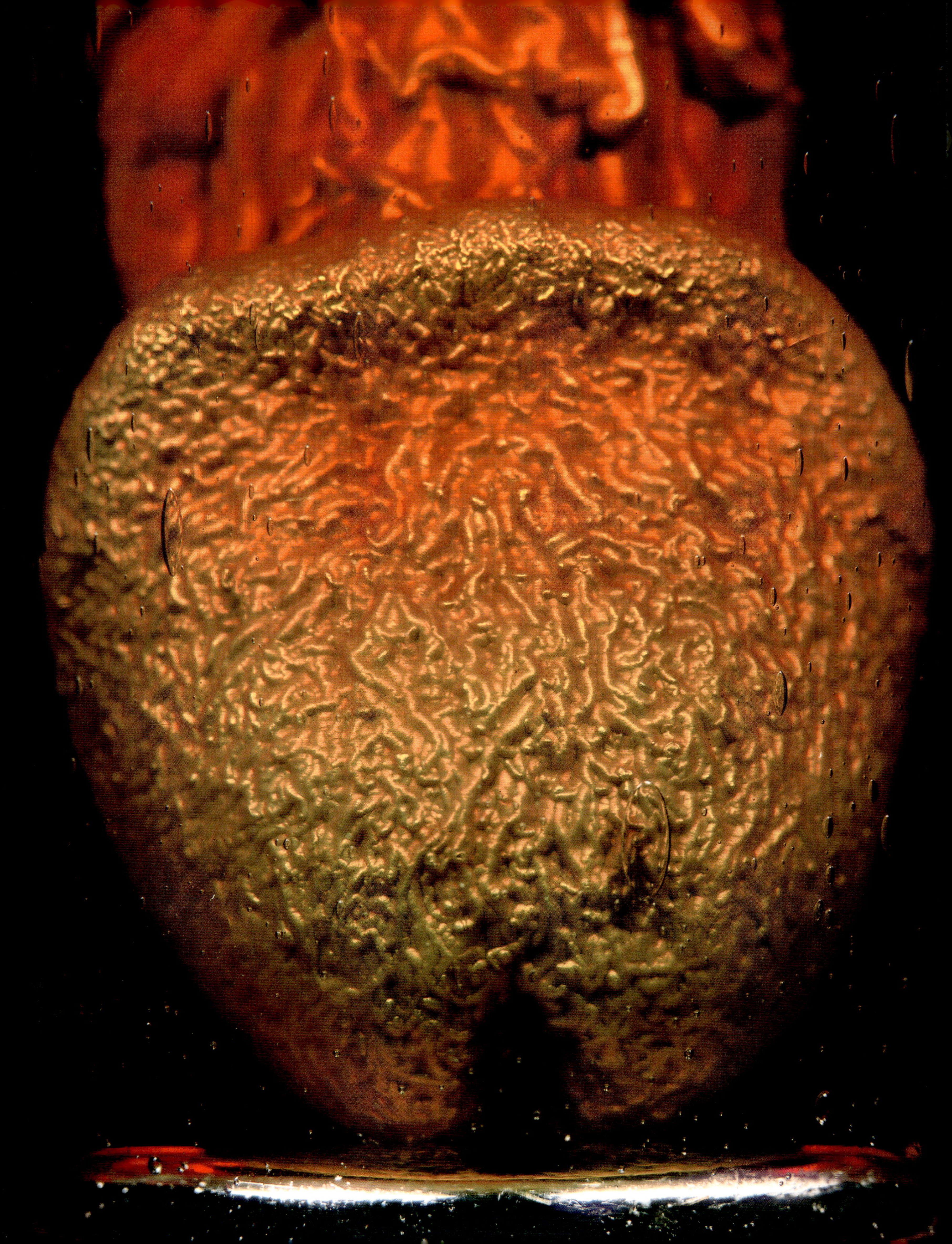

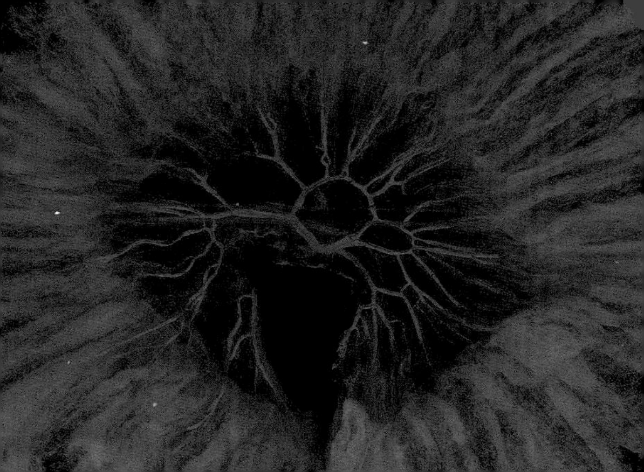

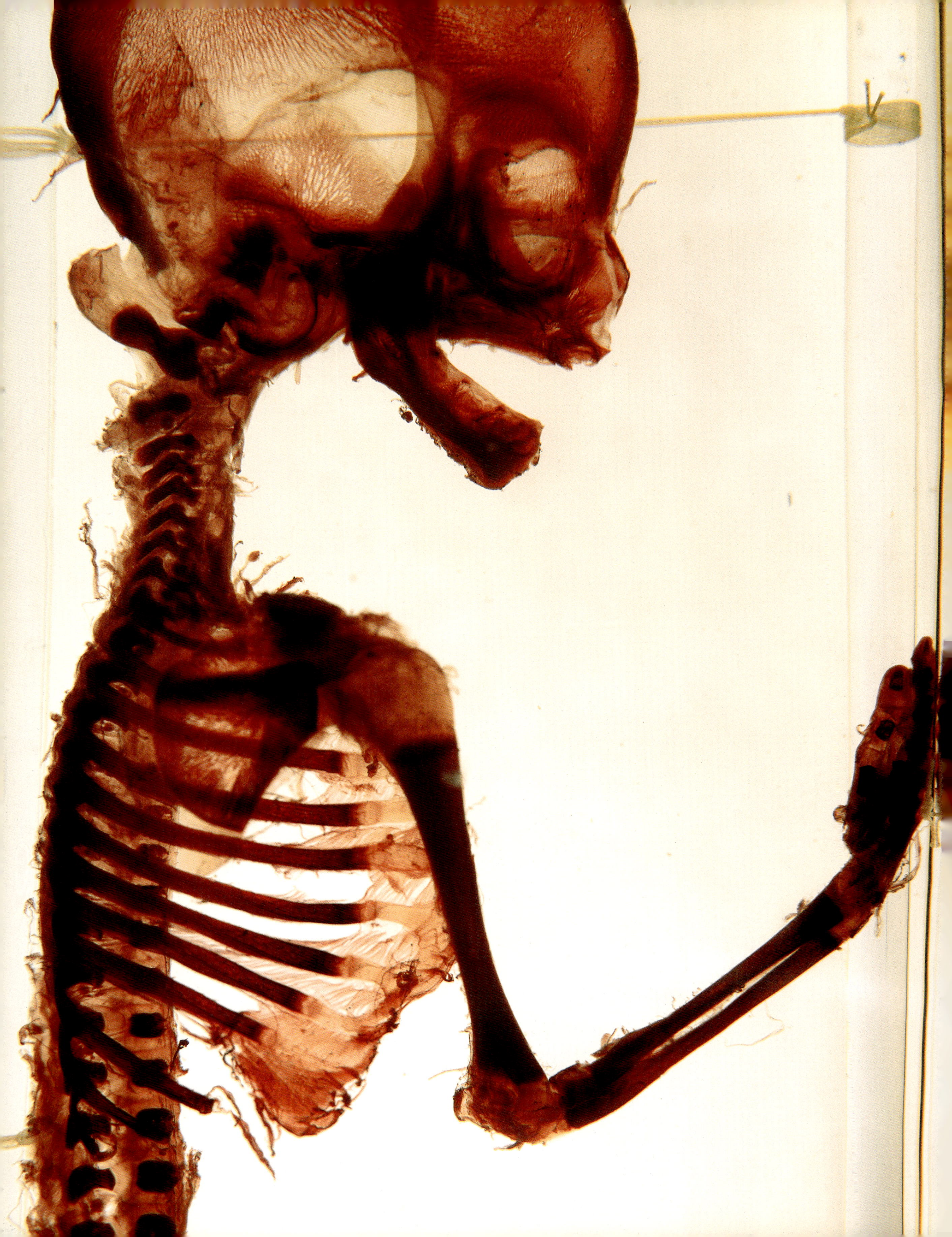

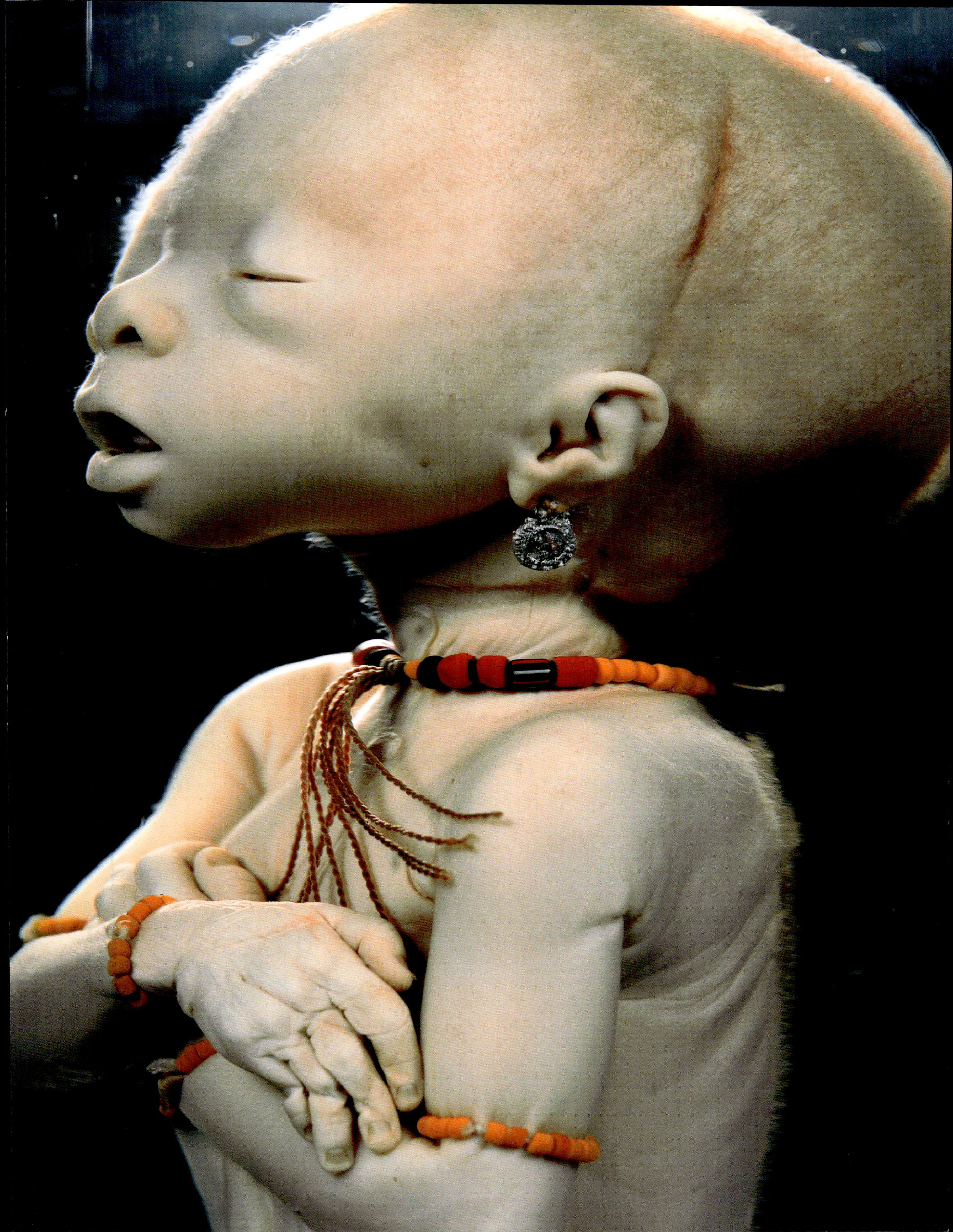

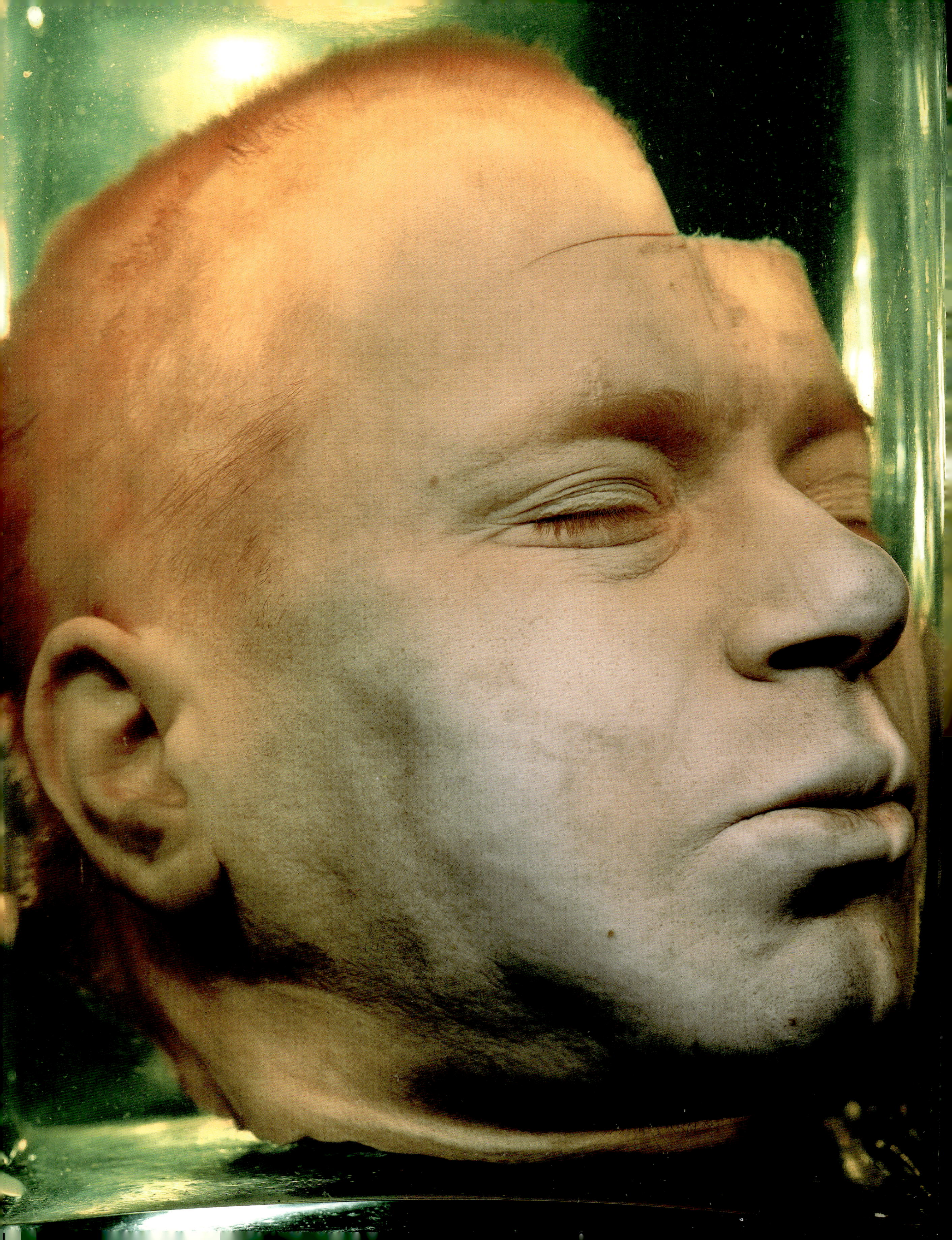

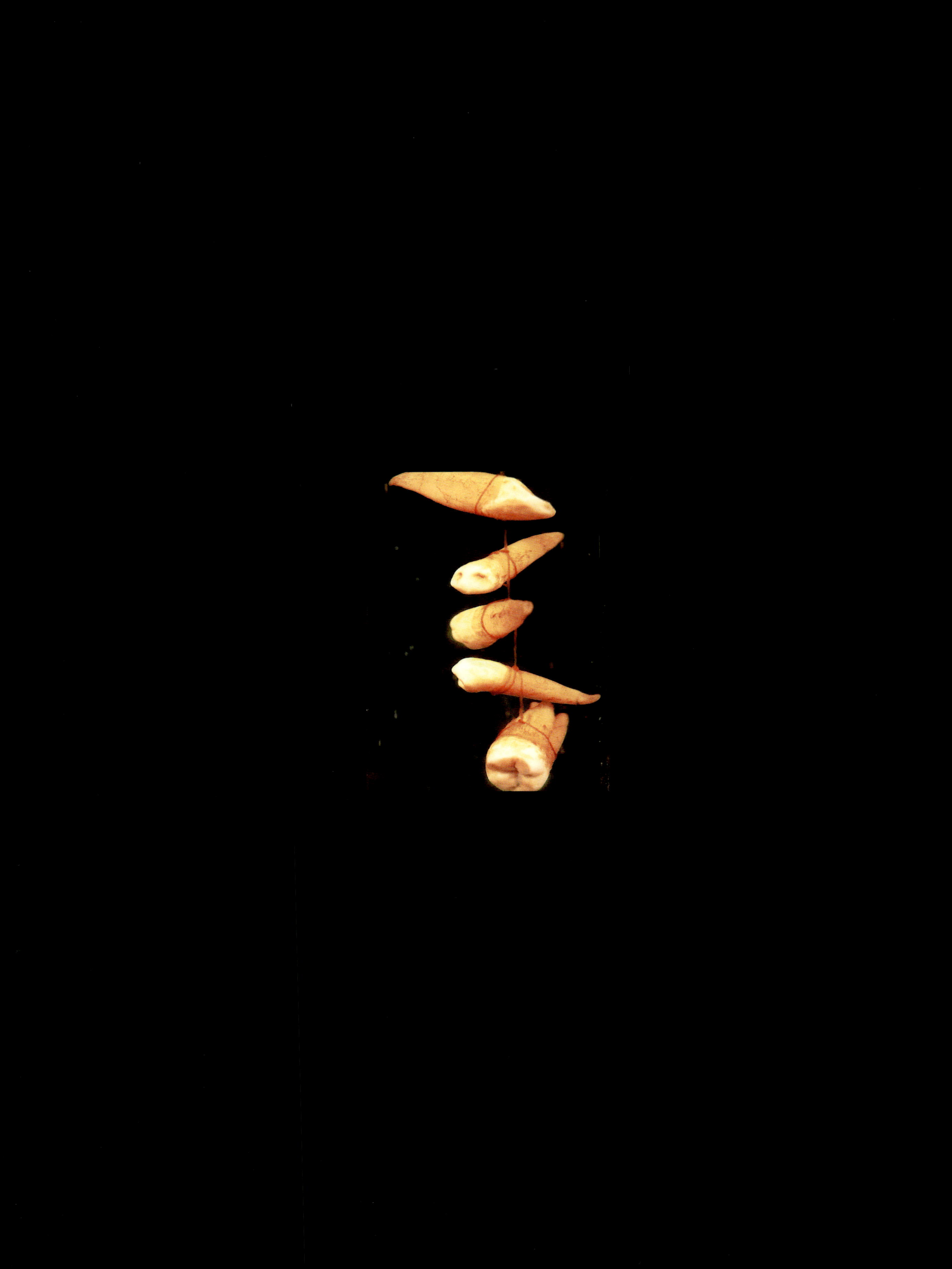

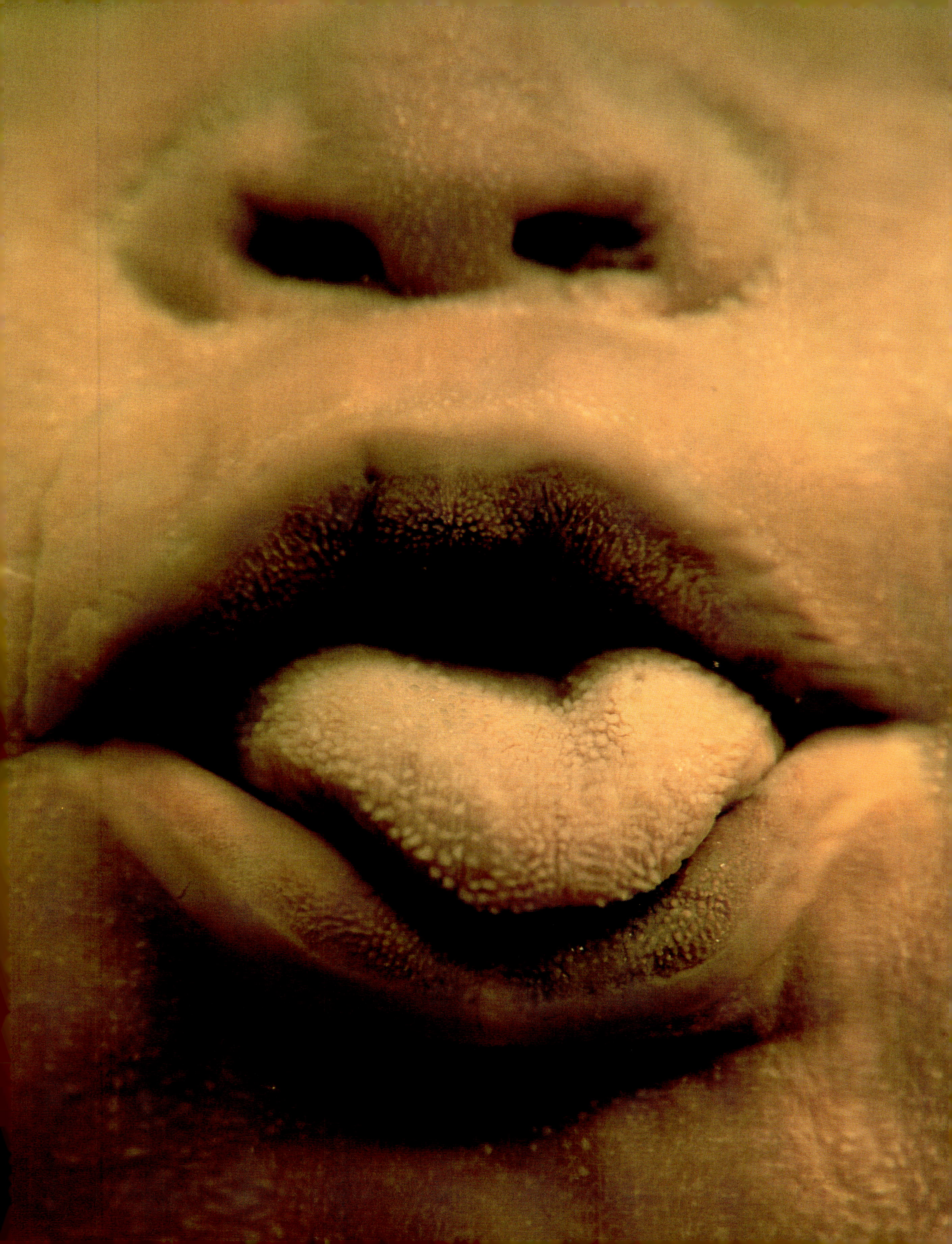

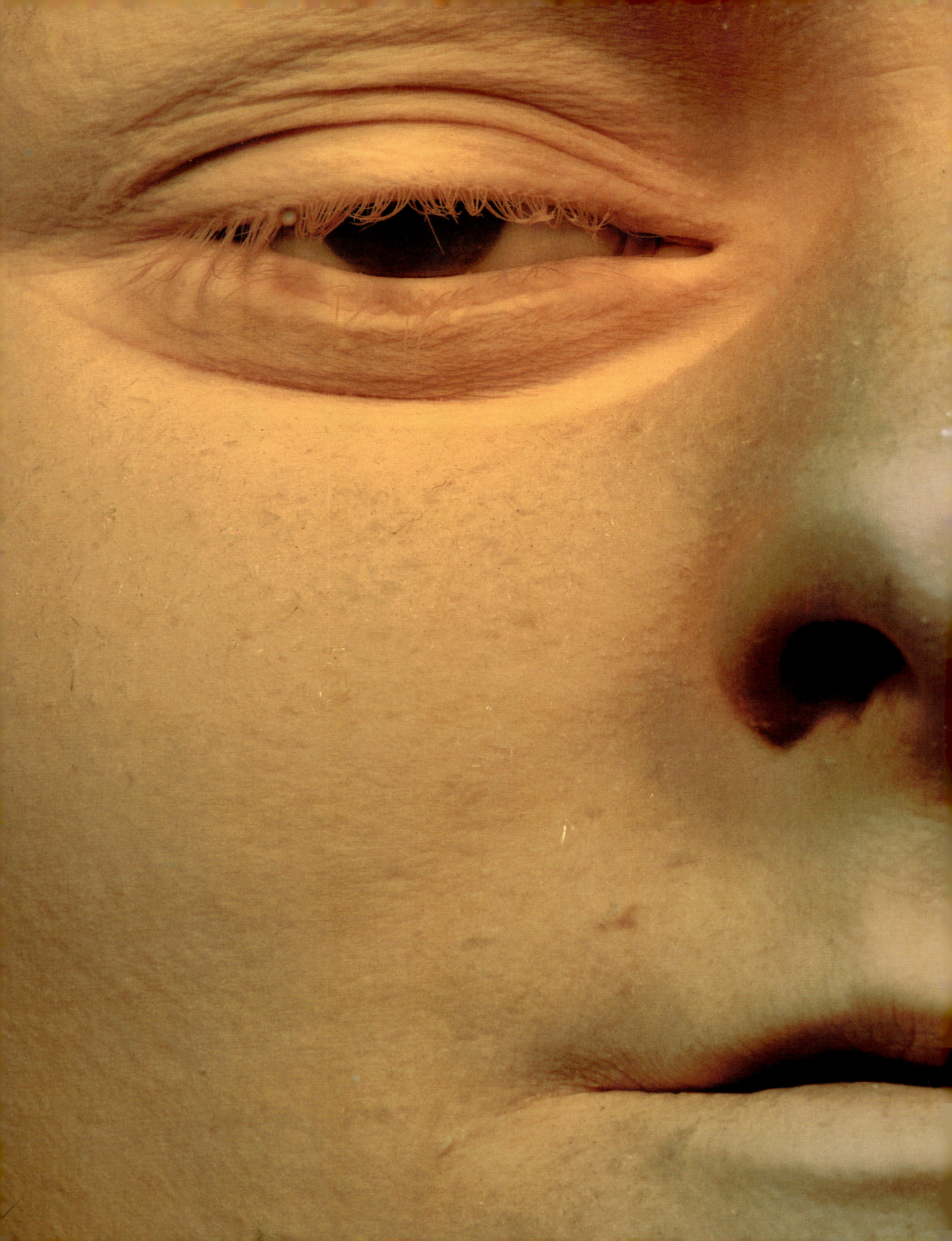

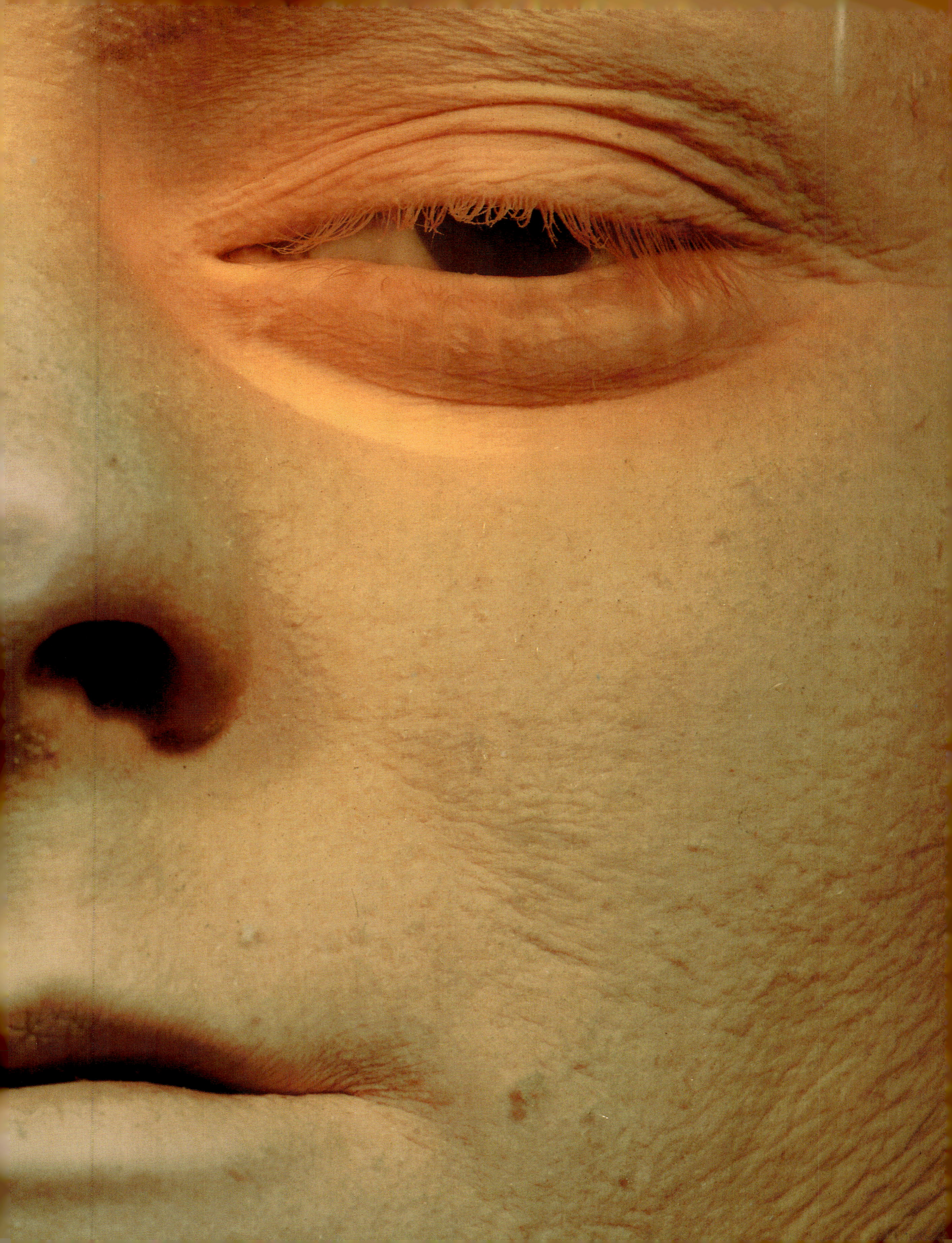

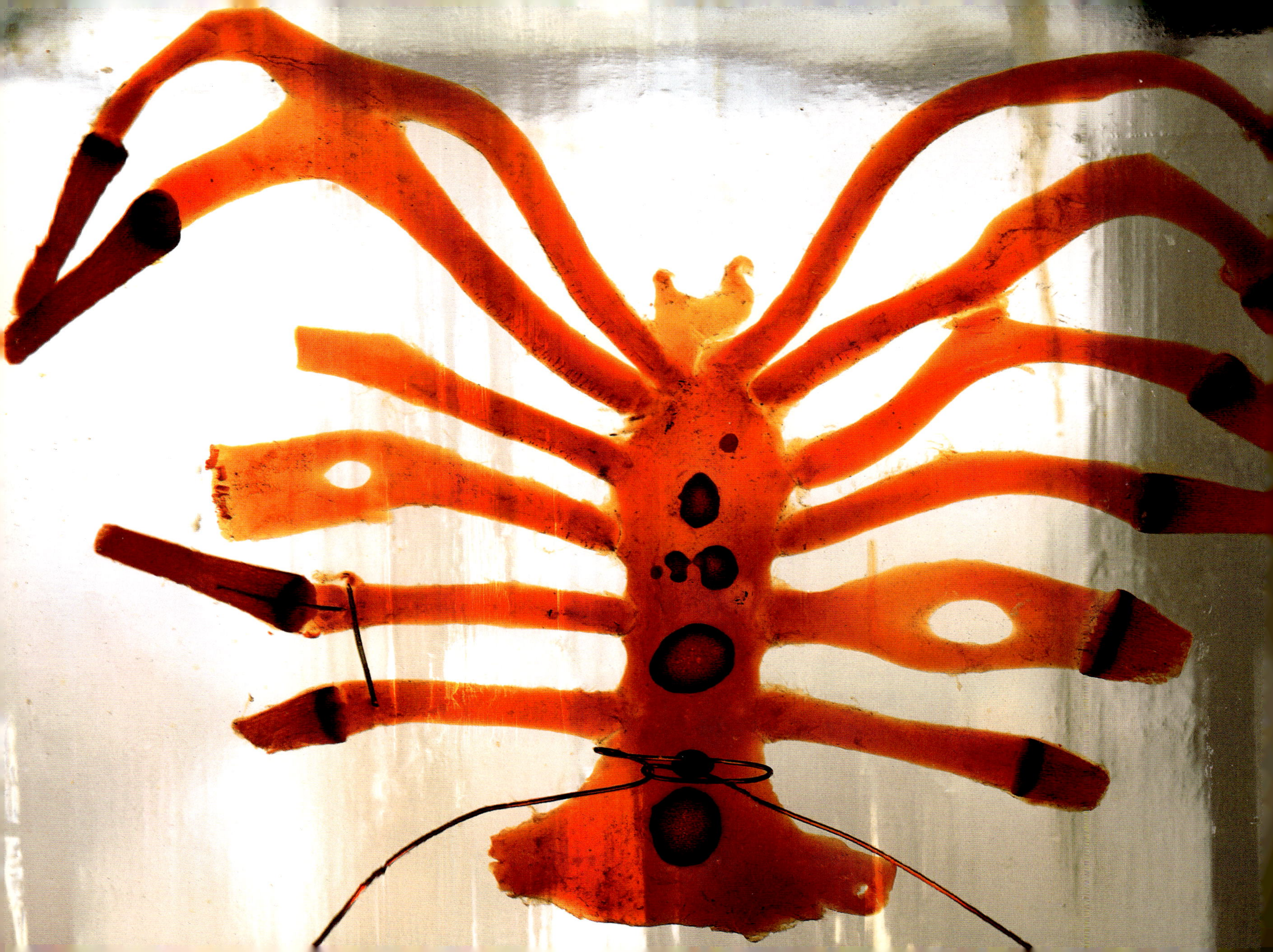

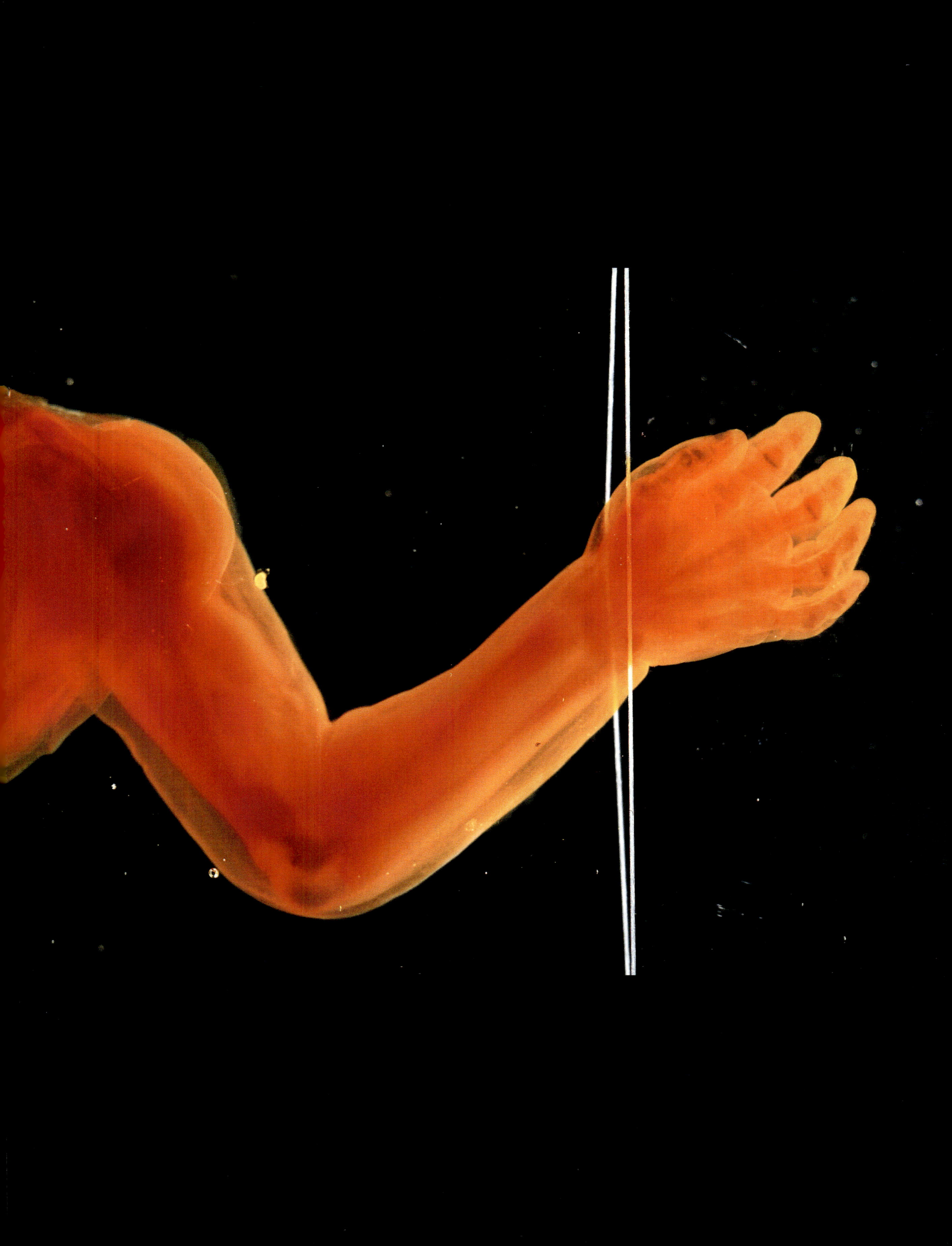

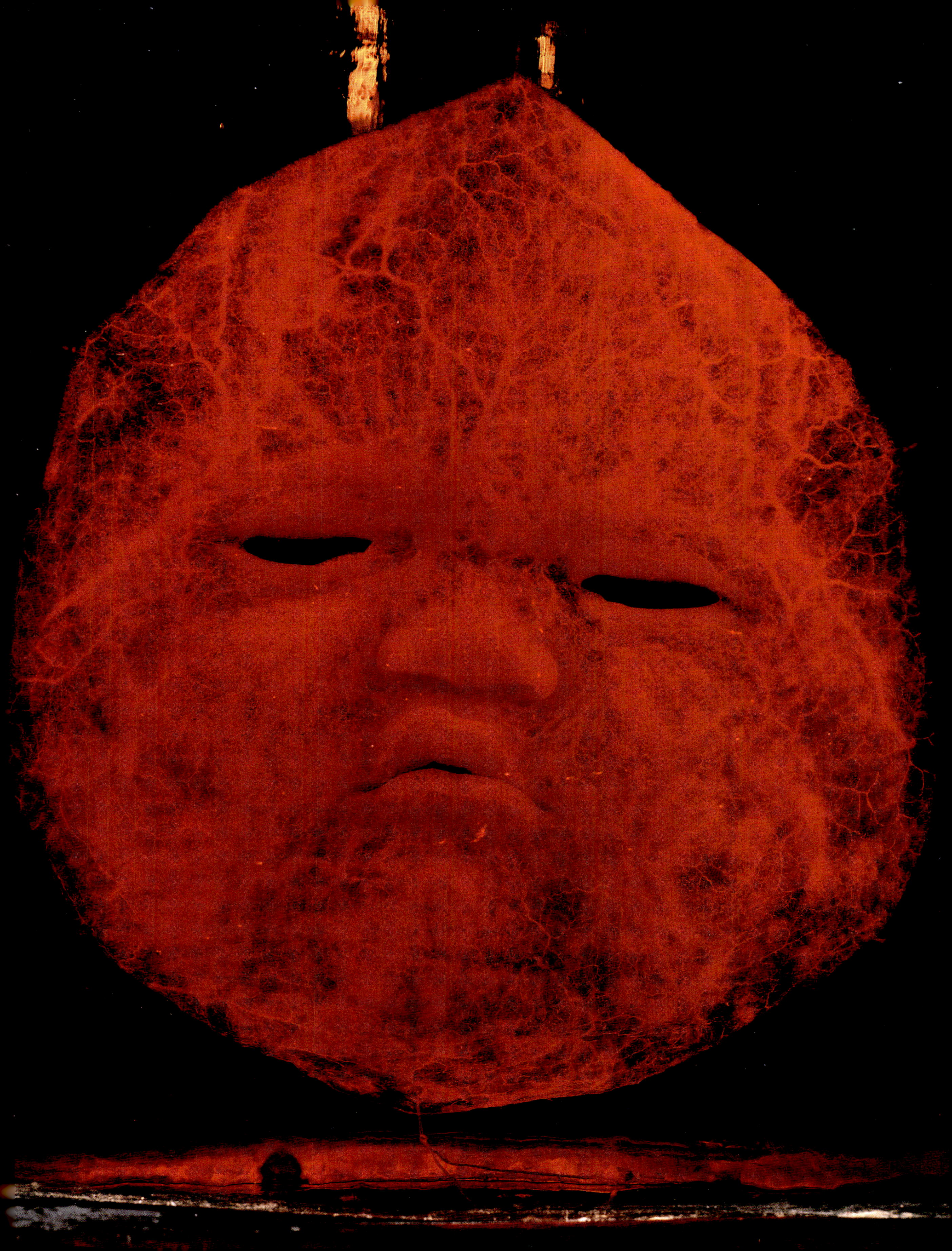

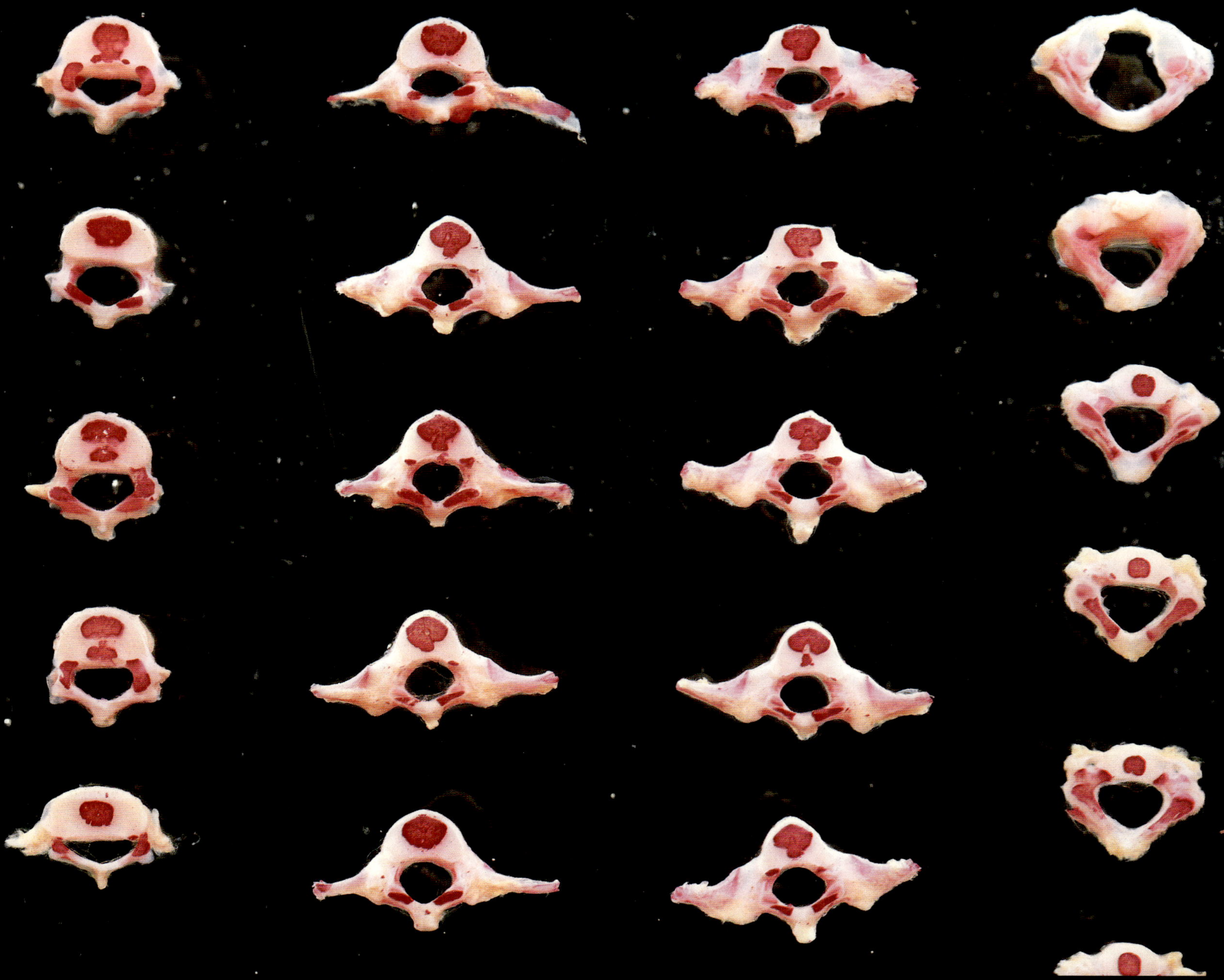

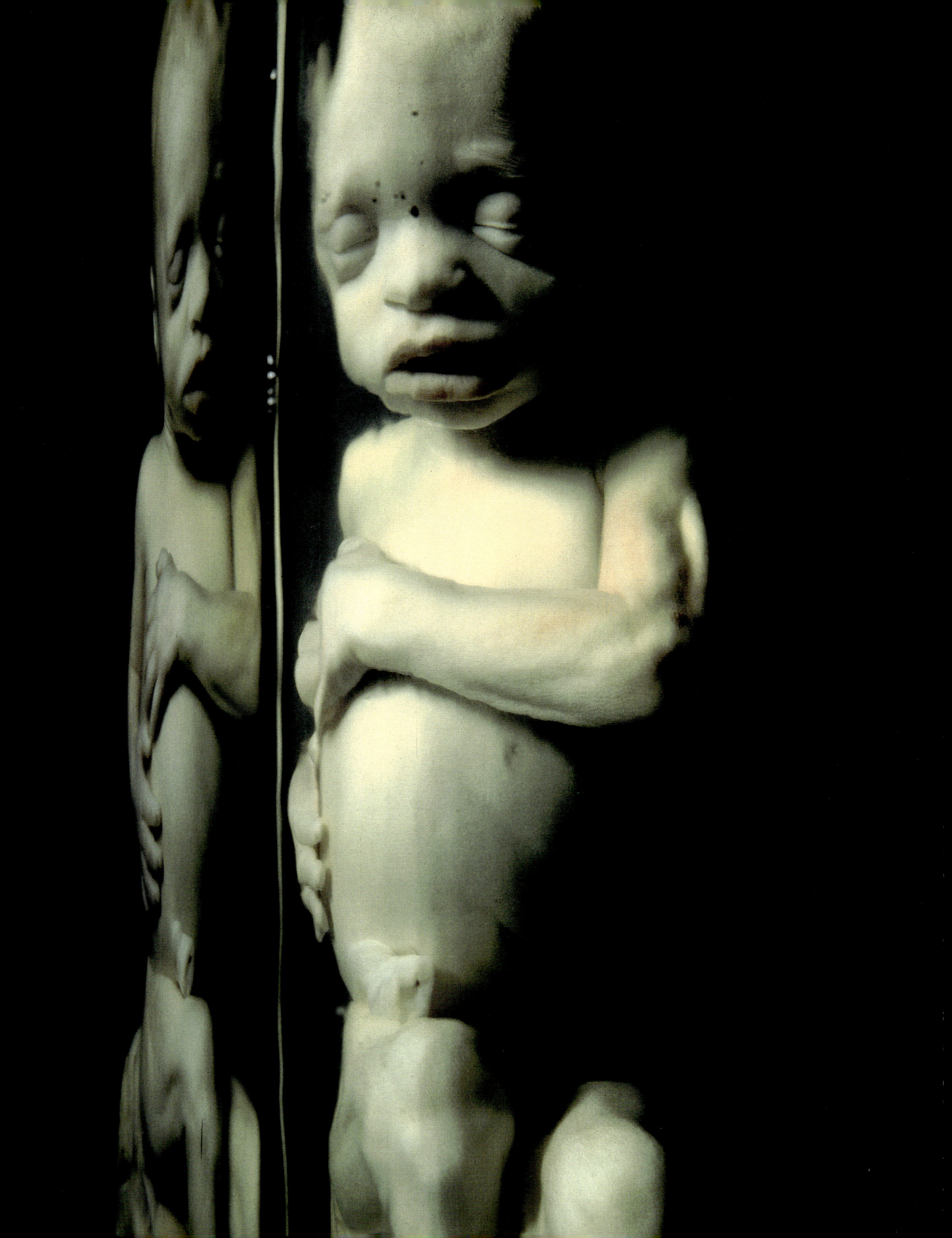

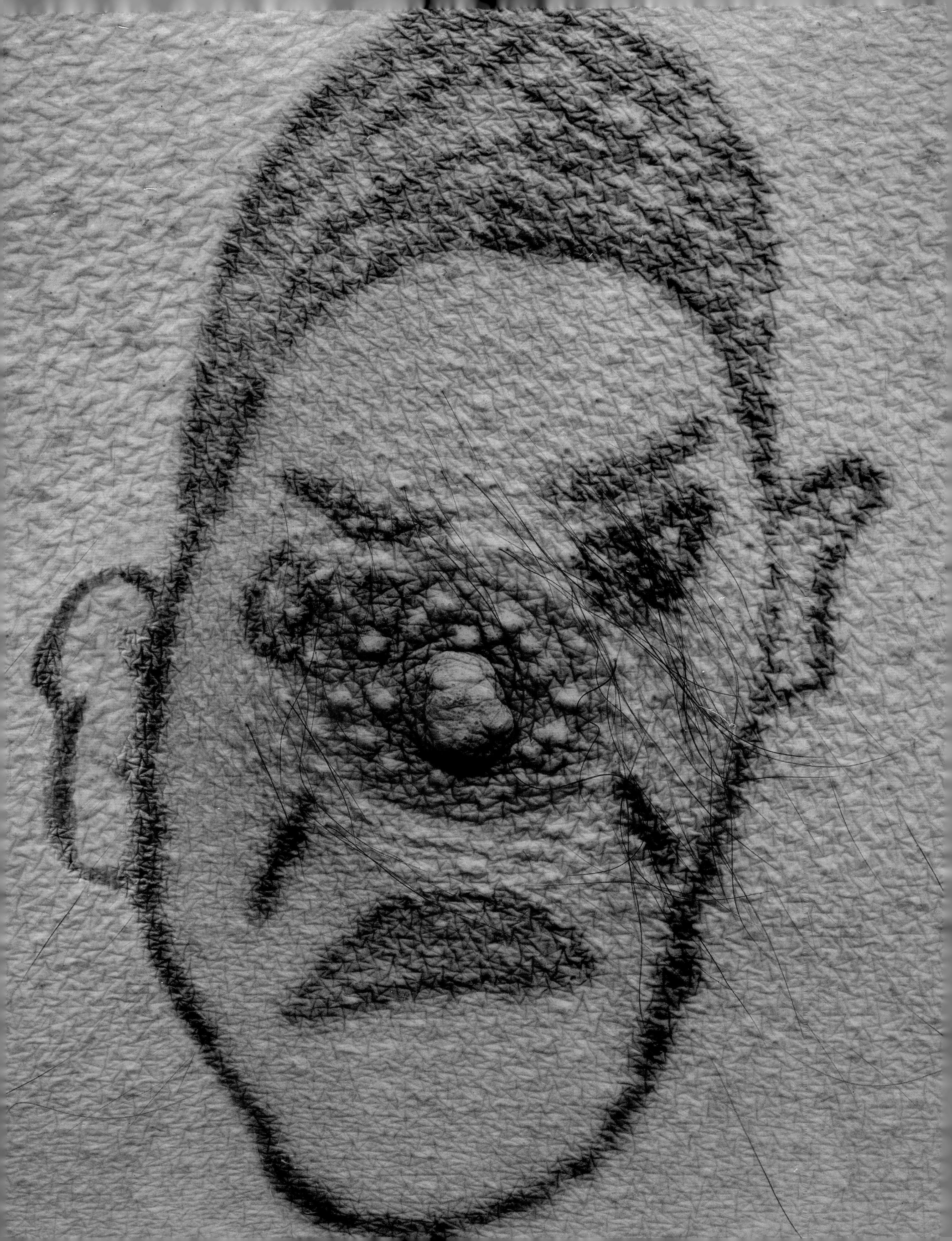

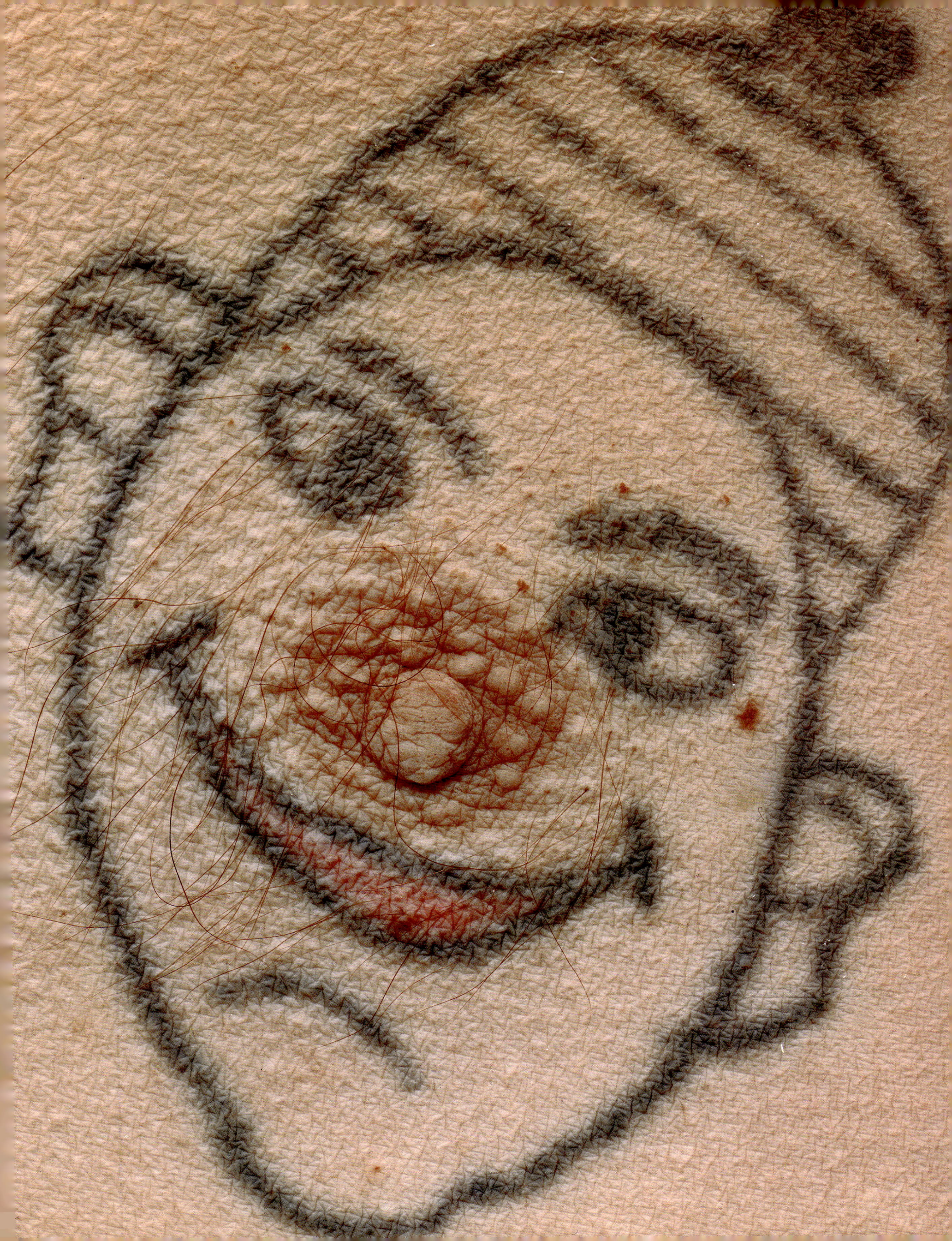

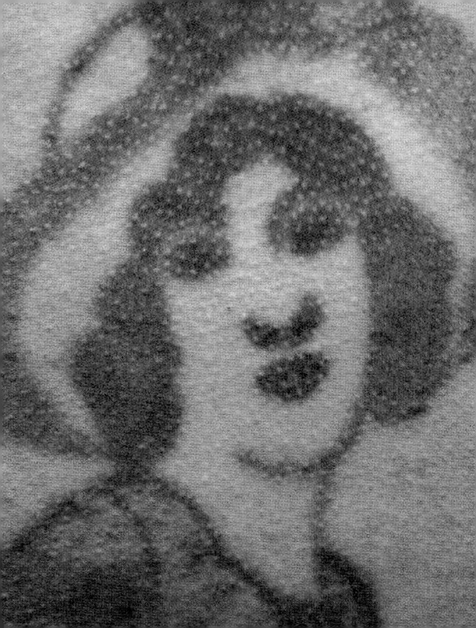

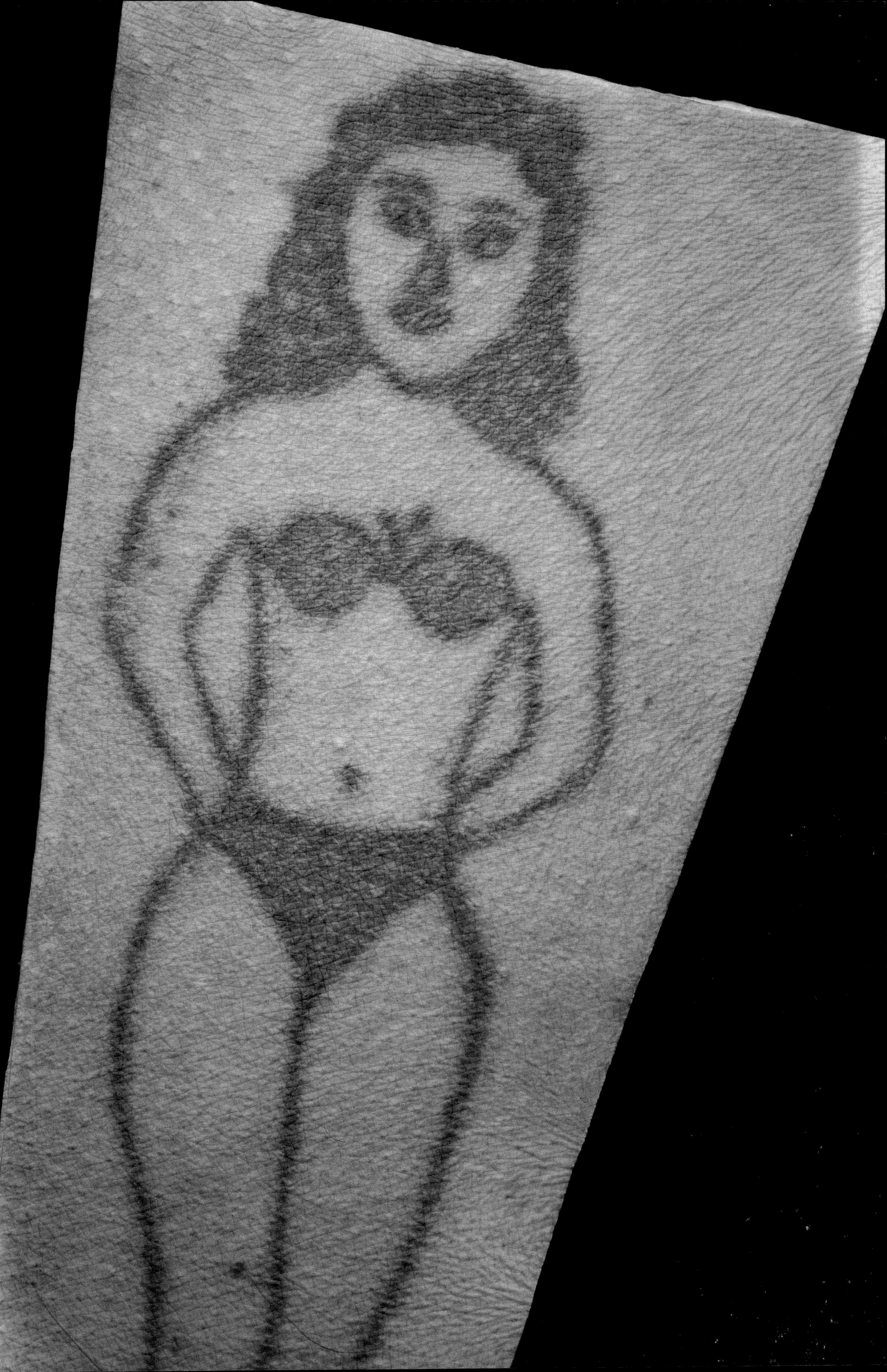

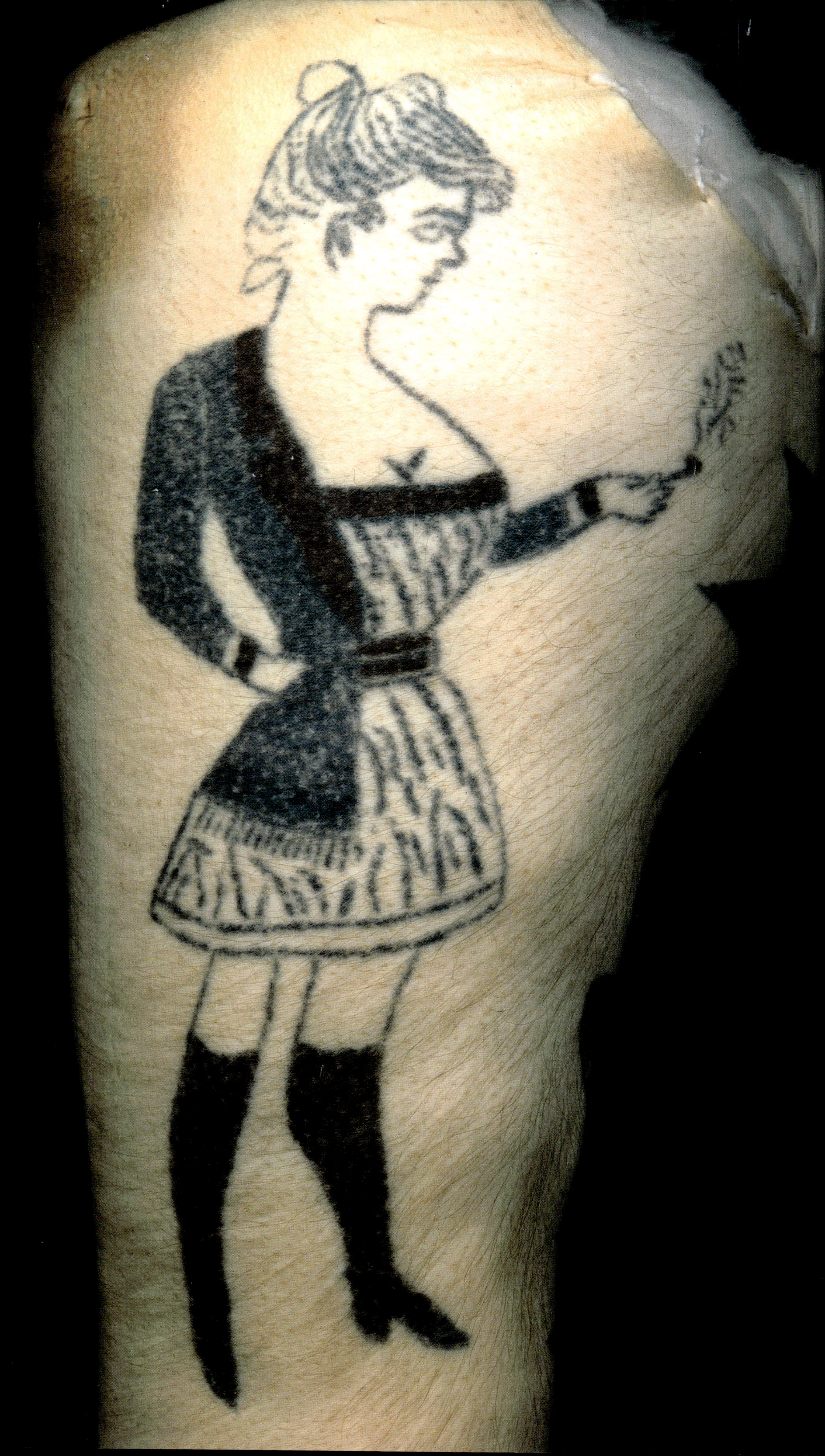

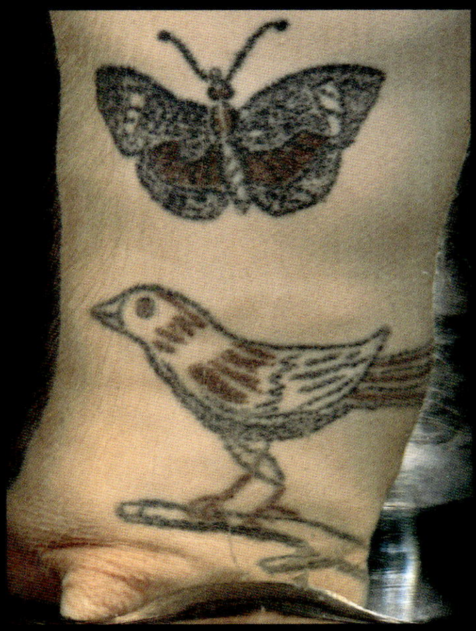

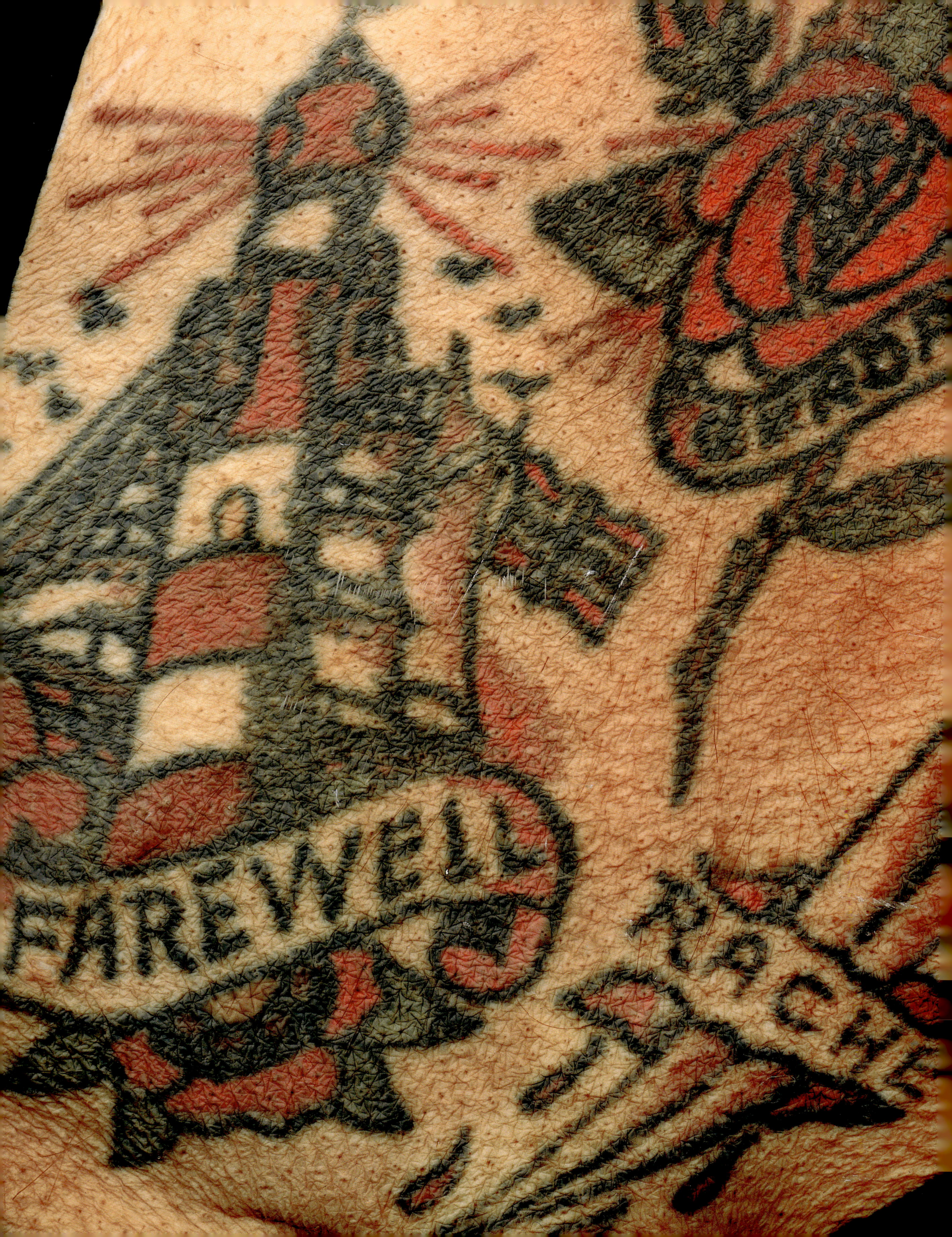
FAREWELL

1808

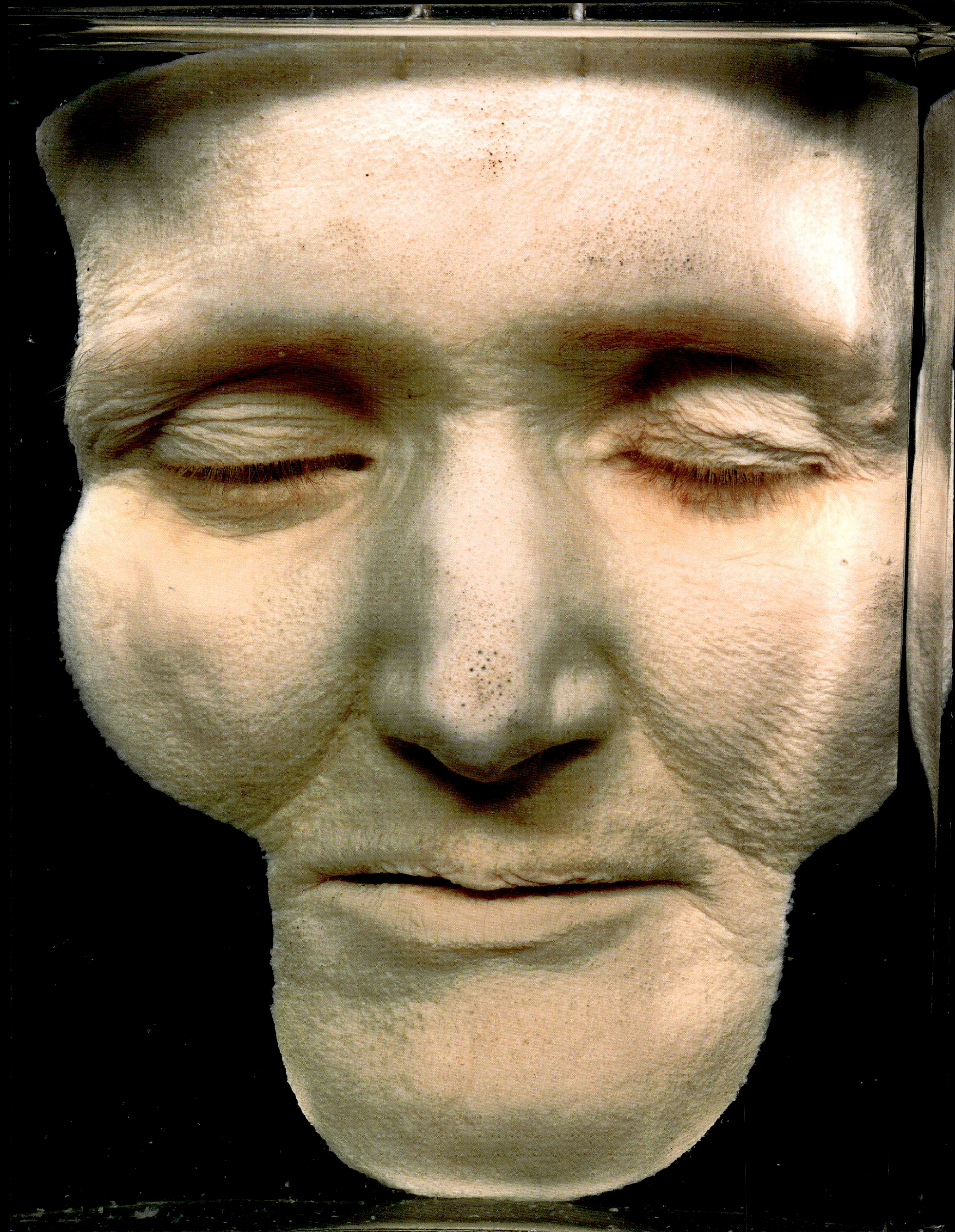

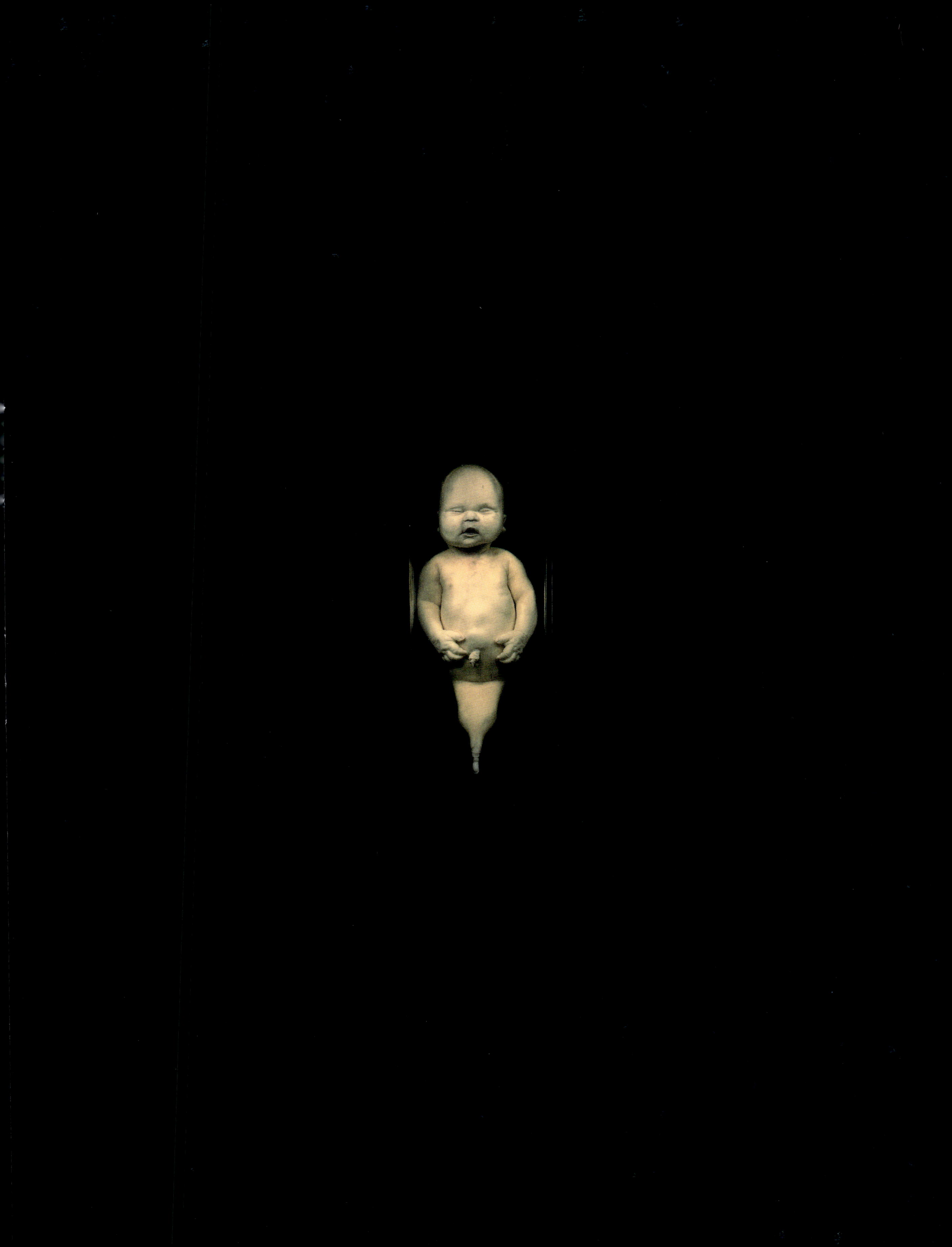

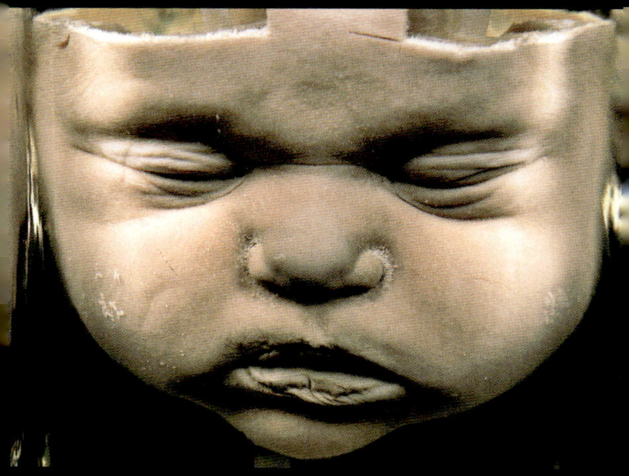

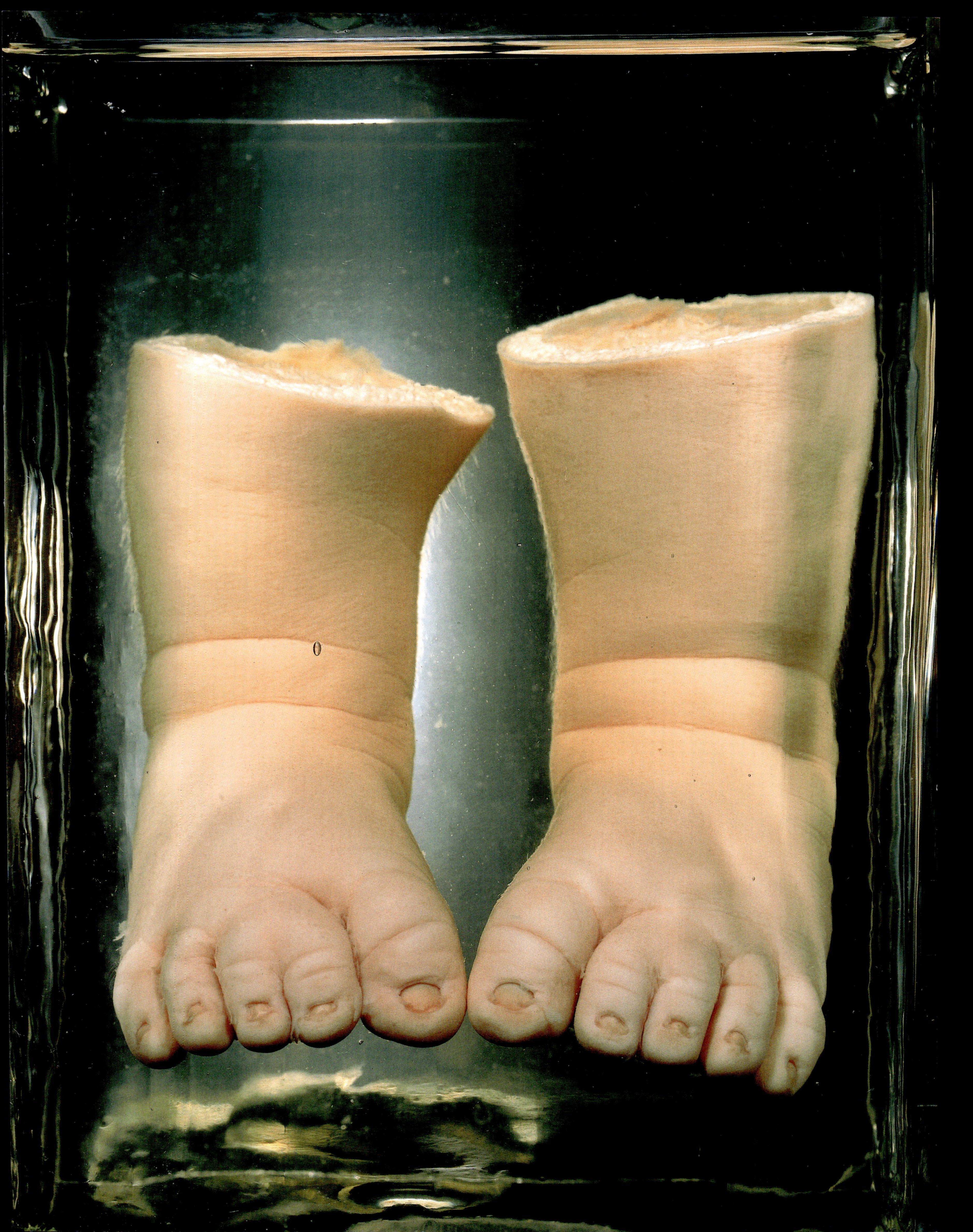

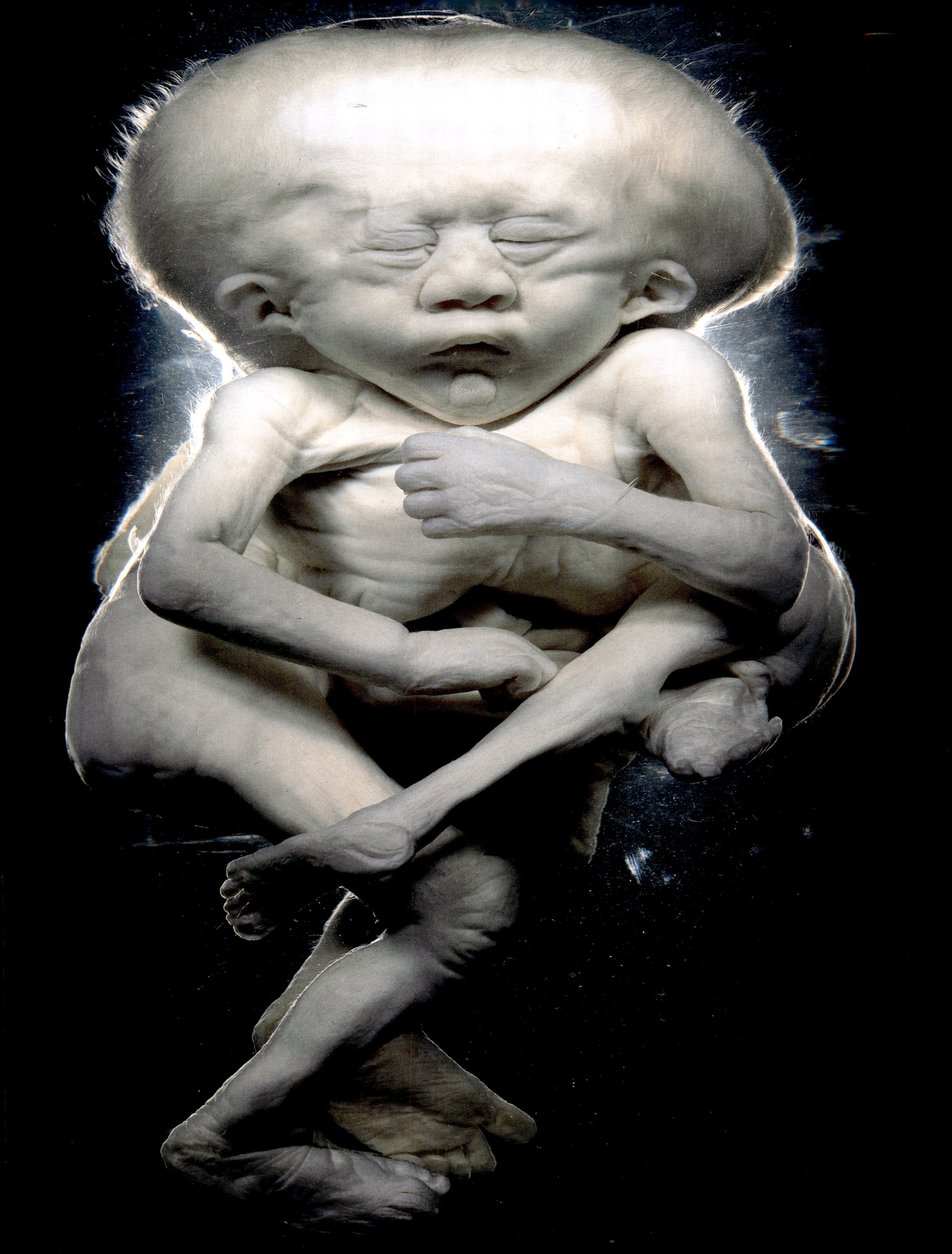

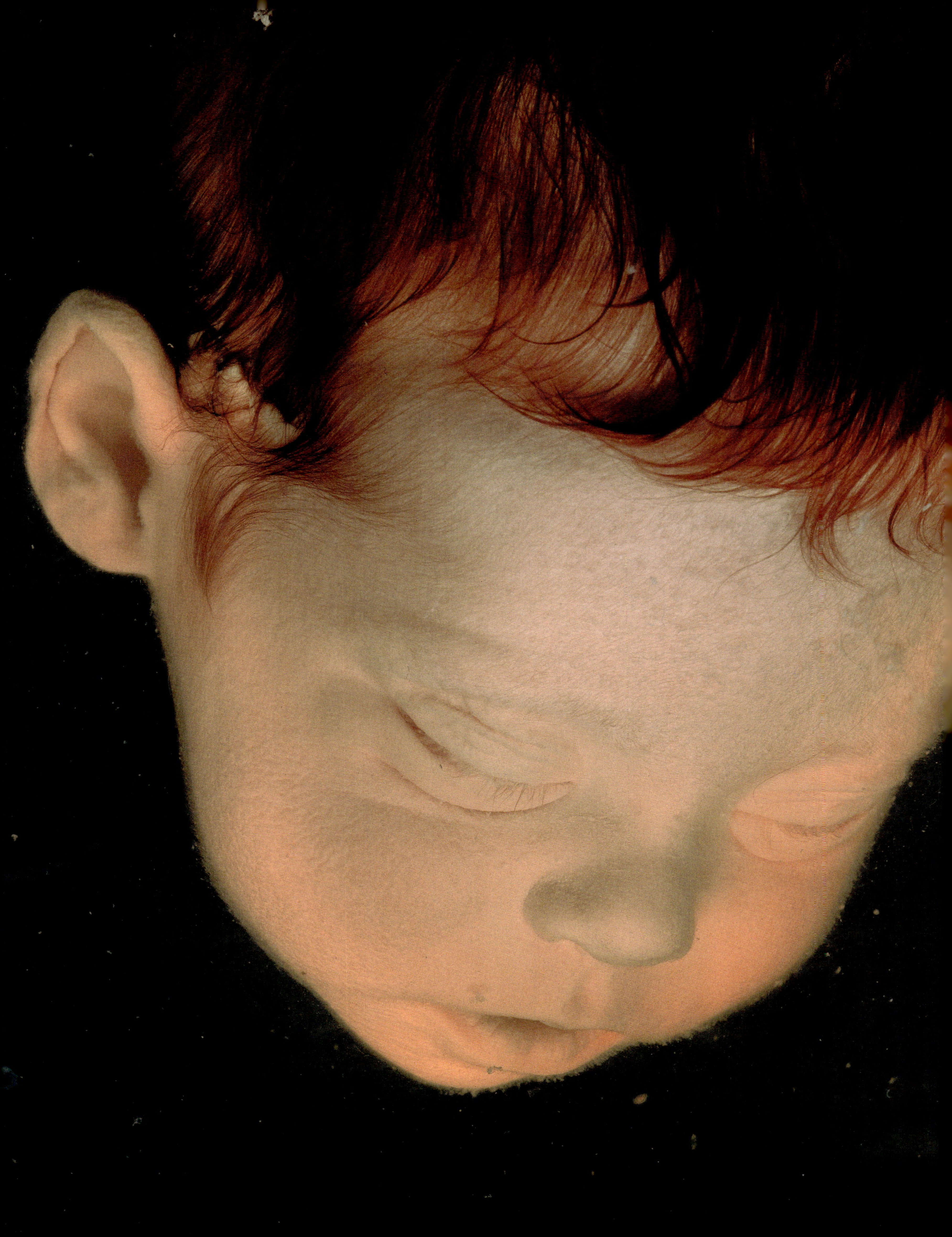

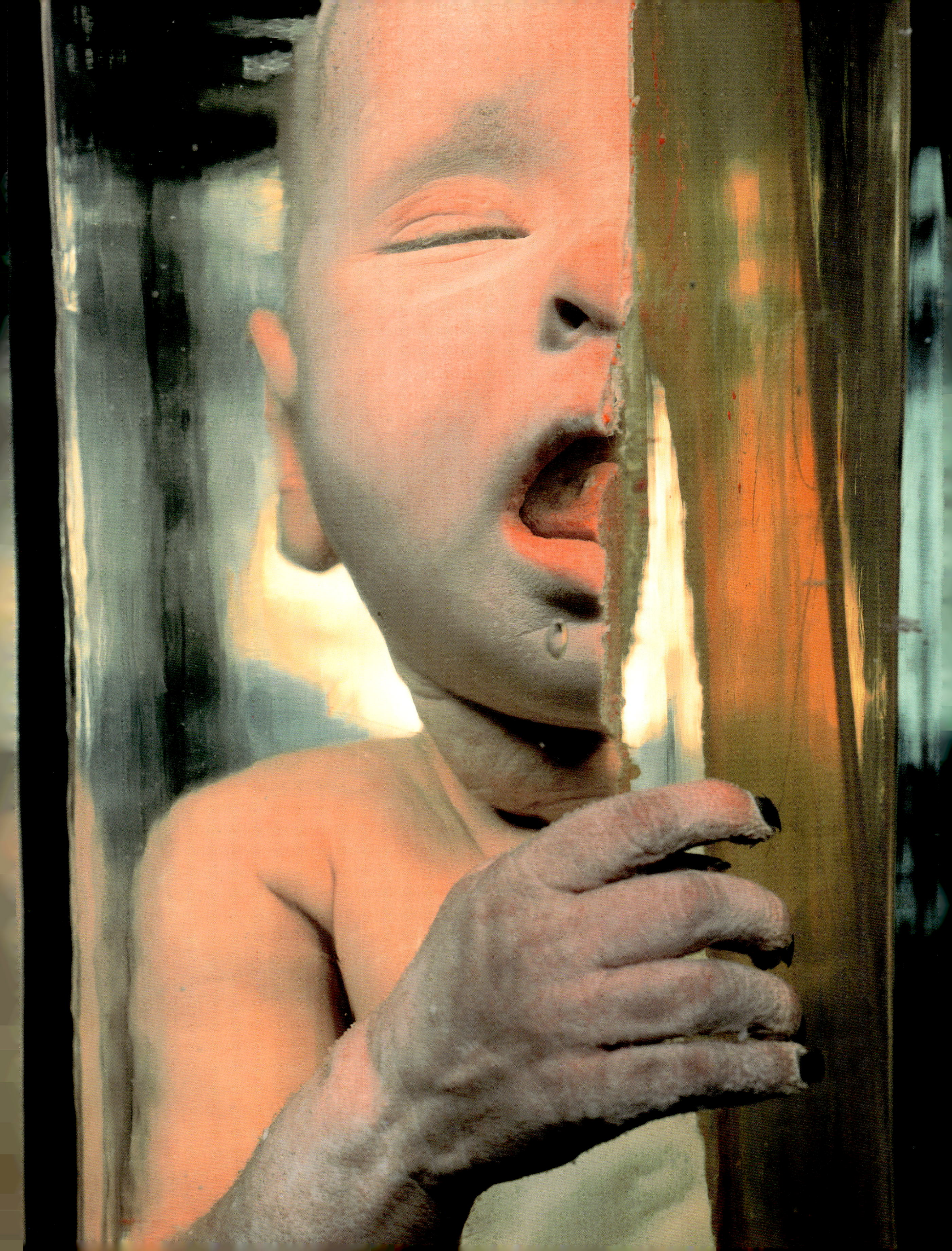

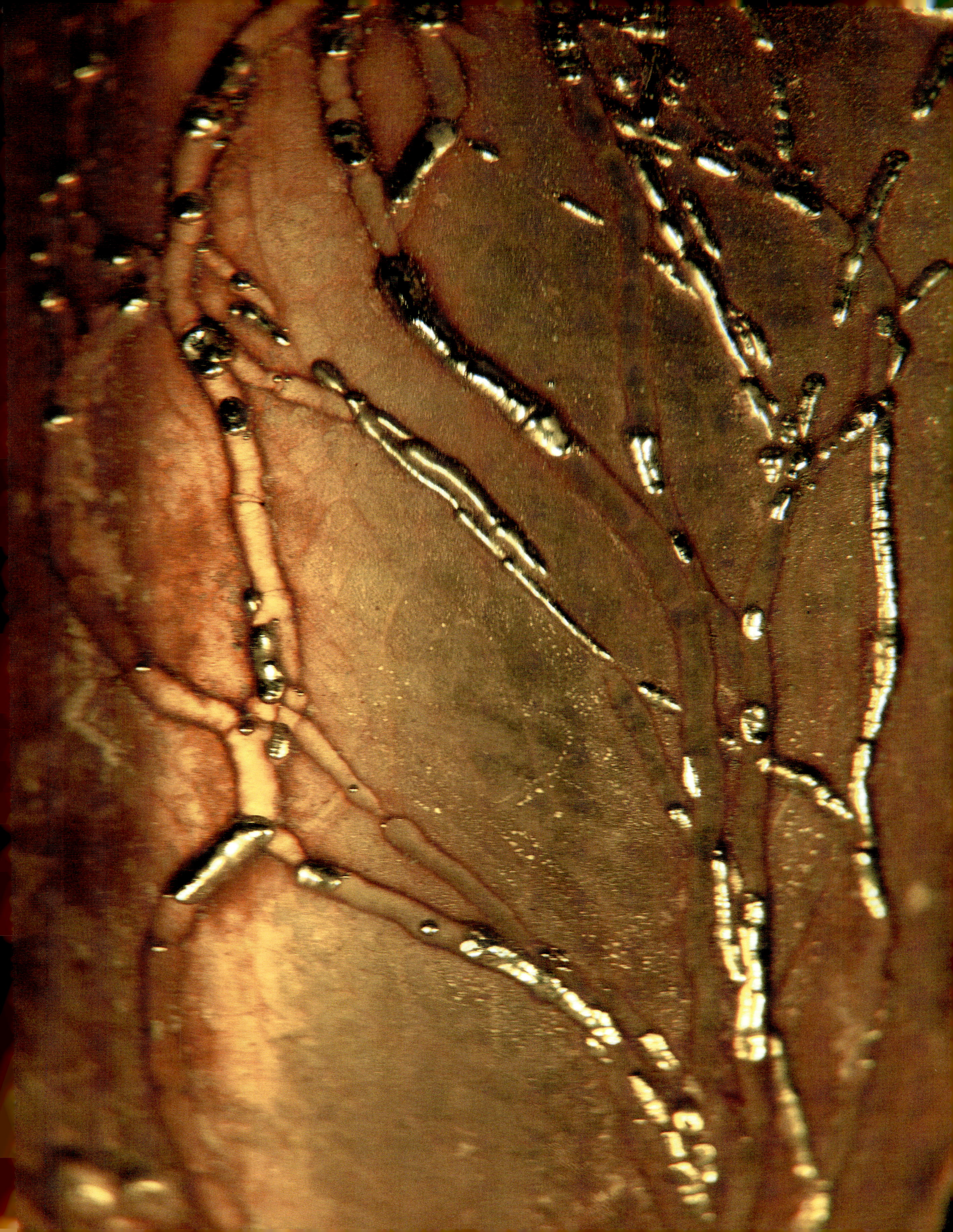

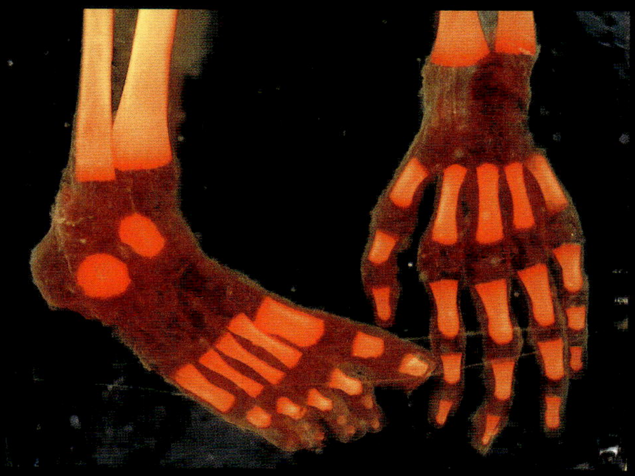

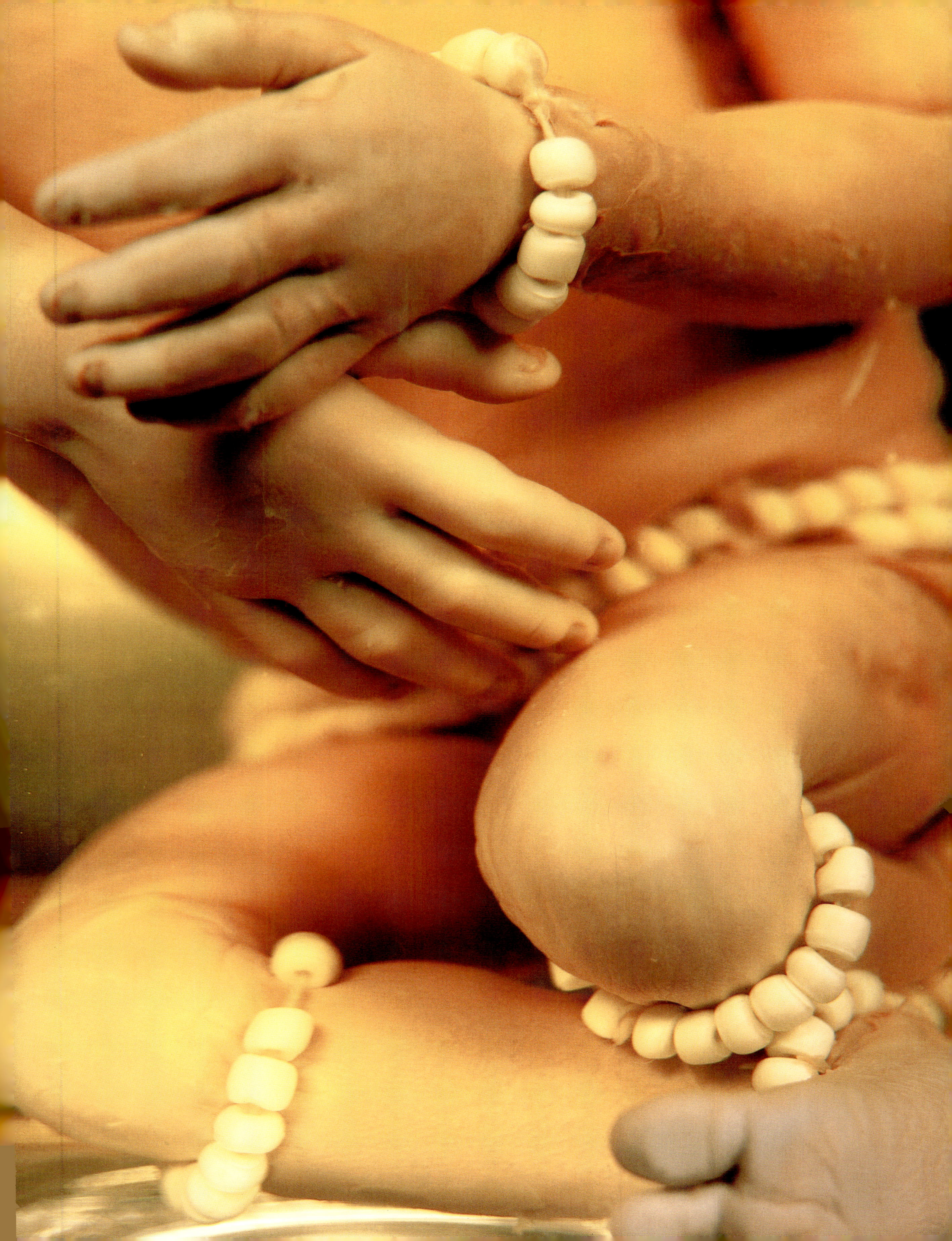

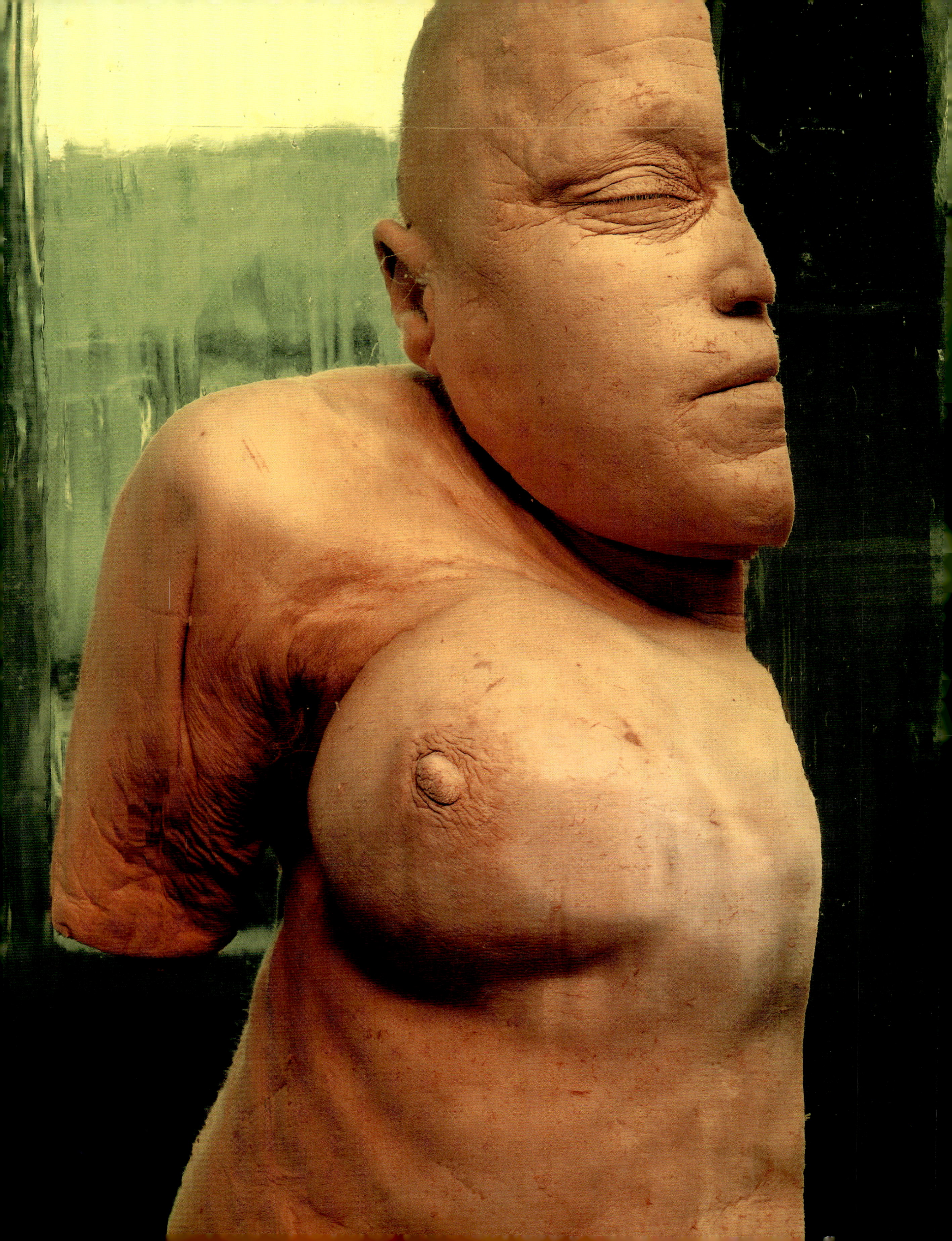

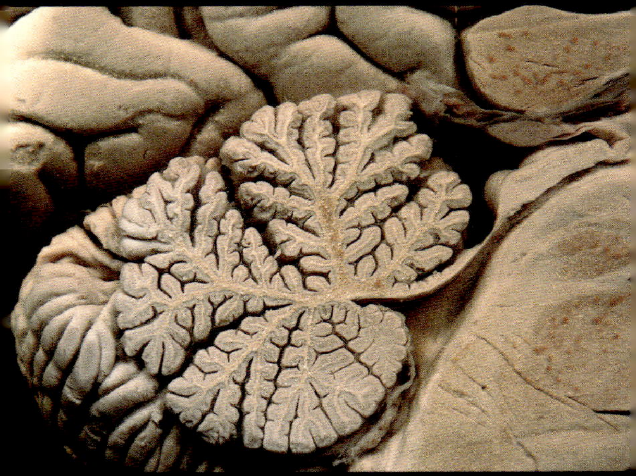

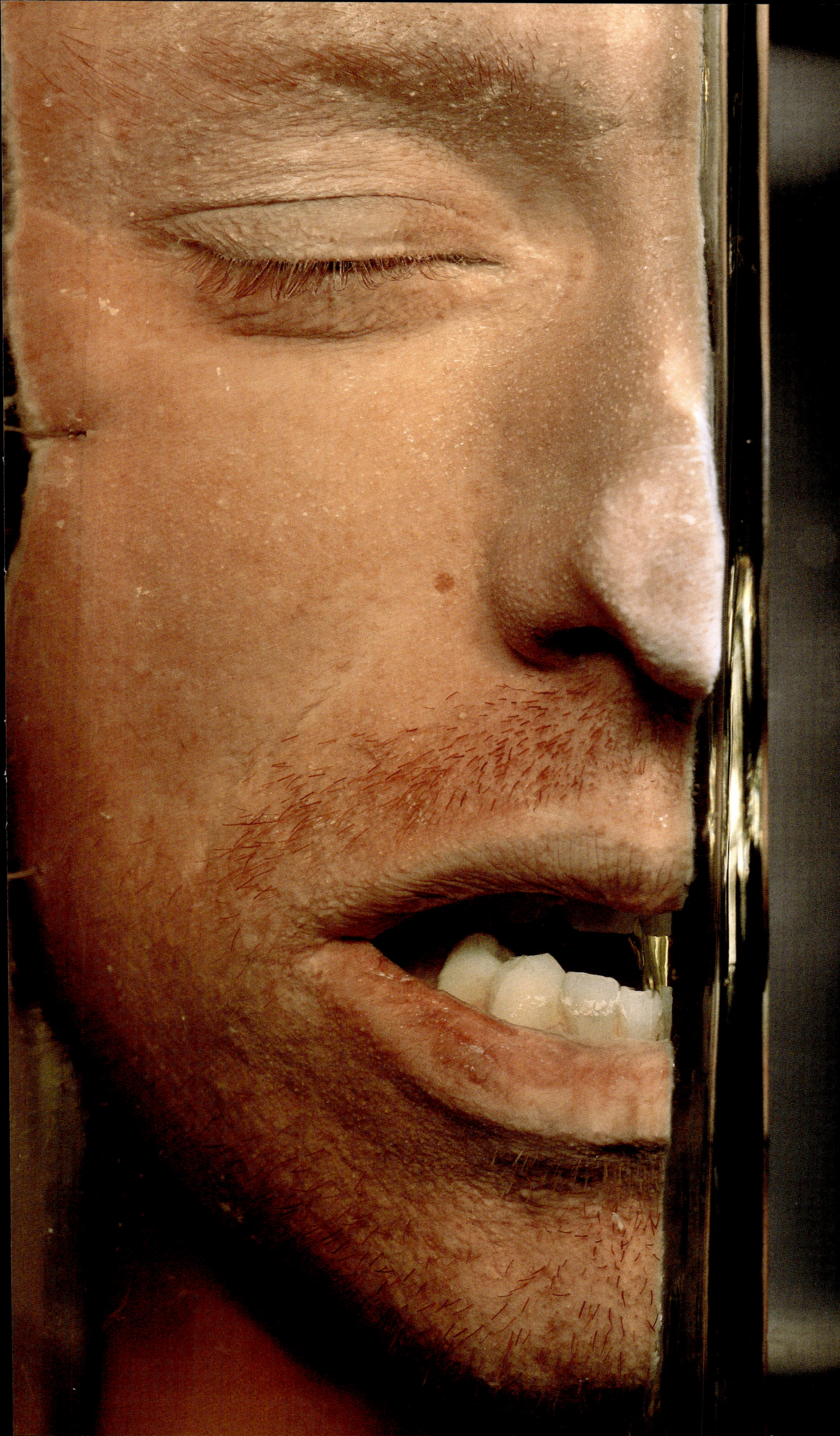

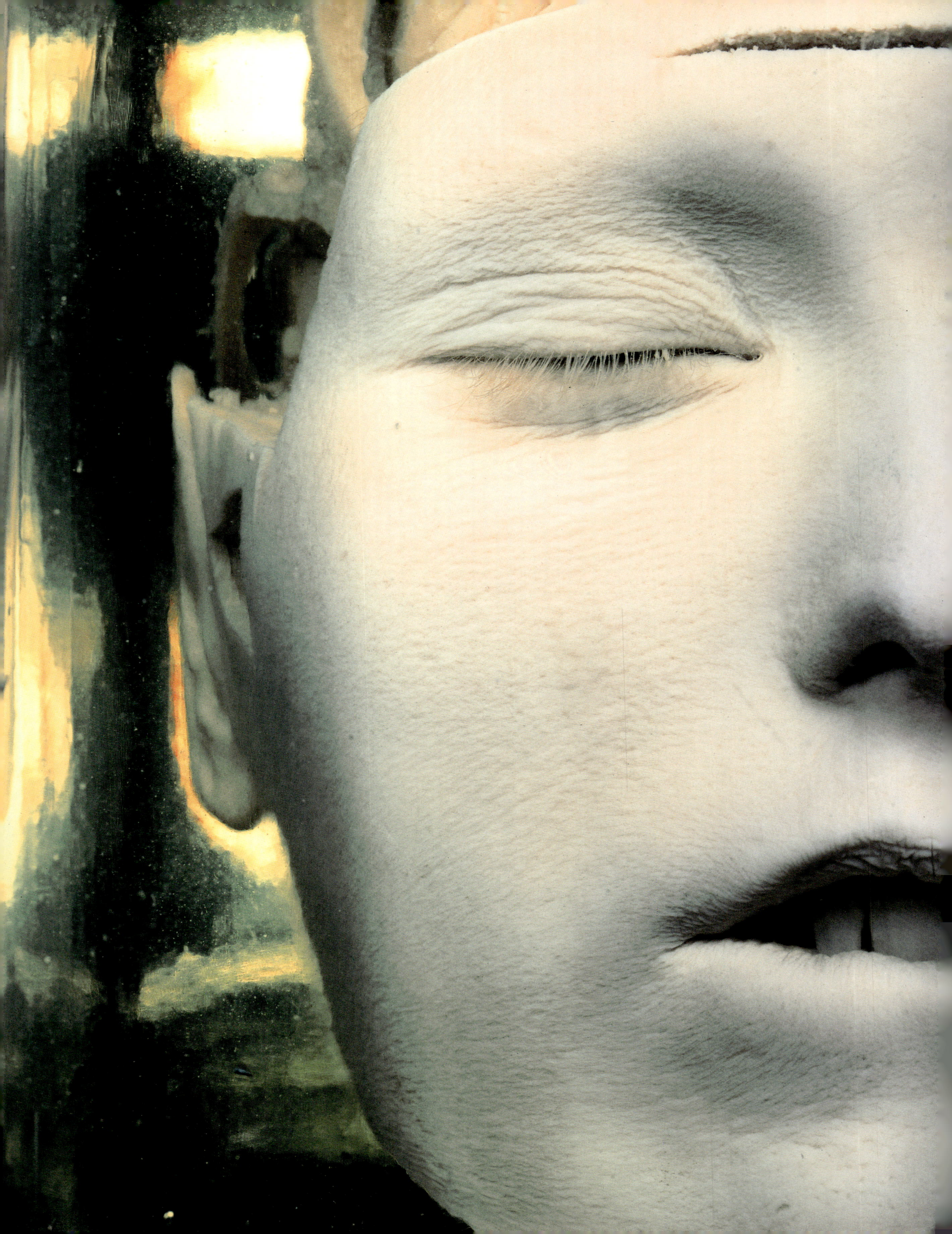

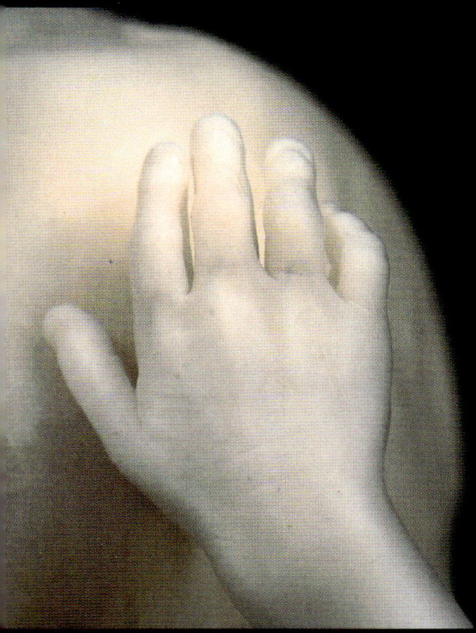

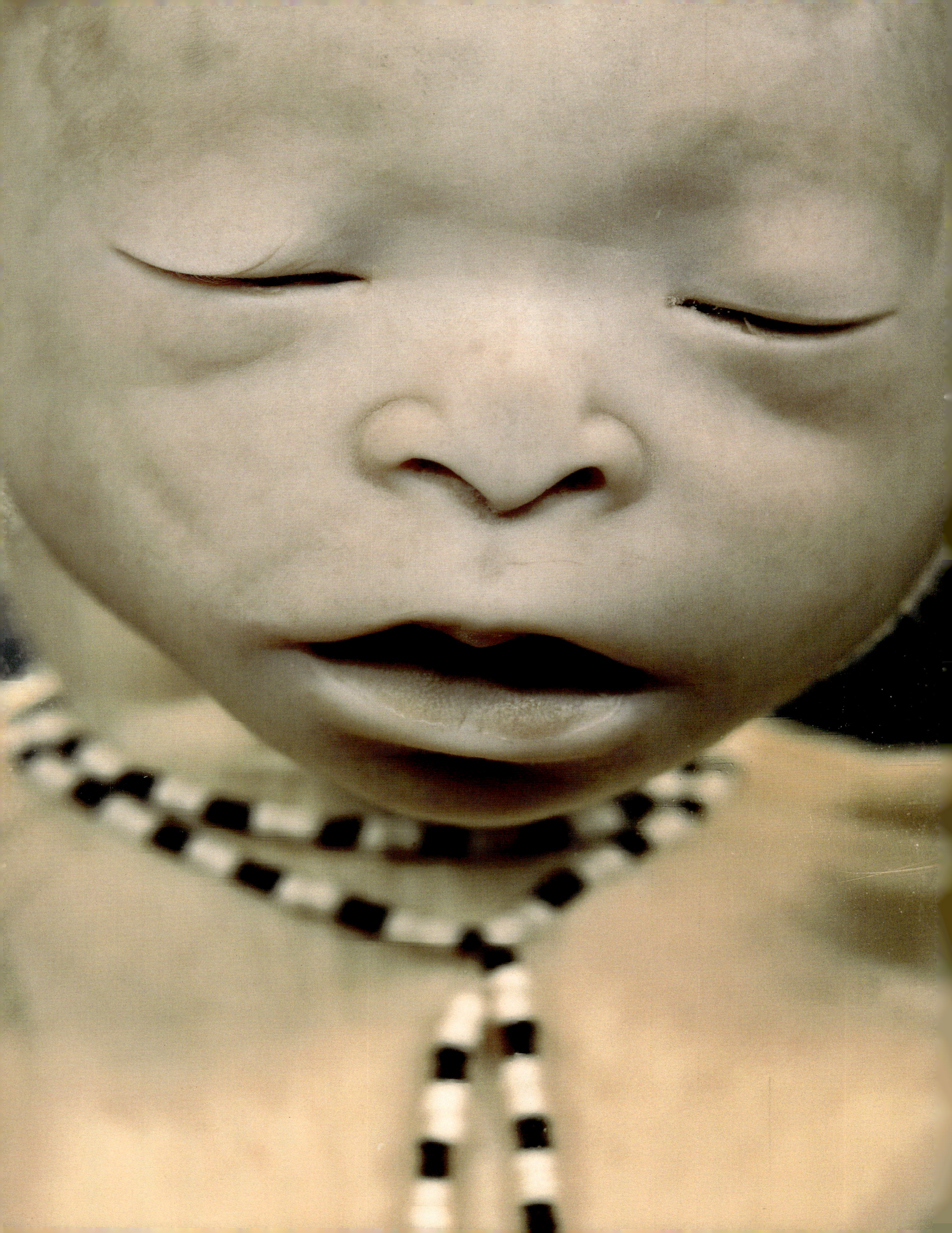

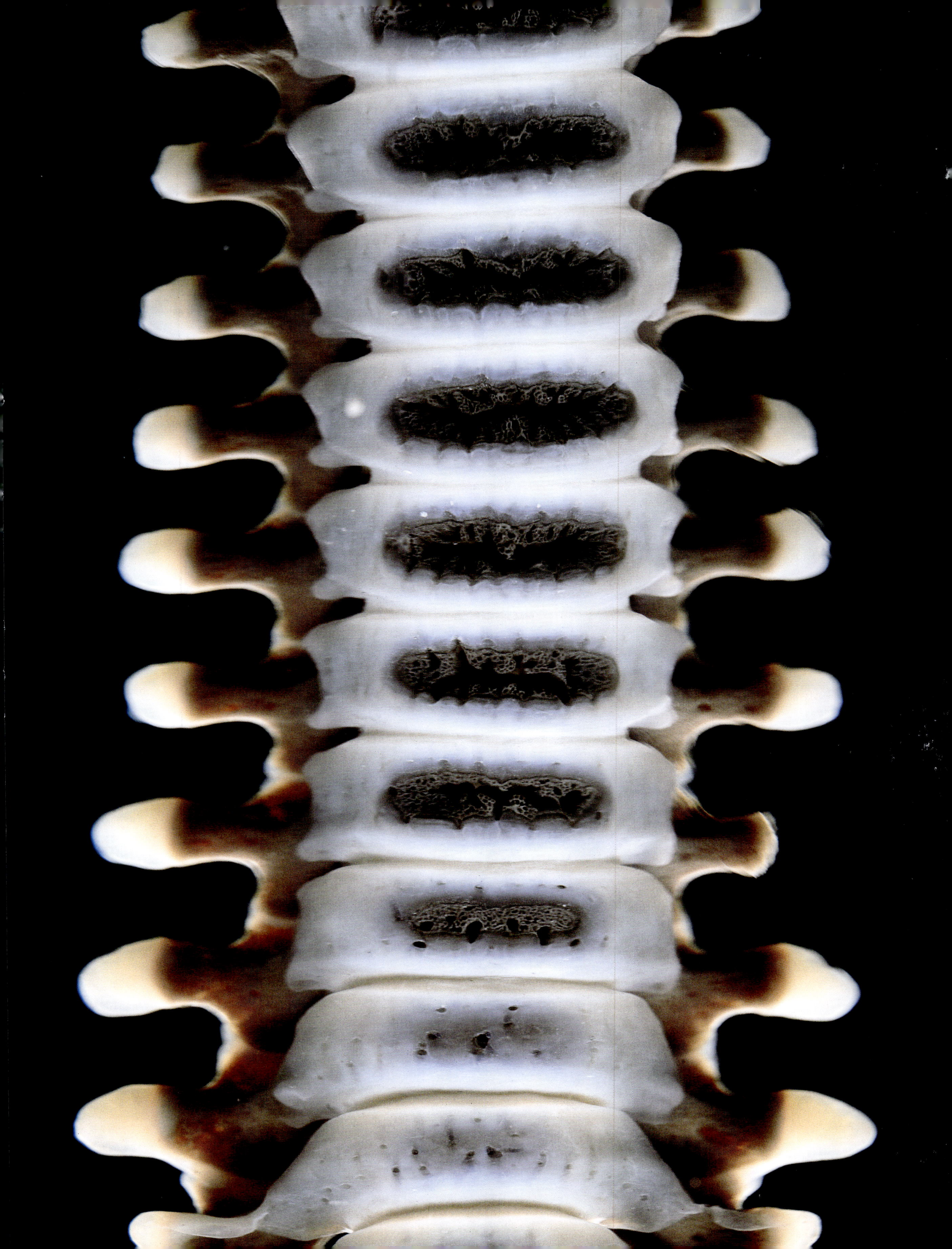

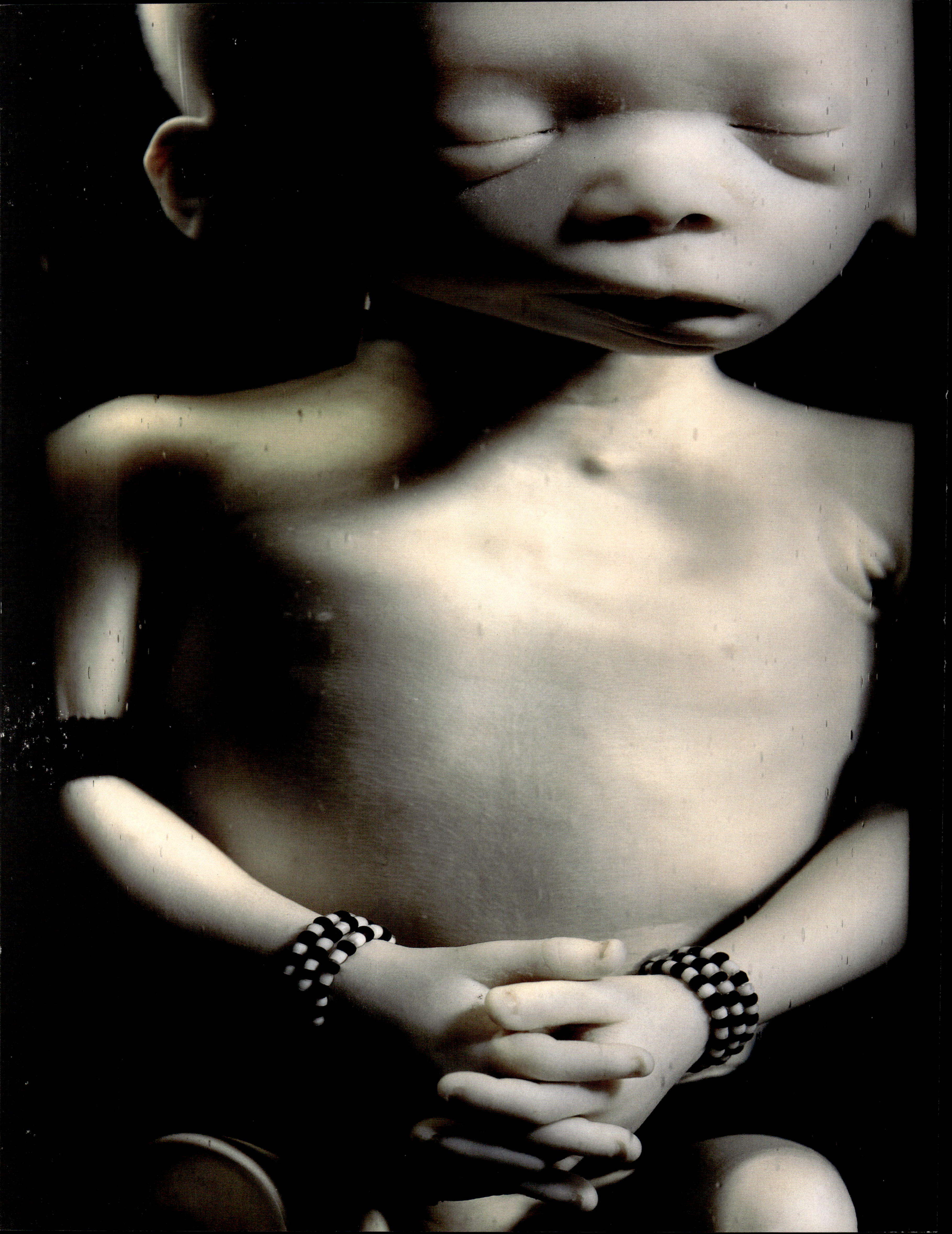

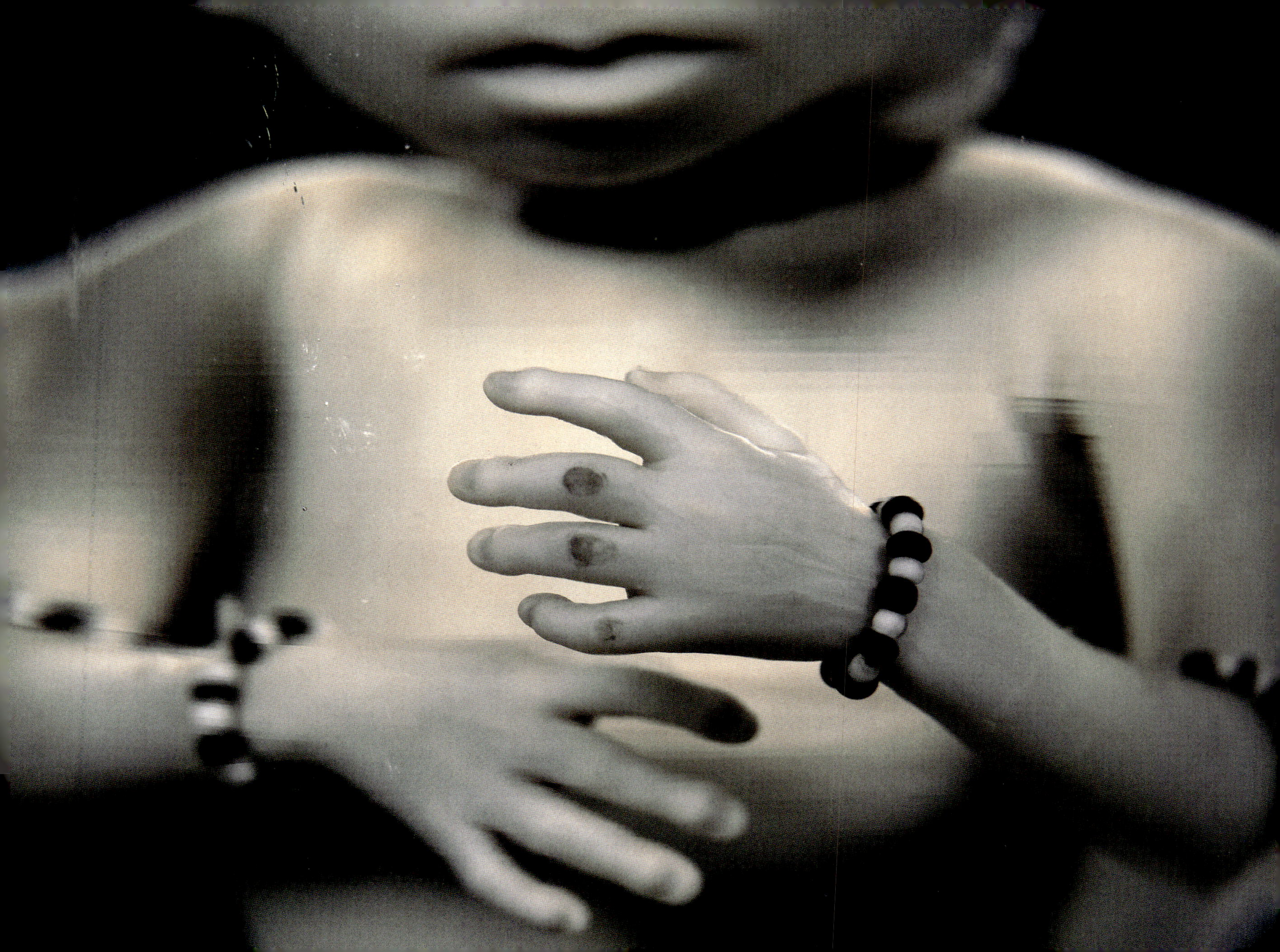

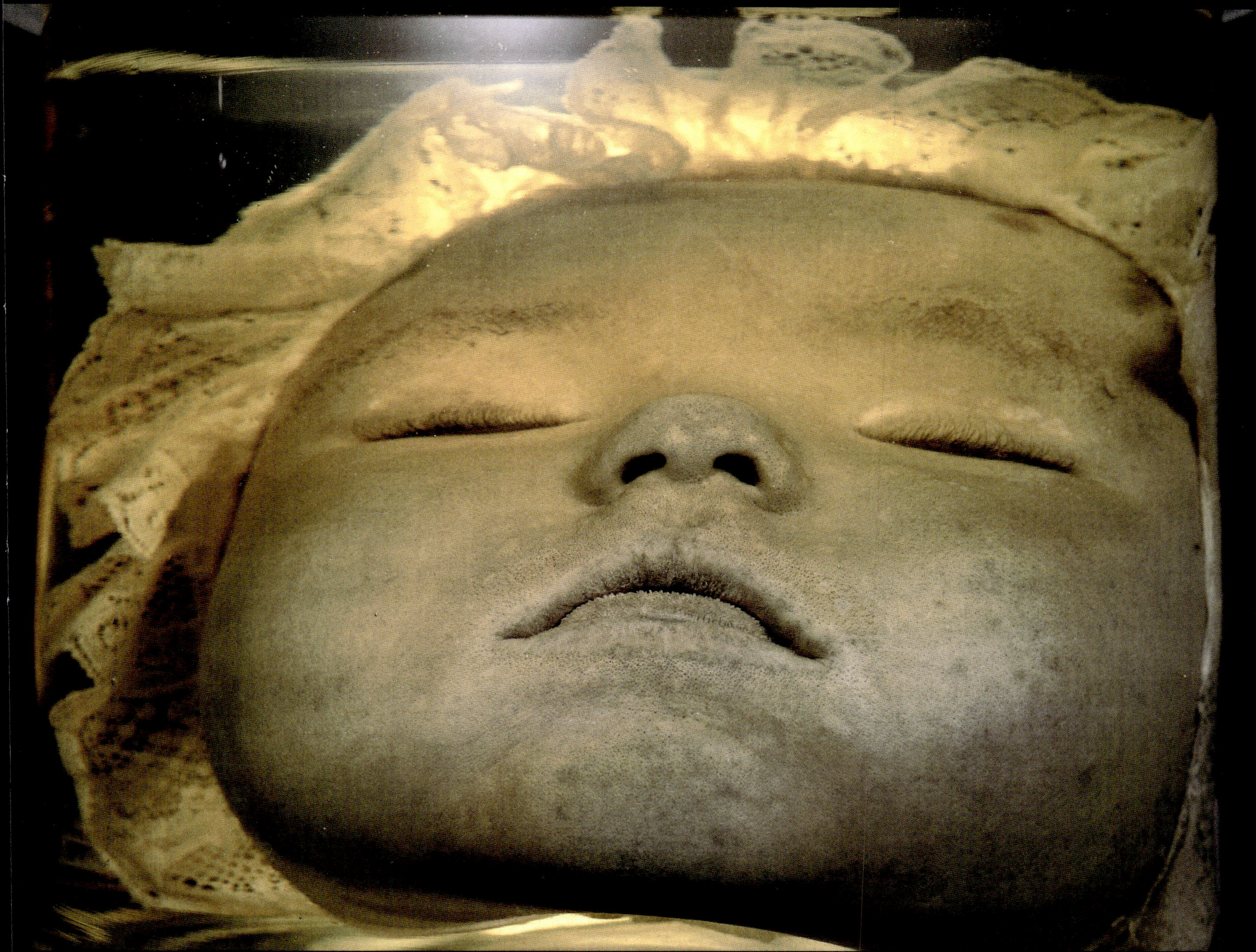

CONSERVING : cycle de Daniel et Geo Fuchs

Les photographies les plus réussies sont celles
sur lesquelles la simultanéité absolue
évoque le cours du temps
qui retient sa respiration.
ADORNO

Un diable écarlate ricane derrière la barrière invisible du verre ; un enfant rêveur, qui n'a jamais vu le jour, ébauche un sourire, comme si le temps ne dérangeait point le sommeil de l'être fugitif ; un poisson, jadis féroce, éternisé dans la transparence de son dernier assaut. Des pattes d'ours, serrées l'une contre l'autre pour conserver la chaleur qui se perd. Un visage de fillette drapé dans de la dentelle, un visage d'il y a trois cents ans figé dans l'intemporel. « Et nos visages, mon cœur, éphémères comme des photographies. » (1)

Tels sont les travaux de Daniel et Geo Fuchs. Des photographies ou reproductions de pièces qu'ils ont découvertes dans les musées d'histoire naturelle. Poissons, animaux, êtres humains, tous conservés. Plus encore, ces œuvres d'art, telles qu'elles sont réalisées, révèlent les prémices d'une amitié entre l'être représenté et les auteurs de sa représentation. Autant de confrontations, de frictions, de frontières franchies. Peut-on surmonter la mort ? Où commencent la dignité et le devoir de l'artiste de la violer sans la détruire ? Les photographies ont d'abord accepté Daniel et Geo Fuchs. Au delà du mur existant entre l'exclu et le contenu, images et artistes se sont trouvés, dans une démarche réciproque.

Ce cheminement de plusieurs années est parsemé de peurs et de transformations. Quiconque s'aventure dans le domaine obscur de la nature morte, précisément, y trouve des fragments de son propre destin transitoire. Si l'intervention de l'Homme permet de prévenir l'ultime déchéance, pouvons-nous, nous aussi, espérer ? Celui qui tient une main conservée pour l'exposer à la lumière devient le témoin muet d'une autre dimension. Il montre. Ce qui a été rendu visible en images nous fait prendre conscience de quelque chose qui se perdrait s'il était montré sans la moindre sensibilité. De même que la signification peut naître non pas de l'addition du visible, mais de sa privation.

Certes, ce sont également les photographies d'une exposition. Conserving Fish. Images exposées aux murs et prêts provenant de musées, à la fois modèles précurseurs et sujets découverts. Poissons protégés dans leur manteau de verre, conservés dans leur milieu liquide, minutieusement analysés et préparés selon des procédés scientifiques. Tous marquent le début de la démarche des deux artistes. Une petite fille qui compare, comprend, apprend et, soudain, voit. Le poisson dans le verre, bien que mort depuis cent ans, vit sur la photographie. Où commence l'« après » ? Avant l'origine ? L'étonnement, le don de voir quelque chose pour la première fois ... les yeux de cette enfant ont vu avec l'âme.

Ce que nous voyons, nous, observateurs observés. La tête pénétrée par la lumière, la main, nue, la délicatesse du profil de l'être qui n'est pas né. Le sommeil profond dans lequel ce visage paisible est plongé. Le duvet, vert tendre, sous le bout du nez. Cette représentation, cette photographie abrite, cache une histoire. Peut-être allons-nous la reconnaître. Rien n'est déterminé.

Il émane de ces « sujets », de ces photographies prises avec et à travers la paroi, aussi transparente soit-elle, le mouvement. Elles racontent le passé qu'elles recréent. Ces histoires sont orientées vers l'avenir. Encore mieux : elles ont retenu le temps que Daniel et Geo Fuchs ont immobilisé. Et nous, observateurs, qui nous en approchons en les regardant, participons.

Pour Maurice Blanchot, l'essence même de la photographie est de se placer à l'extérieur et d'être dépourvue de la moindre intimité, tout en étant plus inaccessible et plus mystérieuse que la représentation intérieure. Dénuée de signification, elle est en même temps un défi lancé à l'impénétrabilité de tout sens.

Il règne ici un défi qui dérange. Outre la virtuosité technique, la légèreté de la lumière, le flou des couleurs qui se mêlent les unes aux autres à l'arrière-plan, l'étrange précision de la perception est frappante : l'œil du poisson, mat et vitreux, dans un état qui ne trompe pas. Un œil mort, à première vue, qui, plus on le regarde longtemps pour l'abstraire de sa matérialité, soudain, s'anime. L'immobilisation, non point d'un moment, mais dans le présent. La sensation précède parfois la connaissance, la prémonition la présence de l'œil. L'énigme résiste à une prépondérance de la raison. Tout cela cache un secret. Le nôtre, peut-être ? L'on peut partir et vouloir oublier ; le temps, la mort, l'histoire. Certaines images restent, parce qu'elles le veulent ; elles ont pénétré l'observateur.

Ces œuvres reposent sur une présence singulière. Sans calculs prétentieux et pleinement conscients de la nature controversée de leur sujet, les artistes rencontrent les êtres représentés avec un respect plein d'étonnement, créent une atmosphère suspendue, une chaleureuse sérénité, et respectent leur grâce et leur dignité. Sans jamais repousser les limites existantes dans le seul but d'obtenir un effet spécial, ils laissent champ libre à l'épanouissement. Les photographies permettent d'ouvrir des espaces, d'accueillir et de susciter des associations. Rien n'agresse ni ne détermine, si nos limites restent préservées.

On percevra toutefois un aspect étrange. Et la solitude. L'être qui s'aventure de l'autre côté d'un mur connait deux réalités : le temps, la mort. Autant d'expériences limites. Même si ces photographies ne nous sont pas étrangères, elles le demeurent et connaissent elles-mêmes la solitude. Si l'on devait extraire ces objets exposés de leur univers de verre pour les plonger sans la moindre protection dans le monde de la réalité, ils se décomposeraient, se désagrégeraient. Nous le savons. Les photographies de cette trilogie sont tout le contraire. Leur milieu auréolé de mystère ignore tout de la mort. La mort est derrière elles et elles sont réinventées.

JANA MARKO

---

(1) Cette citation est la traduction du titre allemand du livre de John Berger *Und unsere Gesichter, mein Herz, vergänglich wie Photos*. Édition Carl Hanser Verlag. Munich et Vienne 1986, page 14

Vie en suspension : Conserving – photographies de Daniel et Geo Fuchs

Pour réaliser leurs photographies, Daniel et Geo Fuchs pénètrent dans des mondes infiniment intimes. Ainsi recherchent-ils des situations, qu'eux-mêmes qualifient de « limites », des situations que d'autres préféreraient éviter, des lieux et des choses normalement dissimulés au public. Ils vont dans les entrepôts des musées et des instituts scientifiques où des animaux et des êtres humains sont conservés à l'intérieur de récipients en verre remplis de formol et d'alcool, mettant en évidence non seulement la tendance institutionnelle à conserver et à collectionner, mais aussi leur propre réaction par rapport à ces objets et ce qu'ils représentent. « Au début des années 1990, nous avons entendu parler des instituts scientifiques où des animaux, poissons, mammifères et reptiles sont conservés dans l'alcool. Nous avons pensé que les poissons notamment devaient, d'une certaine façon, recouvrer ainsi une nouvelle forme de vie, une fois de retour dans leur élément liquide. Lorsque nous nous sommes rendus pour la première fois dans un musée d'histoire naturelle, nous avons remarqué que les poissons avaient conservé une vivacité remarquable qui convenait parfaitement à notre sujet : trouver la vie dans la mort ... À nos yeux, la mort et la conservation vont de pair dans la mesure où nous considérons cette dernière comme une autre forme d'existence. Certains des objets exposés ont plus de trois cents ans. » (1)

Daniel et Geo Fuchs ont intitulé leur travail « Conserving », un titre équivoque, cela va de soi. Les objets photographiés ont eux-mêmes été conservés et préservés. Ce faisant, ces artistes ont créé un autre niveau de documentation et de préservation. Ils ont procuré à ces objets normalement cachés une nouvelle arène en les plaçant dans le cadre sérieux de la photographie d'art et documentaire, soumis à un regard ô combien différent de celui des scientifiques. Cet ouvrage témoigne de la capacité de la photographie à créer un nouveau public pour des sujets et des idées qui, jusqu'ici, restaient obscurs.

À la fin des années 1980 et au début des années 1990, le thème de la mort a suscité un intérêt considérable parmi nombre de photographes d'art européens et américains. Ainsi les portraits de morts de Rudolf Schäfer (réalisés dans une morgue en Allemagne de l'Est en 1987) ont-ils attiré l'attention de façon immédiate, à l'instar des clichés spectaculaires d'Andres Serrano pris dans les instituts mortuaires de New York. Puis le photographe anglais Nick Waplington, connu au début des années 1990 pour ses commentaires abrasifs sur la vie de deux familles habitant dans un lotissement à Nottingham, a produit une série de photographies mises en scène portant sur la mort non naturelle et le suicide. Ce regain d'intérêt pour la photographie de la mort reste inexpliqué et peut-être est-il trop diversifié pour être amalgamé en une théorie cohérente. Il n'en demeure pas moins que depuis les années 1980 les photographes se sont de plus en plus adonnés à cet art, utilisant des situations réelles pour donner une vue du monde des vivants très subjective et dirigée. La photographie de la mort telle qu'elle apparaît à la morgue ou dans les musées a joué un rôle important dans cette nouvelle approche. Ainsi les morts étaient-ils bien de notre monde mais n'en faisaient pas partie, devenus des objets immobiles et silencieux à la disposition de l'artiste ou du photographe, d'une façon unique et inimitable.

Ce nouvel art de photographier la mort a remis en question le statu quo, s'intégrant ainsi parfaitement à la méthodologie postmoderne des années 1980 et 1990. Les paramètres de la photographie étaient en train de changer quand les artistes ont entrepris de l'utiliser non plus comme un moyen de représentation, mais pour explorer la représentation elle-même. Cindy Shermann met en scène un théâtre de masques et d'imaginaire destiné à sonder l'identité féminine, Martin Parr a usé de couleurs vives pour réaliser des documentaires acerbes sur la vie dans les rues de Grande-Bretagne et a révolutionné l'idée du reportage classique. Thomas Ruff a procédé à des innovations retentissantes dans le genre du portrait photographique. Les frontières se sont peu à peu effondrées et l'importance croissante de la théorie photographique (par opposition à l'histoire de la photographie de plus en plus considérée comme rébarbative et obsédée par les processus) a attiré l'attention d'un public nouveau sur les dimensions intellectuelles et politiques de cet art. De plus en plus d'artistes en ont fait un outil pratique et accessible pour exprimer des idées qui, dans les années 1960 et 1970, auraient pu être véhiculées par la peinture, la sculpture ou l'illustration.

L'œuvre de Daniel et Geo Fuchs se place précisément dans ce cadre du postmodernisme ; ils ont considéré comme acquis le droit de travailler dans des domaines qui autrefois ont fait figure de tabous, insistant sur le fait que loin d'être un documentaire, ces photographies constituent un processus très subjectif dans lequel intervient une histoire personnelle et de profondes émotions. Ils semblent nous dire que le monde existe pour être regardé et étudié, accessible à une multitude de regards différents. Leur souhait continu de travailler sur ce qu'ils définissent comme une expérience limite, leur engagement personnel sur ce sujet, la distance qu'ils prennent par rapport à l'aspect sensationnel ou voyeur, tout cela est le résultat d'une démarche extrêmement ambitieuse et fondée, en attente de toute l'attention et de la compréhension de l'observateur. Dans l'ensemble de la trilogie « Conserving », ils nous offrent un travail doté d'un sens esthétique très particulier transformant en objets fascinants des éléments de notre monde que l'on pourrait tenir pour macabres ou repoussants. Les êtres morts qu'ils photographient, immortalisés dans le liquide de conservation, deviennent des icônes, presque des objets de vénération. Bien que morts depuis longtemps, ils font partie du monde contemporain ; ils ont acquis un statut étrange, peut-être même mystique, en suspens entre la mort et la vie.

« Conserving Fish » est sans doute le cycle le plus éthéré et le plus abstrait de la trilogie « Conserving ». Le fait que les poissons flottent dans un liquide les fait évoluer dans un milieu plus « naturel ». Nous les regardons à travers le verre comme s'il s'agissait d'un aquarium et seules les couleurs ternies et l'absence de mouvement rappellent leur état. On entrevoit ça et là un flacon en verre ou une étiquette signalant que nous sommes en présence d'une étude de taxinomie plutôt que d'un poisson fraîchement pêché, mais malgré tout d'un animal sauvage vivant. Si « Conserving Fish » s'avère en apparence le moins « choquant » des trois cycles de la trilogie, il en émane une mélancolie indubitable. Les écailles qui devraient briller sont ternes et sans éclat, tandis que les yeux vivants sont sans vie et les bancs de poissons si rapides sont immobiles, figés à jamais par la mort et le temps.

« Conserving Animals » constitue une expérience d'un genre complètement différent. Un seul regard suffit pour comprendre qu'il s'agit d'une étude de la mort. Des griffes et des pattes ont été détachées des corps et flottent dans le liquide de conservation. Des yeux qui ne voient pas nous regardent fixement sans que nous puissions croiser ce regard. Parfois, des gestes semblent indiquer la présence d'une vie, un frottement des yeux, des jambes ou des bras tendus, une étreinte, qui ne sont que de purs accidents, brefs moments de vigueur bravant la mort. Ces animaux dorment d'un sommeil agité, abandonnés dans leur liquide, hors de leur élément naturel, à la merci des regards des curieux et des experts. Ils sont devenus des sujets, des objets, privés d'amour-propre.

En observant notre propre espèce (dans la série « Conserving Humans »), le renoncement à toute intimité est frappant. L'expression ou la personnalité n'ont pas disparu des visages des morts ; un jeune homme, un bébé, une vieille femme et même un embryon, tous ont jadis fait partie de notre tribu, ont été habités par des pensées et des émotions, ont pris des décisions, forgé des relations et existé au sein de notre société. Pourtant, une fois transformés en spécimens à la disposition de la science ou en pièces d'exposition dans les musées, ils sont exclus de nos rites tribaux, rites de passage. De manière grotesque, à l'instar du Faust de Goethe, ils sont devenus immortels. « Conserving Humans » constitue le point culminant de la trilogie et nous transporte d'un monde abstrait à la réalité. Cet ouvrage sera inéluctablement considéré comme une étude du grotesque, même si là ne fut pas l'intention de Daniel et Geo Fuchs. Ces derniers portent leurs regards sur les crânes déformés, mutilés ou sectionnés, les squelettes et la peau tatouée presque comme les témoins, les archivistes du temps dans lequel nous vivons. Au sujet de « Conserving Humans », ils ont déclaré : « Cet ouvrage a constitué un immense défi, car nous avons voulu montrer ces êtres humains en veillant à ce qu'ils conservent leur beauté, que la vie reste visible, que la connaissance l'emporte sur la mort et qu'au delà de la mort leur dignité soit préservée. » (2)

Le travail fourni par Daniel et Geo Fuchs au cours de la réalisation de « Conserving Humans » les a menés à un vaste projet en cours sur la mort. Ces artistes se sont aventurés au delà de

l'institution scientifique des musées et ont pénétré le monde étrangement silencieux des morgues où rien n'échappe à leurs regards, des corps aux instruments de dissection. Là où le corps humain, bien que toujours identifiable en tant qu'individu, est devenu un objet privé de son intimité. Ils portent leur regard sur les traces que laissent ces corps, un bout de chair ou un cheveux. Une démarche qui n'a rien de morbide en soi, car si la curiosité tend à révéler un aspect grotesque et absurde, le regard de l'artiste finit par s'imposer. Les photographies que nous voyons ne sont ni abruptes ni horribles, mais s'offrent à notre imagination et à notre esprit, prêtes à susciter une pléiade d'interprétations. À propos de leur travail, les artistes font le commentaire suivant : « Après nous être consacrés à une forme plus abstraite de la mort, il était logique que nous nous tournions vers l'étude de la mort réelle. Sans le verre et le liquide protecteurs, ces situations et les photographies qui en résultent sont bien plus directes. La mort et la décadence du corps humain deviennent très naturelles. Celui-ci ressemble simplement à une coquille vide. » (3)
On ressent à quel point leur soif de savoir et d'obtenir des réponses est devenu dévorant et profondément personnel.

Dans chaque partie de la trilogie « Conserving », Daniel et Geo Fuchs sont restés fidèles aux principes qui régissent leur œuvre. En contemplant les poissons, les animaux et les êtres humains qu'ils ont photographiés, on ressent le lien intime qui les unit à eux. Bien que ces artistes se fassent les observateurs de situations inconfortables, il émane de leur regard de la sollicitude, voire de la gentillesse. « L'Homme ressent autant de fascination que de répugnance face à la mort, le dernier mystère de la vie. Nous savons notre mort inévitable et cette pensée ne nous quitte pas. En ce sens, la vie et la mort sont inséparables. Ainsi notre mission est-elle de réunir ces deux dimensions sur nos photographies. » (4)
On ne peut ignorer leur vulnérabilité, le risque qu'ils prennent personnellement en choisissant de consacrer leur carrière artistique à étudier ce territoire caché et souvent controversé. Ils se font antiquaires et explorateurs d'un pays étranger, ils enregistrent fidèlement des vérités qui dérangent, réalisent des photographies qui engendrent la polémique, provoquent des questions et contribuent de façon majeure à nous faire comprendre le rôle de la photographie s'agissant de décrire notre monde.

VAL WILLIAMS

(1) Citation d'une interview des artistes de 1999

(2) Ibid

(3) Citation d'une interview des artistes réalisée par Francesca Alfano Migletti pour Virus Magazine à Milan en Italie

(4) Citation d'une interview des artistes de 1999

CONSERVING – ciclo di Daniel e Geo Fuchs

Le immagini meglio riuscite
sono quelle in cui l'assoluta
contemporaneità risulta come
scorrere del tempo, in cui il
respiro resta immobile.
ADORNO

Un demone scarlatto sogghigna, dietro una parete invisibile di vetro; un bimbo sognante, mai nato, quasi sorridente, come se il tempo non disturbasse il sonno della caducità; un pesce, una volta aggressivo, ora solamente dietro la trasparenza del suo ultimo viaggio. Zampe d'orso che si pressano come a mantenere il calore perduto. Una faccia di ragazza ornata da pizzi drappeggianti. Un viso, lí da trecento anni, esiliato nell'atemporalità. "E le nostre facce, cuore mio, effimere come fotografie." (1)

Le opere di Daniel e Geo Fuchs. Fotografie. Anche immagini di pezzi d'esposizione nelle collezioni di storia naturale. Pesci, animali, persone. Pesci, animali, persone conservati. Più di questo: il lavoro artistico in questo senso si trasforma in una manifestatione di un'amicizia fra il rappresentato e il rappresentante. Contrasto, attrito, superamento dei limiti. Come si può superare la morte? Dove comincia la dignità, dove il dovere dell'artista, di ferirla senza distruggerla? Daniel e Geo Fuchs sono stati i primi ad accettare le immagini. Oltrepassato il muro fra escluso ed incluso, le immagini hanno trovato gli artisti, e gli artisti le immagini.

Un percorso di molti anni. Paure, cambiamenti. Chi si addentra in sfere oscure, trova nella natura morta frammenti della propria caducità. C'è una speranza anche per noi, se la natura morta, attraverso l'intervento dell'uomo, può essere preservata dal decadimento definitivo? Chi tiene in mano una mano conservata per metterla in luce come motivo d'arte, diventa testimone muto di un'altra dimensione. Mostrare. Ciò che è stato reso visibile, nel vero senso della parola, ci rende coscienti di un qualcosa che potrebbe andare perso se venisse dimostrato in maniera insensibile. Come il significato, il quale non nasce attraverso l'apporto del visibile, bensì attraverso la sua sottrazione.

Ovviamente anche: immagini di una mostra. "Conserving fish". Le immagini fotografiche alle pareti. Gli oggetti, precursori e contemporeamente motivi ritrovati, come cose prese in prestito. Corpi di pesci nell'involucro protettivo del vetro, conservati nel liquido, analizzati minuziosamente, preparati scientificamente. Essi segnano l'inizio del diverbio dei due artisti. Una ragazzina che paragona, capisce, impara e all'improvviso vede. Il pesce nel vetro è morto da cent'anni, nella fotografia è vivo. Dove comincia il dopo? Prima dell'origine? Lo stupore, il dono di vedere qualcosa per la prima volta ... l'occhio del bambino ha visto con l'anima.

La questione del vedere. Noi. Osservatori osservati. La testa oltrepassata dalla luce, la mano, soltanto la mano, la linea delicata del profilo di un mai-nato. Il sonno giace profondamente in questo viso tranquillo. Lí, sotto la punta del naso, il verde pallido come peluria. L'immagine nasconde una storia. Forse ci sarà un riconoscimento. Niente è prestabilito.

In queste fotografie, in questi soggetti, ripresi

attraverso una parete, anche se trasparente, si nota movimento. Tramandano il passato mentre si ripropongono nel presente. Le storie mirano al futuro, in avanti. O meglio: il tempo si è imprigionato in loro e le immagini di Daniel e Geo Fuchs lo hanno a sua volta immobilizzato. E noi, che ci avviciniamo con lo sguardo, ci rendiamo partecipi.

Secondo Maurice Blanchot l'essenza delle immagini persiste nel restare al di fuori, senza intimità, restando però più innavvicinabile e più enigmantico di una rappresentazione interna; senza significato ma allo stesso tempo una sfida all'infondatezza di ogni senso possibile.

In queste immagini c'è sfida, tanta da irritare. All'infuori della loro brillantezza tecnica, della leggerezza della luce, del flusso dei colori che si mescolano sullo sfondo, balza allo sguardo una percezione particolarmente limpida: l'occhio del pesce — di un incelabile offuscamento opaco. Al primo sguardo un occhio morto. Poi all'improvviso, quanto più lo sguardo stacca l'oggetto dalla sua oggettività, tanto più nasce il movimento. Non uno stato immobile del momento, ma un essere silenzioso del presente. La sensazione precede a volte la conoscenza, il percepire precede l'esistenza dell'occhio. L'enigma si ribella alla prepotenza della ragione. Dietro di noi, un segreto - il nostro? Si può andare via e voler dimenticare. Il tempo, la morte, la storia. Ci sono immagini, che restano, perché vogliano restare. Si sono intrufolate nell'osservatore.

Una strana presenza si nasconde nelle opere. Senza calcoli pretenziosi e coscienti della natura particolarmente difficile del loro soggetto vanno incontro a ciò che osservano con stupito rispetto, creano l'atmosfera della sospensione, di una pace calda, rispettando grazia e dignità. I limiti prestabiliti non vengono spostati per produrre un effetto qualsiasi, permettendo così uno spiegamento completo. Immagini possono creare spazio, accogliere qualcuno, suscitare associazioni. Niente è aggressivo o determinato, se esso e noi restiamo incolumi.

Ciò nonostante estraneità. Anche: solitudine. Chi si protrae al di là del limite, conosce due realtà. Il tempo, la morte. Esperienze limite. Anche se le immagini non ci sono estranee, rimangono comunque per se stesse estranee - e in solitudine. Dal momento che vengono prese fuori dal loro mondo di vetro, abbandonandole al mondo reale prive di protezione - potrebbero decadere e decomporsi. Noi lo sappiamo. Al contrario di noi le fotografie di questa trilogia, la cui attuale patria segreta non conosce la morte, la quale si trova dietro di loro, mentre loro sono state ricreate.

JANA MARKO

---

(1) Titolo del libro. John Berger. *Und unsere Gesichter, mein Herz, vergänglich wie Fotos.* Carl Hanser Verlag. München und Wien 1986, pg. 14

VITE FLUTTUANTI : CONSERVING — fotografie di Daniel e Geo Fuchs

Per realizzare le loro foto, Daniel e Geo Fuchs penetrano in mondi infinitamente intimi. Si avvicinano a situazioni che loro stessi definiscono "limite" situazioni che altri preferiscono evitare, in luoghi e oggetti normalmente nascosti all'occhio del pubblico. Entrano nei sotterranei dei musei e degli istituti scientifici dove animali e esseri umani vengono conservati in formalina o alchol in recipienti di vetro, in uno stato apparente di vita fluttuante. Non solo si confrontano con la tendenza istituzionale a conservare e a collezionare ma anche con la propria reazione a certi oggetti e a ciò che essi rappresentano. "All'inizio degli anni '90 abbiamo sentito parlare dell'istituto scientifico dove animali, pesci, mammiferi e rettili sono conservati nell'alcool. Ci affascinava soprattutto l'idea che i pesci conservati nel loro elemento — il liquido — dovevano aver ritrovato una vitalità e già la prima volta che visitammo una collezione in un museo di storia naturale constatammo che i pesci possedevano ancora una considerevole vitalità, che era quello che esattamente si adattava al nostro tema, che appunto riguarda — il trovare la vita nella morte. Per noi morte e conservazione vanno di pari passo nella misura in cui consideriamo l'ultima come un'altra forma di vita. Alcuni pezzi stanno lí da più di trecent'anni." (1)

Daniel e Geo Fuchs hanno intitolato il loro progetto "Conserving" — un nome che ovviamente ha un doppio significato. Gli oggetti da loro fotografati sono stati conservati e mantenuti, ma anche con l'atto del fotografare gli artisti hanno creato un altro livello di documentazione e mantenimento. Hanno inoltre procurato una nuova arena per questi oggetti normalmente nascosti, e dato loro un posto nell'area dell'arte impegnata e della fotografia documentaristica dove sono sottoposti a sguardi ben diversi da quelli degli scienziati. "Conserving" dimostra la capacità della fotografia di creare un nuovo pubblico per soggetti e idee che finora erano rimasti in ombra.

Alla fine degli anni '80 e all'inizio dei '90 molti fotografi d'arte europei e americani cominciarono ad interessarsi notevolmente al tema "morte".
I ritratti di morti di Rudolf Schäfer (realizzati in un obitorio della Germania est nel 1987) fecero subito scalpore. Lo stesso vale per l'impressionante serie realizzata da Andres Serrano negli obitori di New York City. Il documentarista inglese Nick Waplington che all'inizio del 1990 fece parlare di sé per i suoi commenti abrasivi sulla vita di due famiglie in un quartiere di Nottingham, produsse una serie di fotografie con attori, riguardanti la morte innaturale e il suicidio. Le ragioni di questo nuovo risvegliato interesse per il fotografare la morte restano inspiegabili e forse sono anche troppo diversificate per poter essere amalgamate in una teoria coerente. Artisti che lavorano con la fotografia hanno sviluppato, dagli anni '80, un forte interesse per gli "objets morts" utilizzando situazioni della "vita reale" per dare una visione del mondo dei vivi molto soggettiva e mirata. Le fotografie dei morti, come si possono vedere negli obitori o nei musei, hanno avuto in questo nuovo approccio un ruolo molto importante. I morti erano parte del nostro mondo eppure fuori di esso. Si presentavano al fotografo/artista come immobili, oggetti silenziosi a disposizione in un modo unico e inimitabile.

Questo nuovo modo di mettersi a confronto con il

tema morte, ha rappresentato una sfida per lo status quo, e si è introdotto nella metodologia postmoderna degli anni '80 e '90. I parametri della fotografia slittarono quando gli artisti man mano cominciarono a non utilizzare più la foto come rappresentazione ma ad esplorare la rappresentazione stessa. Cindy Sherman ha costruito un teatro di maschere e fantasia, allo scopo di farsi un'idea sull'identità femminile, Martin Parr si è servito di foto a colori vivaci per realizzare delle documentazioni amare sulla vita nelle strade della Gran Bretagna, e con ciò ha rivoluzionato l'idea del documentario classico. Thomas Ruff ha percorso nuove esaltanti vie portando idee innovative per quanto concerne il ritratto fotografico. I limiti caratteristici dell'arte fotografica sono caduti e l'importanza crescente della teoria fotografica (in opposizione alla storia della fotografia considerata sempre di più come sterile e ossessionata dalle procedure) ha attirato l'attenzione di un pubblico nuovo sulle dimensioni intellettuali e politiche dell'immagine fotografica. In numero sempre maggiore, cominciarono, artisti a utilizzare la fotografia come un mezzo accessibile e leggero, per esprimere idee, che negli anni '60 e '70 probabilmente avrebbero affidato ai mezzi della pittura, illustrazione, e scultura.

Le opere di Daniel e Geo Fuchs sono da inquardrare in questo contesto postmoderno. Si prendono la libertà di farlo in aree che prima forse erano considerate tabù. Il loro lavoro non vuole essere assolutamente un documentario ma molto più un processo molto soggettivo, caratterizzato da una forte partecipazione personale e profonde emozioni. Sembrano voler dire che il mondo è lí per essere osservato e studiato accessibile a innumerevoli sguardi differenti. Il loro intento continuo è di lavorare a qualcosa che loro stessi definiscono "esperienza limite". L'impegno personale, con cui si dedicano al tema, la loro distanza dal sensazionalismo e dal voierismo è il risultato di una determinazione ambiziosa e fondata, che richiede l'attenzione e la comprensione dell'osservatore. Le immagini della trilogia "Conserving" sono impregnate tutte di un estetica particolare. Elementi del nostro mondo, che normalmente vengono considerati macabri o nauseanti, vengono trasformati in oggetti affascinanti. Le creature o gli umani morti, fotografati da loro e resi immortali dal liquido che li contiene, divengono icone, quasi oggetto di venerazione. Morti da tempo e lo stesso parte del nostro mondo attuale, si trovano in uno strano stato fluttuante, da qualche parte tra la vita e la morte e forse hanno raggiunto uno stato mistico.

"Conserving fish" è probabilmente la parte più eterea e astratta delle 3 serie di "Conserving". Il nuotare nel liquido permette ai pesci di ottenere uno stato che si può definire più "naturale". Li vediamo attraverso il vetro come attraverso le pareti di un acquario, e solo una certa opacità dei colori, una mancanza di vitalità fanno pensare al loro strano stato fluttuante. Qua e là un flacone, un'etichetta ci segnalano che siamo in presenza di uno studio tassonomico, piuttosto che di animali prigionieri ma pur sempre vivi. Nonostante "Conserving fish" sia il meno scioccante della trilogia, è impregnato di evidente malinconia. Squame che dovrebbero brillare risultano opache e prive di lucentezza, occhi vivaci senza vitalità, branchie mobilissime restano silenti, rese immobili dalla morte e dal tempo.

"Conserving animals" è tutt'altra esperienza. Già dal primo sguardo si capisce che questo è uno studio sulla morte. Artigli e zampe sono staccati dal corpo e galleggiano nel liquido. Occhi senza sguardo fissano attraverso il liquido e noi non riusciamo a guardarli. Qualche volta c'è un gesto che ricorda la vita, uno sfregamento degli occhi, gambe o braccia tese, un abbraccio, ma sono solo gesti occasionali, attimi fuggenti di una vitalità che si ribella alla morte. Questi animali non dormono un sonno tranquillo, condannati al liquido, strappati dal loro elemento, consegnati agli sguardi dei curiosi e degli esperti. Sono diventati oggetti, cose, senza il minimo senso di sé stessi.

Quando osserviamo la nostra propria specie in ("Conserving humans") la mancanza di sfera privata diventa sensibile. I visi dei morti non sono privi di espressione o di personalità. Un ragazzo, una vecchia donna, un neonato, perfino un embrione appartenevano, una volta, al nostro genere, avevano pensieri e sentimenti, prendevano decisioni, avevano relazioni, facevano parte della nostra società. Ma qui, dove fungono da protoptipi per gli scienziati, o da oggetti da esposizione, sono stati esclusi dai riti tribali dai riti di passaggio. Come Faust di Goethe sono stati resi immortali in modo grottesco. "Conserving humans" è l'apice della trilogia "Conserving" che ci porta dall'astratto al reale. Irrefutabilmente si considererà "Conserving humans" come studio del grottesco, ma Daniel e Geo Fuchs hanno poco interesse al lato spettacolare del grottesco. Guardano le mutilazioni, crani spaccati, scheletri e pelle tatuata, diventano quasi testimoni, cronisti del tempo in cui viviamo. Hanno scritto a proposito di Conserving humans, che la produzione di questa serie si è rivelata una sfida, "poiché volevamo mostrare immagini di persone che comunicavano bellezza, dove la vita resta visibile e la sapienza vince la morte e la dignità umana va oltre la morte." (2.)

I lavori per "Conserving humans", hanno portato a un progetto fotografico più ampio e duraturo sul tema morte. Gli artisti si sono addentrati nei reparti scientifici dei musei e nel mondo strano e silente delle sale d'autopsia, dove niente sfugge al loro sguardo, dai cadaveri agli strumenti per tagliare e sezionare. Qui dove il corpo umano, sebbene ancora riconoscibile come l'individuo che era una volta, è diventato oggetto, derubato della sua intimità. Si occupano delle tracce che lascia il corpo, un pezzo di carne, un capellIo. Il loro interesse a questi resti non è morboso, perchè anche quando la curiosità spinge a scoprire il grottesco e l'assurdo, l'occhio dell'artista si impone. Le immagini che vediamo non sono drastiche o nauseanti, ma si offrono alla nostra immaginazione e al nostro intelletto e con ciò a innumerevoli interpretazioni. "Dopo esserci occupati intensivamente delle forme più astratte della morte, era logico per noi come prossimo passo, guardare in faccia la morte reale. Senza il vetro e il liquido proteggente, la nostra situazione, e di conseguenza le immagini che ne risultano sono molto più dirette...La morte e la decadenza del corpo umano diventano qualcosa di molto naturale. Il corpo sembra essere una conchiglia vuota..." (3.) Si capisce che la loro sete di sapere, la pretesa di risposte è diventata una fissazione profondamente personale.

Nelle tre serie che compongono "Conserving", sono

rimasti fedeli ai principi che distinguono la loro opera. Quando guardiamo i pesci, i mammiferi, e gli umani, da loro fotografati, ci accorgiamo del loro legame intimo con i soggetti. Nonostante osservino situazioni spiacevoli, sentiamo che il loro sguardo è preoccupato perfino benevolo: " L'uomo è affascinato e allo stesso tempo disgustato dall'idea della morte — l'ultimo segreto della vita. Siccome siamo sempre accompagnati dalla coscienza che la nostra morte è sempre presente, la morte e la vita si tengono a braccetto. Questo è il notro scopo, mostrarle entrambi nelle nostre fotografie." (4.) Possiamo percepire la loro vulnerabilità, il rischio personale a cui sono andati incontro, dal momento che hanno deciso di dedicarsi a questo terreno nascosto e spesso discusso. Si sono fatti antiquari, ricercatori in terra straniera, che scoprono verità scomode e creano fotografie, che provocano discussioni, propongono domande e contribuiscono molto di più a farci capire il ruolo della fotografia.

Val Williams

---

1. Citazione di un'intervista con gli artisti nel 1999.
2. Ibid.
3. Citazione di un'intervista di Francesca Alfano Miglietti per *Virus Magazine*, Milano, Italia.
4. Citazione di un'intervista con gli artisti nel 1999.

Index

The notes beside the small photos below were taken by the artists based on information written on the labels of the conserved specimens. The "Conserving" trilogy is intended to be viewed in an artistic context; the photographers' notes are not meant to be exhaustively complete.

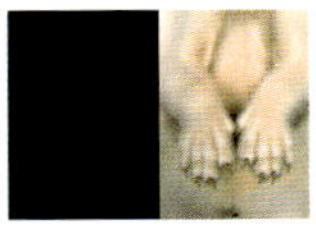
Siamkatze, Leiden
Siamese cat, Leiden

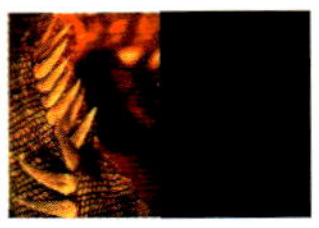
von 1928
from 1928

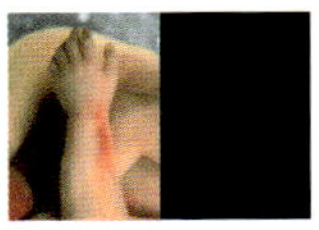
Fetus, Zwillinge
fetus, twins

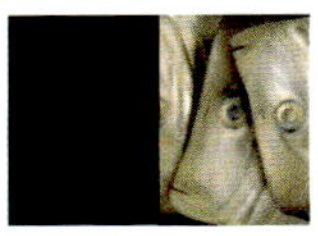
Selene vomer
Meerbarsch
lookdown

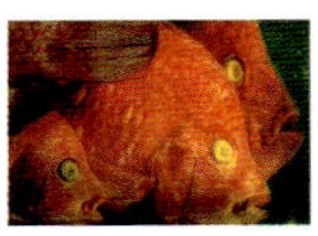
Herichthys spec.

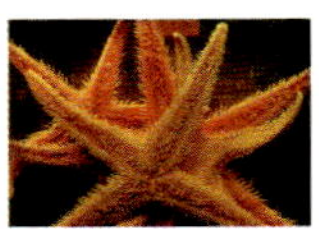
Seestern
starfish

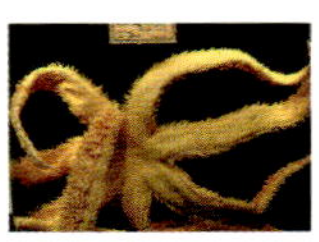
Seestern
starfish

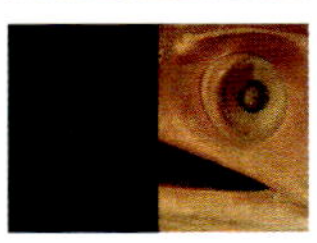

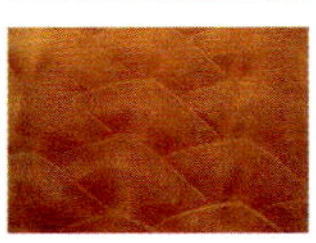
Cyprinus carpio
Karpfen
carp

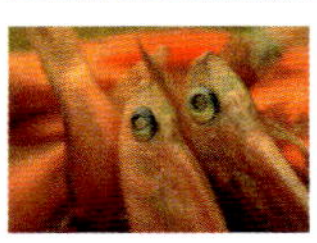

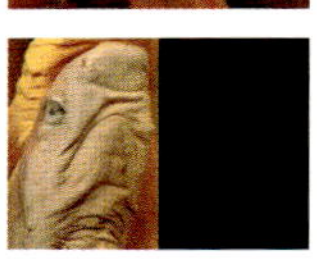
Haifisch
shark

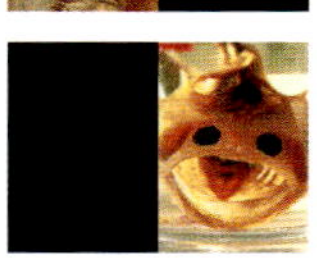

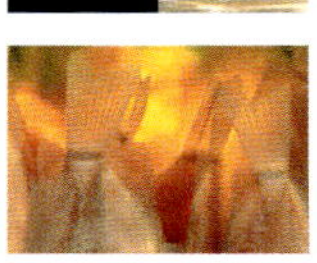

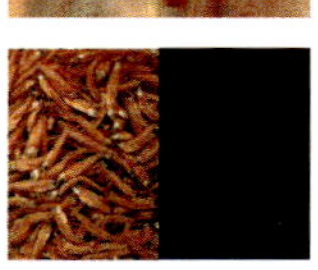

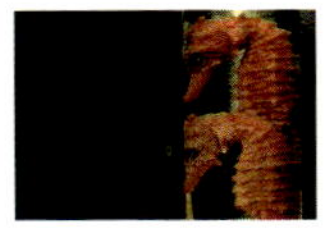
Seepferdchen
seahorse

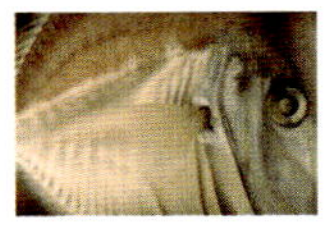
Selene vomer
Meerbarsch
lookdown

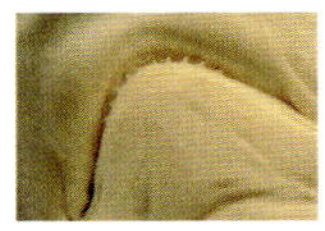
Haifisch
shark

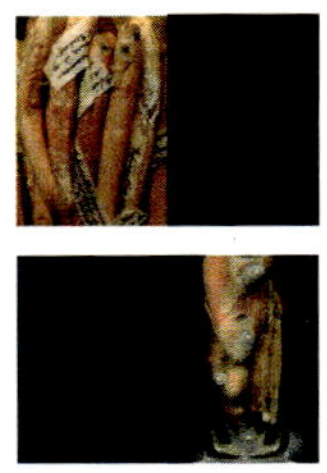

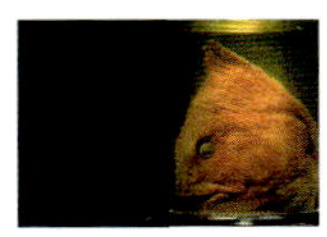
Anarhichas denticulatus
Seewolf
wolf fish

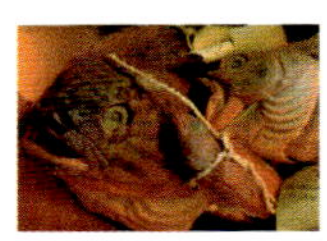

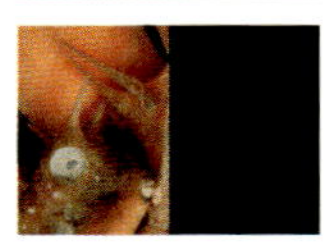
Kofferfisch
trunkfisch

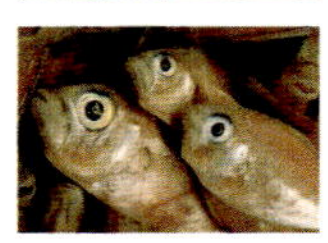

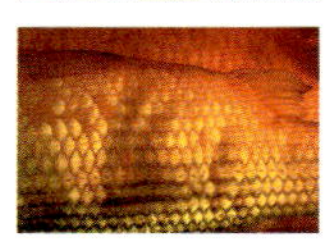

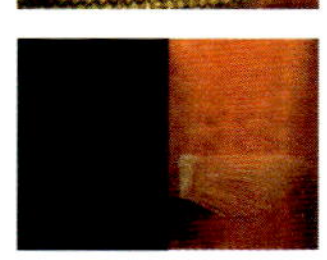

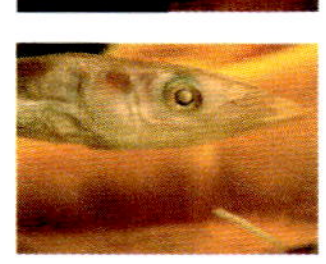

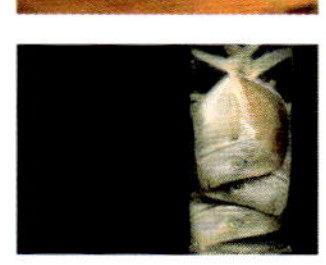
Selene vomer
Meerbarsch
lookdown

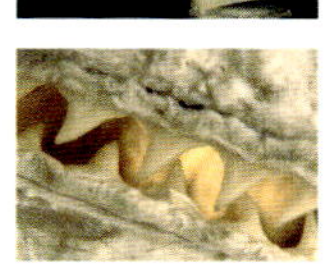
Haifisch
shark

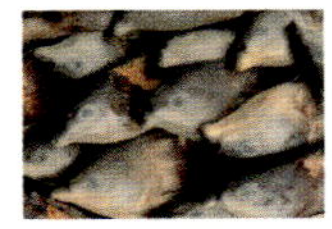

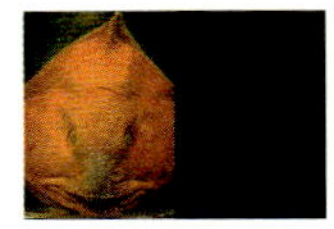
Anarhichas denticulatus
Seewolf
wolf fish

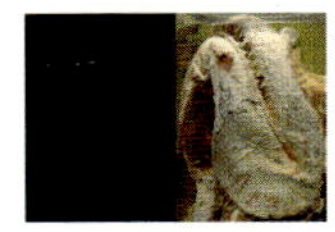
Haifisch
shark

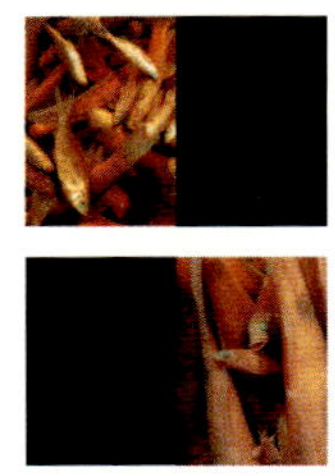

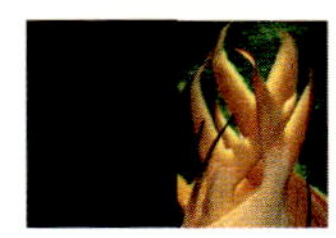

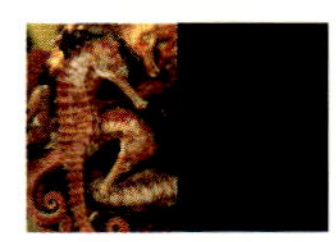
Seepferdchen
seahorse

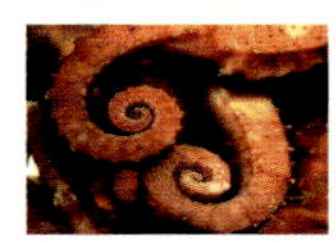
Seepferdchen
seahorse

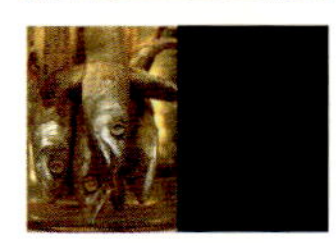

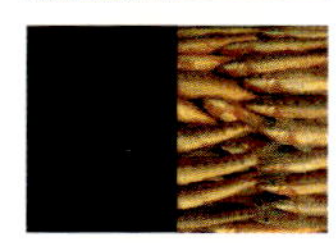

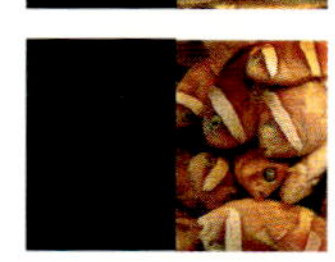

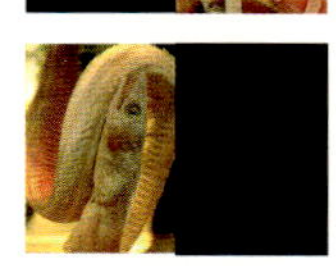
Moräne
moray eel

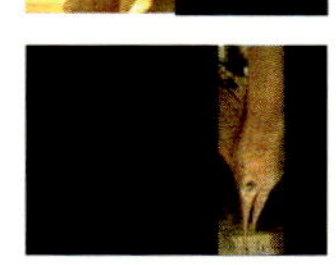

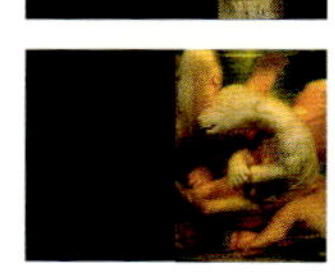
Eisbären
polar bears

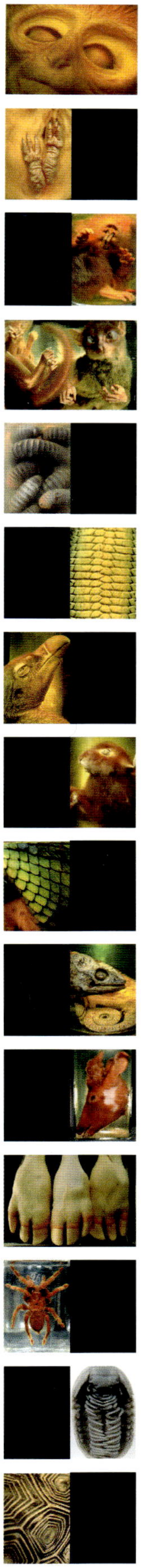

Zwergmeerkatze
dwarf guenon

Meerschweinchen
guinea pig

Sibergrauer Erdbohrer
mole rat

Fischgabun
Demidoff's bushbaby

Riesen-Tausendfüßler
giant millipede

Smaragdeidechse, 1953
emerald lizard, 1953

Geier
vulture

Schneehase, 1910
mountain hare, 1910

Schlange
snake

Chamäleon
chameleon

Reh
deer

Füße eines Rindes
feet of a cow

Tarantel
tarantula

Riesenkugler
sphaerotherids

Schildkröte, 1932
turtle, 1932

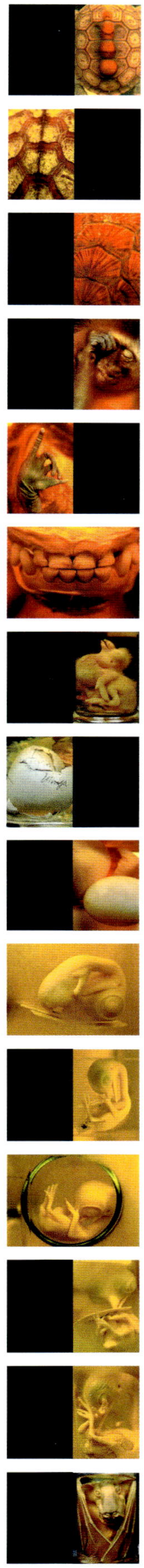

Diamantschildkröte, 1879
diamondback terrapin

Schildkröte
turtle

Schildkröte
turtle

Meerkatze
guenon

Helmkopfkröte, 1960
toad, 1960

Gebiss eines
Schimpansen,1914
teeth of a chimpanzee

Haustauben-Embryo
domestic pigeon embryo

Vogelei
bird's egg

Echsenei
lizard's egg

Nachtreiher-Embryo
night heron embryo

Seidenreiher-Embryo
little egret embryo

Seidenreiher-Embryo
little egret embryo

Schwarzer-Milan-Embryo
black kite embryo

Nachtreiher-Embryo
night heron embryo

Hammerkopf-Flughund
hammer-headed fruit bat

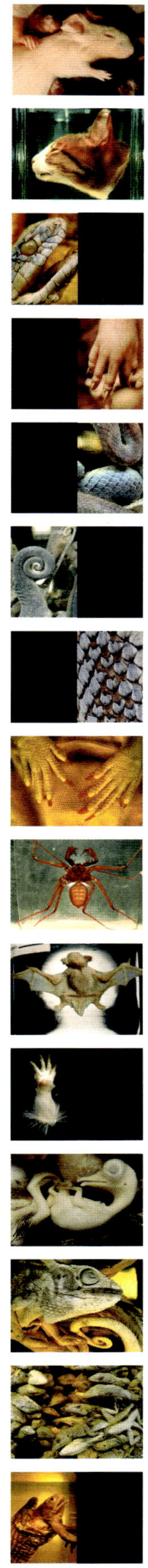

Mäuse, 1972
mice, 1972

Hauskatze
domestic cat

Eidechsennatter
colubrid snake

Fuß eines Opossums
foot of an opossum

Weißlippen-Bambusotter
asiatic lance head snake

Bergchamäleon
mountains chameleon

Eidechsennatter
colubrid snake

Ägyptischer Dornschwanz
Egyptian spiny-tailed lizard,
1828

Fledermaus
bat

Vogel-Embryo
bird embryo

Riesenchamäleon
Oustalet's chameleon

Eidechsen
lizards

Schlange frisst einen Frosch
snake eating a frog

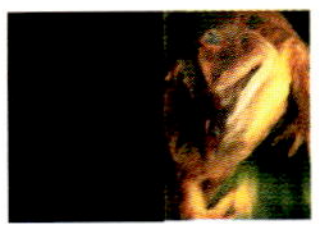
Südamerikanischer Hornfrosch frisst Maus
South American horned frog eating a mouse

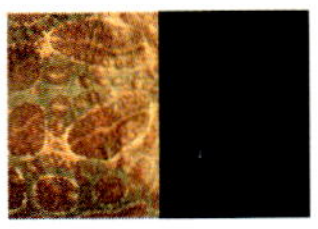
Schmuck-Hornfrosch, 1893
frog, 1893

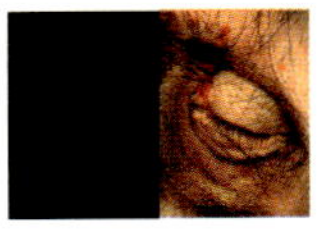
Auge einer Meerkatze
eye of a guenon

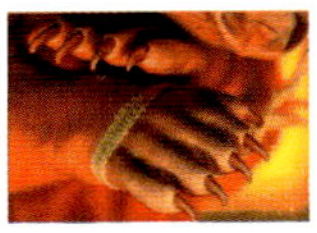
Bärentatzen
feet of a bear

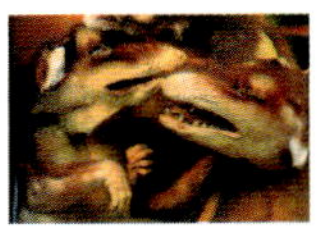
Opossum
opossum

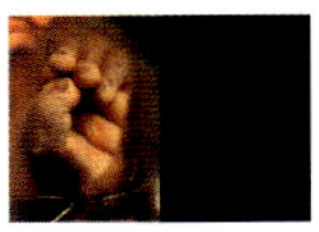
Hand eines Plumplori
hand of a slow loris

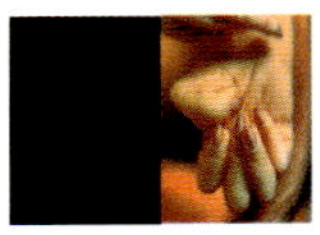
Hand eines Affen
hand of a monkey

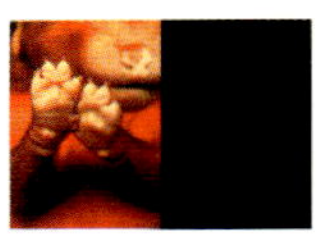
Fuchs
fox

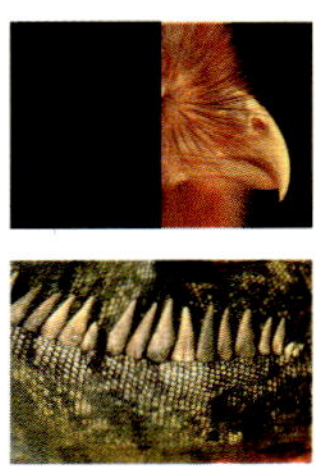
Waldkauz
tawny owl

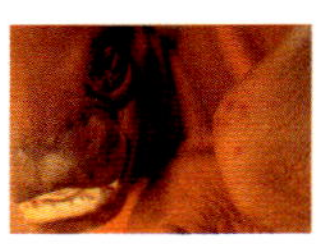
Zwergflusspferd
pigmy hippopotamus

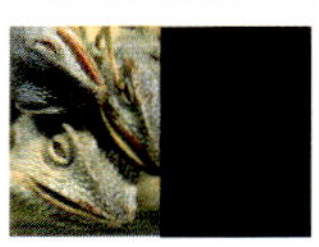
Schönechsen, 1889
lizard 1889

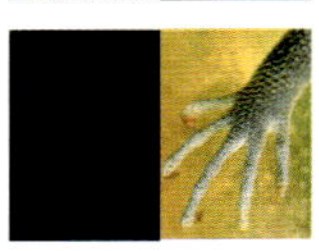
Kurzkammleguan, 1963
iguanid, 1963

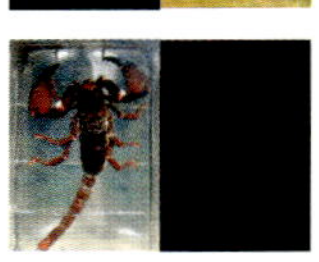
Riesenskorpion
giant scorpion

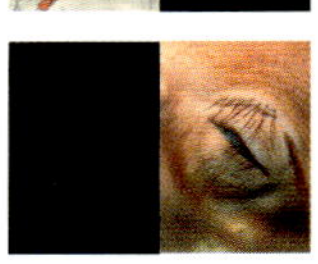
Buschschwein
bush pig

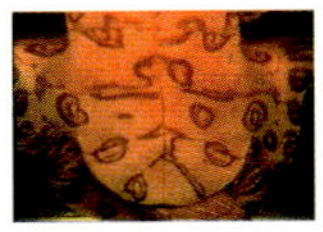
Sumpfschildkröte
fresh-water turtle

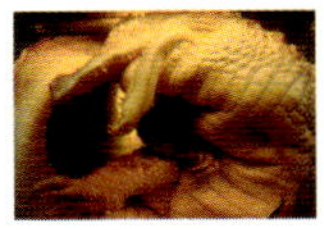
Mississippi Alligator
american alligator

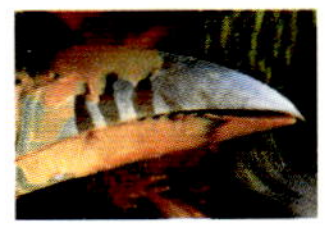
Nashornvogel
hornbill bird

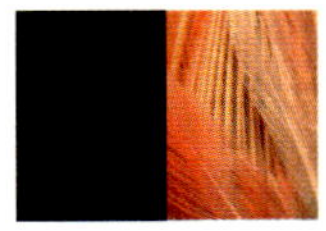
Scharlachsichler (Rotibis)
scarlet ibis

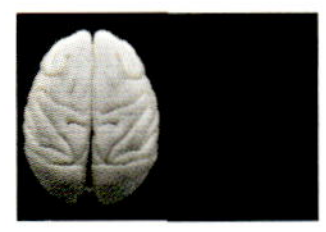
Gehirn eines Rhesusaffen
brain of a rhesus ape

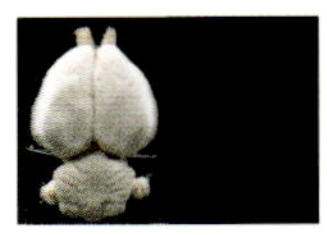
Gehirn eines Langschwanz-Chinchilla
brain of a chinchilla

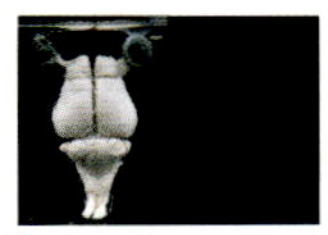
Gehirn eines Igels
brain of a European hedgehog

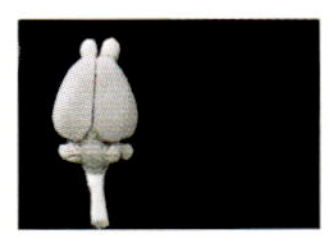
Gehirn eines gewöhnlichen Spitzhörnchens
brain of a common tree-shrew

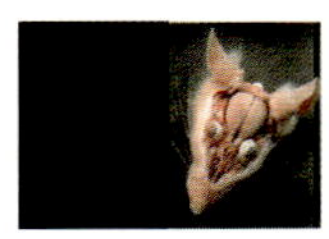
Gehirn einer Ratte
brain of a rat

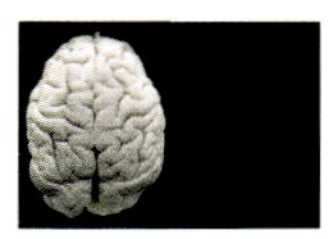
Gehirn eines Pavians
brain of a baboon

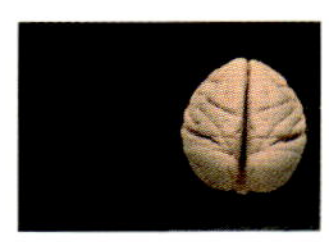
Gehirn eines gehaupten Kapuzineraffens
brain of a brown capuchin

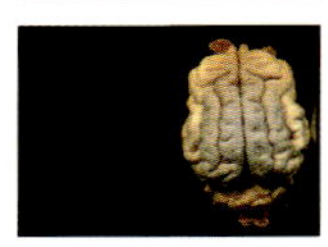
Gehirn eines Tigers
brain of a tiger

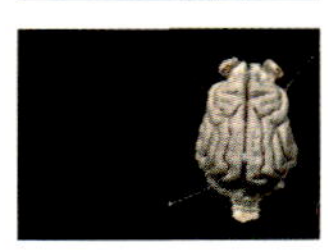
Gehirn eines Hundes
brain of a dog

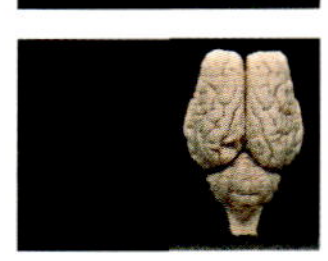
Gehirn eines Schafes
brain of a sheep

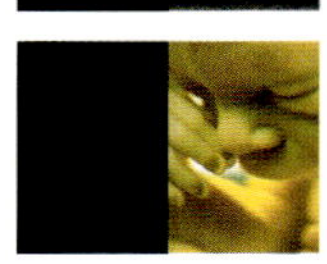
Fetus, ca. 5. Monat
fetus, ca. 5 months old

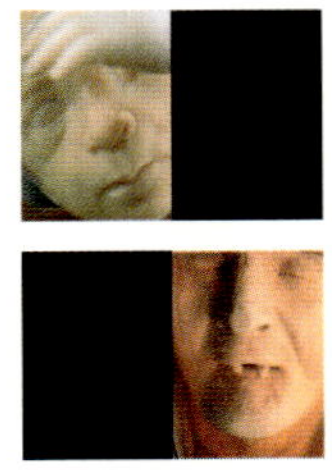
Fetus
fetus

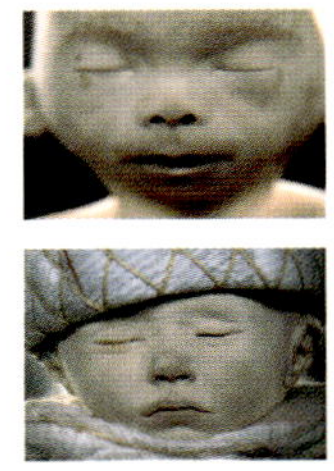
Hottentotte, Sammlung Bonn, 18.Jhd., Leiden
Hottentot, Bonn Collection, 18th century, Leiden

Kinderkopf mit türkischer Mütze, 18.Jhd., Utrecht
child's head with Turkish cap, 18th century, Utrecht

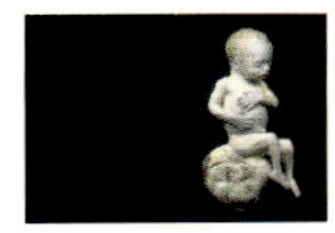
Fetus, 8. Monat, Steißteratom
fetus, 8th month, with tumor

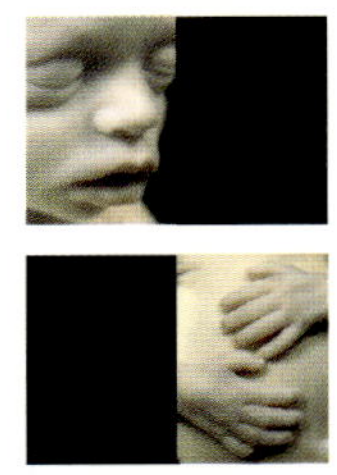
Fetus
fetus

Fetus
fetus

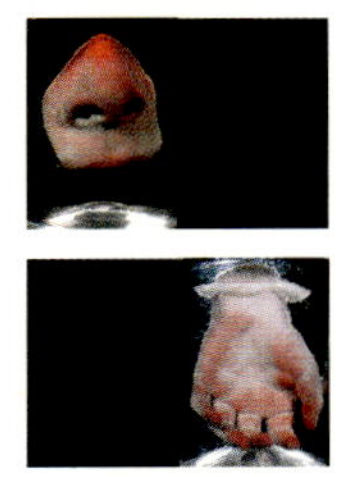
Nase, Sammlung Albinus, Anf. 18.Jhd., Leiden
nose, Albinus Collection, early 18th century, Leiden

Hand, Sammlung Albinus, Anf. 18.Jhd., Leiden
hand, Albinus Collection, early 18th century, Leiden

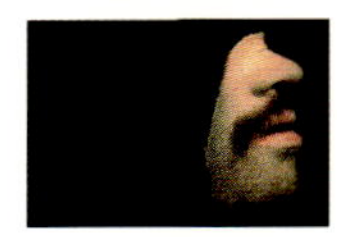
Sammlung Bleuland, Anf. 19.Jhd., Utrecht
Bleuland Collection, early 19th century, Utrecht

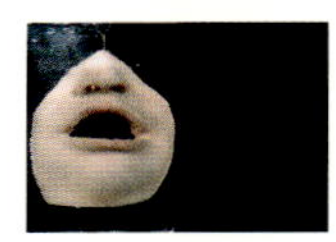
Mund eines Kindes, 18.Jhd., Leiden
mouth of a child, 18th century, Leiden

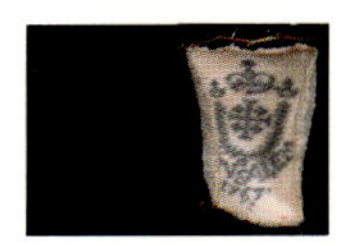
Tätowierung, Sammlung Bonn, 18.Jhd., Leiden
Tatoo, Bonn Collection, 18th century, Leiden

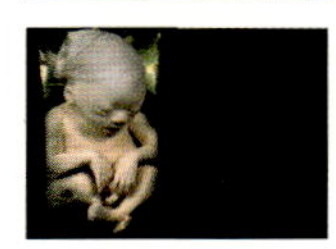
Fetus, 18.Jhd., Leiden
fetus, 18th century, Leiden

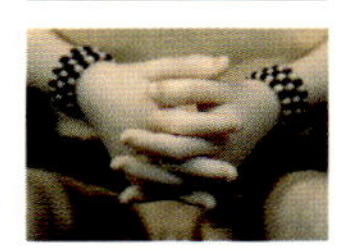
Fetus, 18.Jhd, Leiden
fetus, 18 century, Leiden

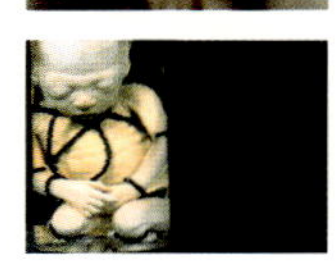
Fetus, Sammlung Bonn, 18. Jhd., Leiden
fetus, Bonn Collection, 18th century, Leiden

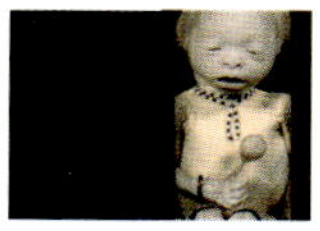
Fetus, 7. Monat, 18.Jhd., Leiden
fetus, 7 month old, 18th century, Leiden

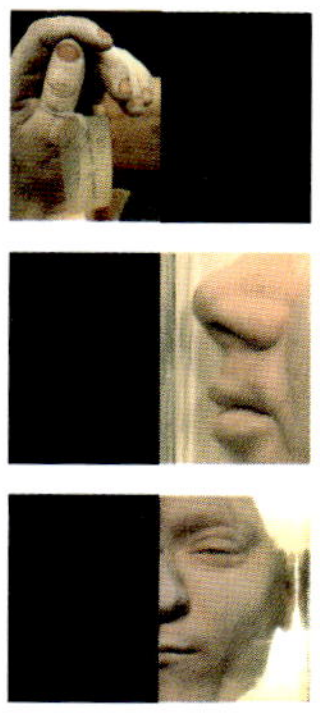
Hand, 1923
hand, 1923

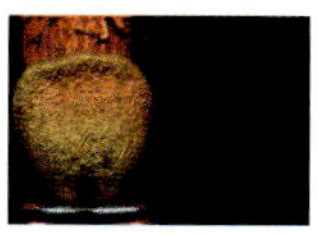
Penis, mit Quecksilber injiziert, Anfang 19.Jhd.
penis, injected with mercury, early 19th century

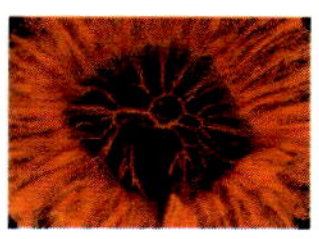
Dünndarm, Gefäßinjektion
small intestine, arterial injection

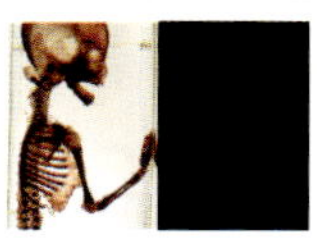
Fetus, Medianschnitt, Aufhellungspräparat
fetus, medial section, bleached specimen

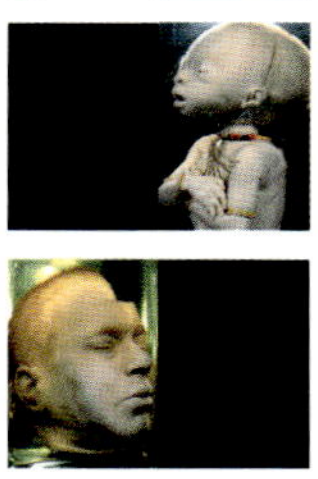
Fetus, 5. Monat, 18.Jhd., Leiden
fetus, 5 month old, 18th century, Leiden

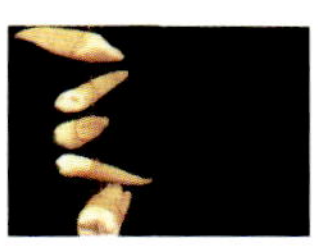
Zähne, Samml. Bleuland, Anfang 19.Jhd., Utrecht
teeth, Bleuland Collection, early 19th century, Utrecht

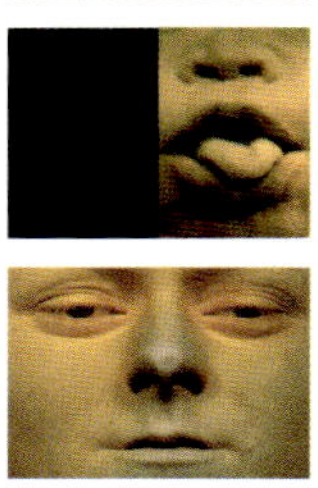
Fetus, Missbildung: Anencephalus
fetus with anencephalic deformity

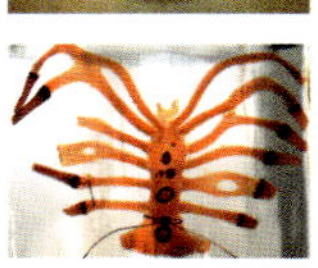
Brustbein, Aufhellungspräparat
breastbone, bleached specimen

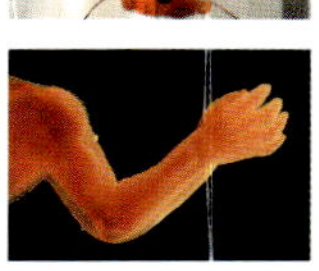
Arm eines Fetus, Aufhellungspräparat
arm of a fetus, bleached specimen

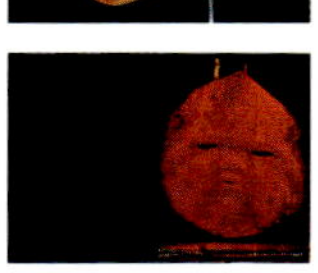
Injizierte Gesichtshaut eines Neugeborenen, 1912
injected facial skin of a new born baby, 1912

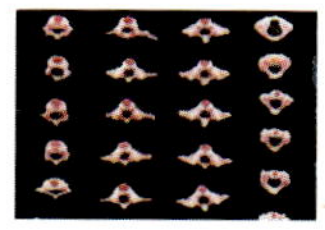
Wirbelsäule eines 19. Wochen alten Fetus
spine of a 19-week-old fetus

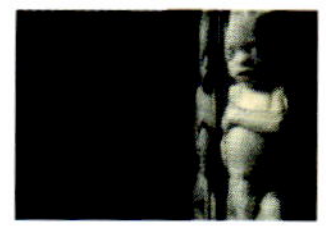
Fetus, 8. Monat
fetus, 8 month old

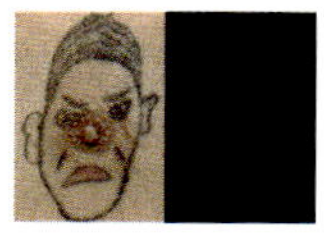
Tätowierung auf der Brust
tatooed chest

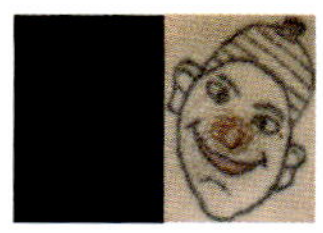
Tätowierung auf der Brust
tatooed chest

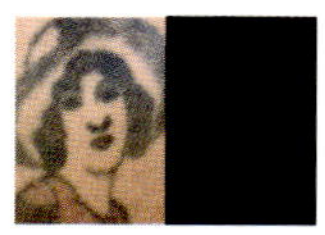
Tätowierung
tatoo

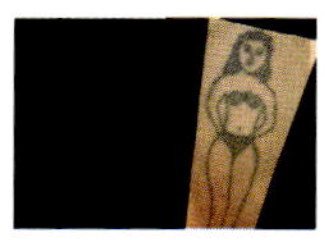
Tätowierung
tatoo

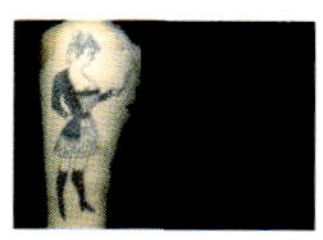
Tätowierung, Staatsprüfung, 1901
tatoo, state examination, 1901

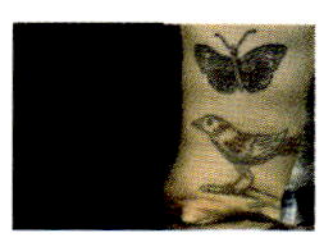
Tätowierung, Staatsprüfung, 1901
tatoo, state examination, 1901

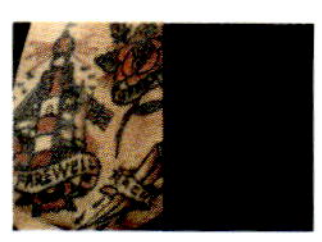
Tätowierung, 1958
tatoo, 1958

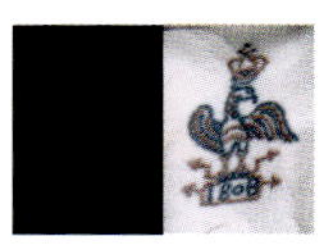
Tätowierung, Anfang 19.Jhd., Utrecht
tatoo, early 19th century, Utrecht

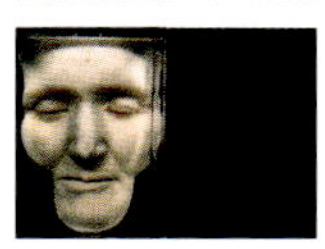
Oberes Gesicht
face, upper portion

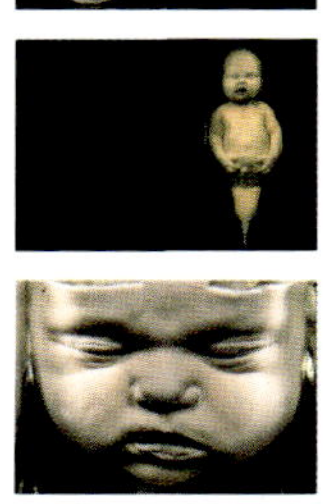
Sirenoide Missbildung
sirenoid deformity

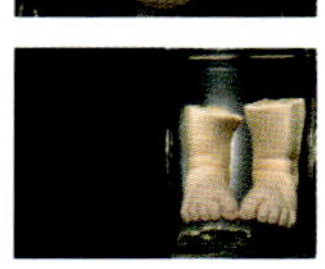
Füße eines Kindes mit zusammengewachsenen Zehen
child's feet with fused toes

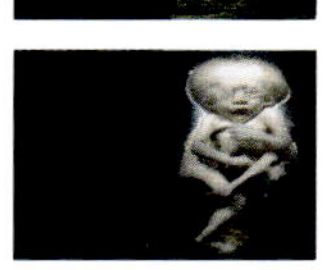
Janus
janus

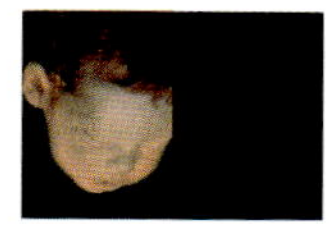
Kopf eines 10 Tage alten Säuglings
head of a 10-day-old infant

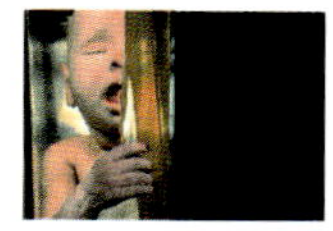
Kopf eines Neugeborenen, Medianschnitt
head of a newborn, medial section

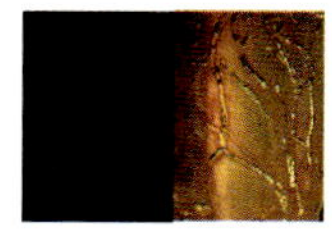
Lymphgefäße, mit Quecksilber injiziert. Lymph vessels injected with mercury, early 19th century, Utrecht

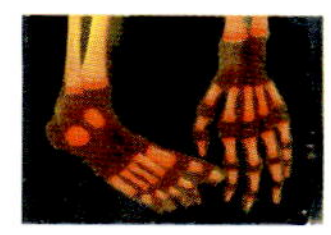
Unterarm und Fuß eines Kleinkindes
forearm and shin of a small child

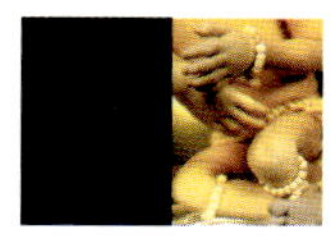
Fetus, Utrecht

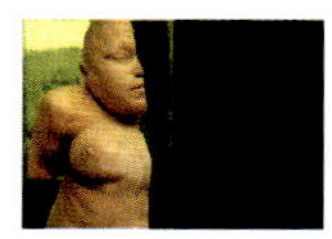
Marburger Lenchen, Sagittalschnitt, 1890
"Marburg's Lenchen", sagittal section, 1890

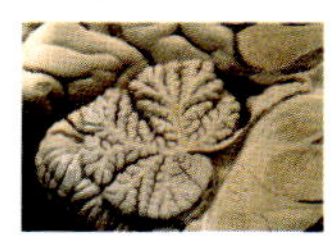
Kleinhirn-Lebensbaum
cerebellum

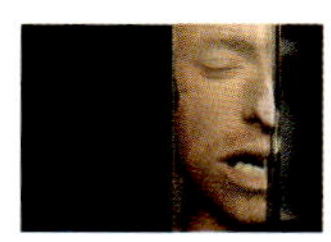
Kopf eines Erwachsenen, Sagittalschnitt
head of a adult, sagittal section

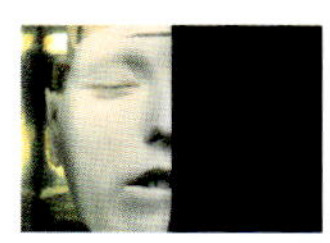
Kopf eines Erwachsenen, Sagittalschnitt
head of a adult, sagittal section

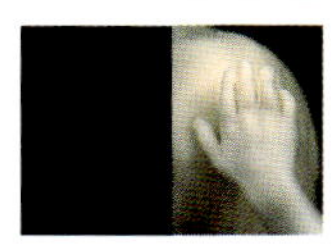
Fetus
fetus

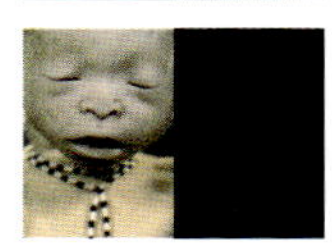
Fetus, 18.Jhd., Leiden
fetus, 18th century, Leiden

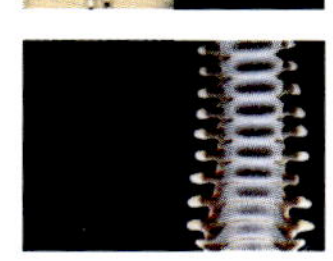
Rückgrat eines Neugeborenen
Spine of a newborn baby, van Doeveren Collection, 18th century, Leiden

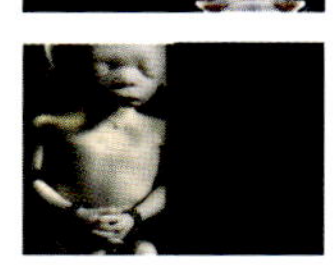
Fetus, 18. Jhd., Leiden
fetus, 18th century, Leiden

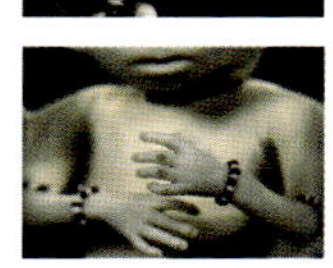
Fetus, 18.Jhd., Leiden
fetus, 18th century, Leiden

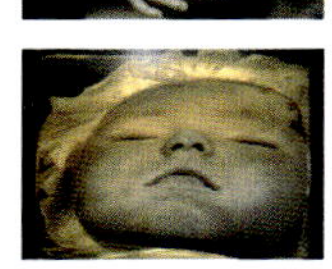
Kinderkopf, Sammlung Ruysch, 1690
child's head, Ruysch Collection, 1690

Artistic Biography

Daniel Fuchs, born in 1966
Geo Fuchs, born in 1969

Solo exhibitions of "Conserving"

"Conserving"
Rencontres Internationales de la photographie Arles, France, 04.07.–20.08.2000
Fotografie Forum international, Frankfurt am Main, Germany, 12.05.–18.06.2000
Stephen Bulger gallery, Toronto, Canada, 28.10.–25.11.2000
Galerie Le Réverbère 2, Lyon, France, December 2000 – February 2001

"Conserving Animals"
Naturhistorisches Museum Basel, Switzerland, 04.05.–27.08.2000
Barlach HALLE K, Hamburg, Germany, 09.03.–22.04.2000

"Conserving Fish"
Galerie Nei Liicht, Dudelange, Luxemburg, 31.03.–24.04.2000
Barlach HALLE K, Hamburg, Germany 16.04.–24.05.1998

Group exhibitions of "Conserving"

"New Natural History" (exhibition and catalog)
Hasselblad Center, Göteborg Museum of Art, Sweden, 22.01.–27.02.2000
National Museum of Photography, Bradford, England, 17.09.–14.11.1999

"Still in motion," Galerie Lipanjepuntin, Trieste, Italy, (exhibition and catalog), 02.10.–13.11.1999
Art Cologne, Galerie Lipanjepuntin, Trieste, Italy, 07.–14.11.1999
Art Brüssel, Galerie Lipanjepuntin, Trieste, Italy, 23.–27.04.1999

Selected solo exhibitions

"Im falschen Körper," photo installation, galerie & edition m, Leipzig, Germany, 12.03.–25.04.1998

"Espada – in den Armenvierteln Portugals"
Triennale der Photographie Hamburg, Kunsthaus, Germany, 31.05.–11.07.1999
Internationale Photoszene Köln, Germany, 18.09.–04.10.1998
Galerie Nei Liicht, Dudelange, Luxemburg, 03.07.–26.07.1998
Schirn Kunsthalle Frankfurt am Main, Germany, 12.10.–09.11.1997

"Menschen in der Psychiatrie"
Grundbuchhalle Ziviljustizgebäude Hamburg, Germany, 18.04.–30.05.1997

"Transsexuelle Menschen in Deutschland – Im falschen Körper"
Haus am Kleistpark and Neue Gesellschaft für bildende Kunst Berlin, Germany, 27.09.–02.11.1997
Paul-Gerhardt-Kirche, Köln, Germany, 10.01.–30.01.1997
Halle K3, Kampnagelfabrik Hamburg, Germany, 16.08.–01.09.1996
Gasteig München, Germany, 23.09.–01.10.1995
Römerhalle, Römer Frankfurt, Germany, 17.03.–26.03.1995
Publication of the book: *Transsexuelle Menschen in Deutschland – Im falschen Körper*,
preface by Peter Weiermair, August 1995, Verlag Martina Rueger, Wiesbaden

"Die Unbehausten"
Schauspiel Frankfurt am Main, Germany, 02.12.1994 –08.01.1995
Kreuzweg in der Weißfrauenkirche Frankfurt am Main, Germany, February 1995

Selected group exhibitions

"Kinder haben Rechte" (exhibition and catalog)
Franckesche Stiftungen, Halle/Saale, Germany, 07.03.–19.06.2000
Airport Gallery, Frankfurt am Main, Germany, 16.09.–31.10.1999

Prometheus – Menschenbilder – Visionen
Elektrizitätswerk Reading, Tel-Aviv, Goethe-Institut Tel-Aviv, Israel, 12.10.1999 –15.01.2000
Alte Völklinger Hütte, Gebläsehaus, Deutsches Historisches Museum Berlin, 06.09.–08.11.1998

Art Frankfurt, galerie & edition m, Leipzig, Germany, 19.03.–23.03.1998

Participation in exhibitions of the "Deutsche Fototage," Airport Gallery, Frankfurt am Main
"Deutsche sehen Schweizer," October 1998,
"Die Farbe Grün," September 1998
"Zeitgenössische Automobilfotografie, " 11.03.–13.04.1998
"Zeitgenössische Modefotografie," 02.10.–02.11.1997

"Sie und Er – Frauenmacht und Männerherrschaft im Kulturvergleich" (exhibition and catalog),
Rautenstrauch-Joest Museum für Völkerkunde, Köln, Germany, 24.11.97–08.03.1998

We especially wish to thank the institutions and collections whose interest in and understanding for our "Conserving" project were essential to its realization.

Sektion Ichthyologie, Forschungsinstitut und Naturmuseum Senckenberg Frankfurt am Main:
Dr. Friedhelm Krupp, Horst Zetzsche, Dr. Uwe Zajonz, Kornelia Jentoch

Zoologische Abteilung, Naturhistorisches Museum Basel:
Dr. Ambros Hänggi, Jolanda Riedi, Edi Stöckli

Zoologische Abteilung, Hessisches Landesmuseum Darmstadt:
Dr. Wolfgang Schneider, Michaela Kurbel, Dr. Ina Busch

Sektion Herpetologie, Forschungsinstitut und Naturmuseum Senckenberg Frankfurt am Main:
Dr. Gunther Köhler, Monika Laudahn

Institut für systematische Zoologie, Museum für Naturkunde der Humboldt-Universität zu Berlin:
Prof. Ulrich Zeller, Dr. Manfred Ade, Dr. Sylke Frahnert, Holger Hackmann

Dr. Senckenbergische Anatomie, Klinikum der J.W.Goethe Universität Frankfurt am Main:
Dr. Helmut Wicht, Prof. Horst-Werner Korf

Institut für Anatomie, Westfälische Wilhelms-Universität Münster:
Prof. Reinhard Hildebrand, Peter Callam

Museum Anatomicum, Institut für Anatomie und Zellbiologie der Philipps-Universität Marburg:
Prof. Gerhard Aumüller, Jens Cordes

Berliner Medizinhistorisches Museum, Universitätsklinikums Charité der Humboldt-Universität zu Berlin:
Dr. Petra Lennig, Dr. Kristina Schubert, Navena Widulin, Prof. Manfred Dietel

Anatomisch Museum Leiden (see index for list of specimens)
Dries van Dam, Prof. Harm Beukers

Anatomisch Instituut, Universiteit Utrecht (see index for list of specimens)
Prof. Berend Hillen, Peter Heller, Dr. Ronald Bleys

Institut für Rechtsmedizin, Heinrich Heine Universität Düsseldorf:
Prof. Wolfgang Bonte, Dr. Peter Gabriel

Institut für Anatomie, Universitätsklinikum Charité der Humboldt-Universität zu Berlin:
Prof. Gottfried Bogusch, Dr. Wolfgang Kreutz, Günter Wilcke

Institut für Anatomie und Zellbiologie, Martin-Luther-Universität Halle / Wittenberg zu Halle (Saale):
Prof. Rüdiger Schultka, Egbert Steinicke, Mario Frommann

Anatomische Sammlung des Institutes für Anatomie 1, Friedrich-Schiller-Universität Jena:
Prof. Werner Linß, Dr. Rosemarie Fröber, Simone Freund

Special thanks for their many years of encouragement are due to:
Jana Marko, Hans Barlach and Barlach Halle K, Hamburg

For their sensitive and fantastic translation of our "Conserving" project into book form, we especially wish to thank Matthias Reuss, Annette Nellessen and Wolfgang Metzger.

We also wish to thank:

Gilles Mora, Gudrun Muschalla, Christoph Herold, Young & Rubicam Frankfurt, Alexander Gérard, Galerie Lipanjepuntin Trieste, Italien, Cristina Lipanje, Marco Puntin, Val Williams, Greg Hobson, Patrick Henry, National Museum for Photography, Bradford, England, Luciano Bobba, Maurizio Galimberti, Grazia Neri, Lello Piazza, Mariateresa Cerretelli, Maria Büscher, Georgine Knöll, Sabine Cuntz, Philip Riefenstahl, Tina Schellhorn, Hans-Eberhard Hess, Josefine Geier, Mathias Reichelt, Erwin Knöll, Fotografie Forum international, Frankfurt am Main, Celina Lunsford, Galerie Nei Liicht Dudelange, Luxemburg, Danielle Igniti, Rencontres Internationales de la photographie, Arles, France, Oschatz Visuelle Medien, Wiesbaden, Daniel Oschatz, Sandra Gilles, Anke Betz, Galerie Le Réverbère 2 Lyon, Frankreich, Catherine Derioz, Jacques Damez, Stephen Bulger gallery Toronto, Canada, Karin and Achim Lahrem, Gertrud and Dieter Fuchs, Claus Mewes, Irene Kern, Sabine Simon, Andrea and Christian Aeby, Otfried Lütt, Dr. Gerhard Storch, Alejandro De Nardi, Peter Teschler, Martin Booth.

Röhm GmbH Darmstadt, Roland Münkel and Jürgen Jourdan.

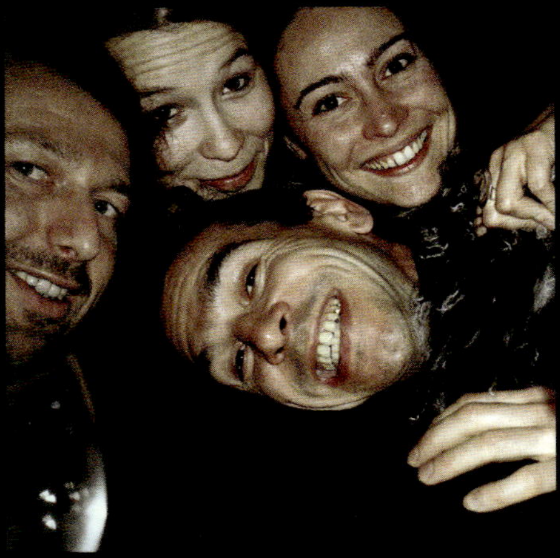

D&G Fuchs, A. Nellessen, M. Reuss
„Einer für alle, alle für einen!"

English translation: Howard Fine
French translation: Sophie Vieillard
Italian translation: Fiorella Palini

Lithography and printing in six-color (hexachrome) process
by Metzger-Druck, Obrigheim, Germany

Edition Reuss Munich, Germany; FAX + 49 (0) 89 50.13.95

ISBN 3-934020-01-1